ANCIENT PHILOSOPHY,
MYSTERY, AND MAGIC

Ancient Philosophy, Mystery, and Magic

Empedocles and Pythagorean Tradition

*

PETER KINGSLEY

CLARENDON PRESS · OXFORD

Oxford University Press, Walton Street, Oxford OX2 6DP
*Oxford New York
Athens Auckland Bangkok Bombay
Calcutta Cape Town Dar es Salaam Delhi
Florence Hong Kong Istanbul Karachi
Kuala Lumpur Madras Madrid Melbourne
Mexico City Nairobi Paris Singapore
Taipei Tokyo Toronto
and associated companies in
Berlin Ibadan*

Oxford is a trade mark of Oxford University Press

*Published in the United States
by Oxford University Press Inc., New York*

© *Peter Kingsley 1995*

*All rights reserved. No part of this publication may be reproduced,
stored in a retrieval system, or transmitted, in any form or by any means,
without the prior permission in writing of Oxford University Press.
Within the UK, exceptions are allowed in respect of any fair dealing for the
purpose of research or private study, or criticism or review, as permitted
under the Copyright, Designs and Patents Act, 1988, or in the case of
reprographic reproduction in accordance with the terms of the licences
issued by the Copyright Licensing Agency. Enquiries concerning
reproduction outside these terms and in other countries should be
sent to the Rights Department, Oxford University Press,
at the address above*

*British Library Cataloguing in Publication Data
Data available*

*Library of Congress Cataloging in Publication Data
Ancient philosophy, mystery, and magic : Empedocles and
Pythagorean tradition / Peter Kingsley.
Includes bibliographical references and index.
1. Empedocles. 2. Pythagoras and Pythagorean school. I. Title.
B218.Z7K56 1995 182'.2–dc20 95-5090
ISBN 0-19-814988-3*

3 5 7 9 10 8 6 4 2

*Printed in Great Britain on acid-free paper by
The Ipswich Book Co. Ltd., Suffolk*

CONTENTS

*

Map of Sicily and southern Italy	viii
Map of Egypt	ix
Introduction	1

I Philosophy

1. Back to the Roots	13
2. *Aither*	15
3. *Aer*	24
4. The Riddle	36
5. The Sun	49

II Mystery

6. An Introduction to Sicily	71
7. The *Phaedo* Myth: The Geography	79
8. The *Phaedo* Myth: The Sources	88
9. The *Phaedo* Myth: The Structure	96
10. Plato and Orpheus	112
11. The Mixing-Bowl	133
12. 'Wise Men and Women'	149
13. Central Fire	172
14. A History of Errors	195

III Magic

15. The Magus	217
16. From Sicily to Egypt	233
17. The Hero	250

18. Death on Etna	278
19. Sandals of Bronze and Thighs of Gold	289
20. Pythagoreans and Neopythagoreans	317
21. 'Not to Teach but to Heal'	335
22. Nestis	348
23. 'Conceal My Words in Your Breast'	359
24. From Empedocles to the Sufis: 'The Pythagorean Leaven'	371

Appendices

I. Parmenides and Babylon	392
II. Nergal and Heracles	394
III. Empedocles and the Ismāʿīlīs	395
Abbreviations	397
Bibliography	403
Index	417

PREFACE

*

THIS book has essentially been written in the form of a narrative; there are certain stories, however ancient, which still deserve to be told. For the sake of the general reader, Greek words and phrases have been either translated or explained; so with other ancient languages. Technical matters have as a rule been confined to the footnotes; the occasional exceptions are where a certain issue has a wider significance or helps to illustrate a principle. The style of reference adopted in the footnotes has been simplified as far as possible. For an explanation of references given in the form of initials, or abbreviations, see the list of abbreviations at the end of the book; fuller details of works cited either by the author's name alone, or by author and date, will be found in the bibliography.

This is an appropriate place to offer my gratitude to all who have helped in various ways in connection with the book. I would like to mention in particular Sir John Boardman, Archbishop Norair Bogharian, Mary Boyce, Walter Burkert, Michel Chodkiewicz, Stella Corbin, John Creed, Luc Deitz, Jill Kraye, Geoffrey Lloyd, Alain Martin, James Morris, Vrej Nersessian, David Sedley, Bob Sharples, Anne Sheppard, Malcolm Willcock, and Fritz Zimmermann. I owe a special debt to Charles Burnett, Stephanie Dalley, Sara Sviri, and Robin Waterfield; to all the library and teaching staff at the Warburg Institute, London; and to Hilary O'Shea at the Press. But above all my thanks go to Martin West, for always being there ready to help at every stage in the writing of this book.

The Kimbell Art Museum, Texas, has given permission to reproduce on the cover the detail from Salvator Rosa's *Pythagoras Emerging from the Underworld* (1662). For Rosa's own comments on the painting, see *Lettere inedite di Salvator Rosa a G. B. Ricciardi*, ed. A. de Rinaldis (Rome, 1939), 141.

viii   *Maps*

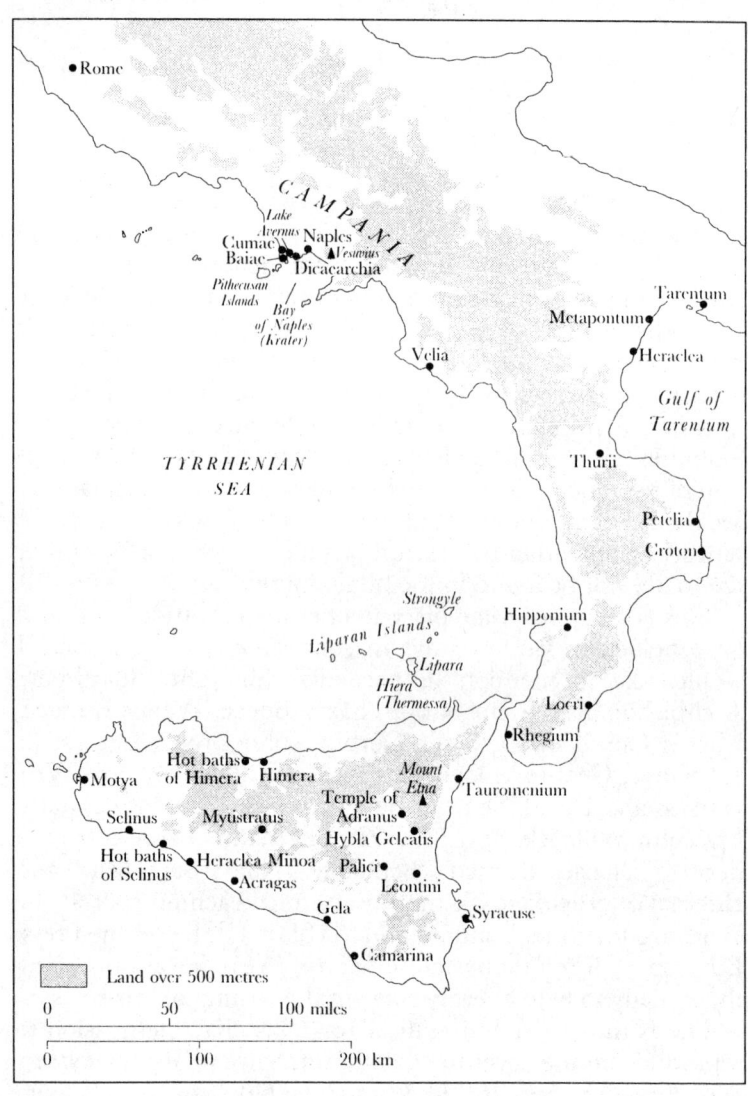

Map of Sicily and southern Italy

Maps

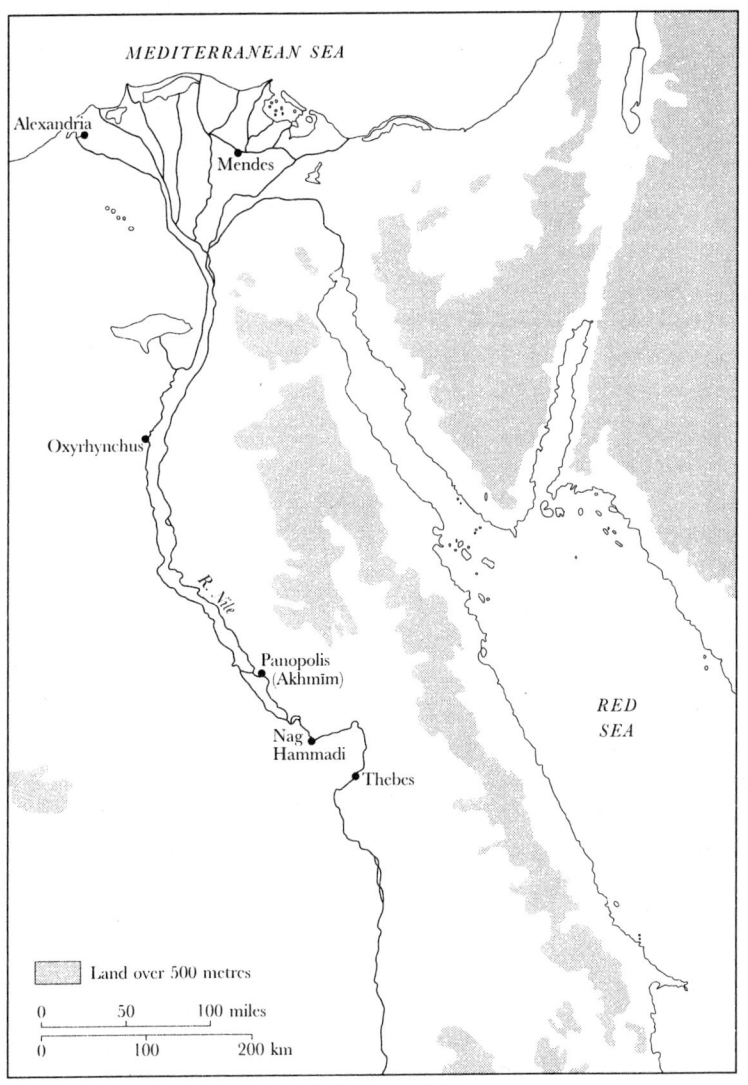

Map of Egypt

Introduction

*

THIS book covers a wide area in space and time, but takes as its starting-point one man who lived well over two thousand years ago. That man was called Empedocles.

Empedocles was probably born around the start of the fifth century BC.[1] He was from the Greek colony of Acragas—modern Agrigento—on the south-west coast of Sicily; but he appears to have spent much of his time travelling, as one would anyway expect from 'seers' of his type in the world of the ancient Mediterranean and Near East.[2] About when he died, or where, or how, we know nothing.[3] And yet this same man, whose life remains such a mystery to us, was to perform an unparalleled role in seeding the subsequent growth of western culture. Formulated in terms of the later trend for defining separate fields of interest or expertise, his influence made itself felt in philosophy, rhetoric, medicine, chemistry, biology, astronomy, cosmology, psychology, mysticism, and religion. The immensely influential theory of four elements, first presented in western literature by Empedocles, is just one very obvious example.[4]

---

[1] Wright 3–5; KRS 280–1.

[2] Regarding his wanderings see DK 31 B112 with Zuntz 189 and e.g. D.L. 8.67, where Empedocles' return to Acragas is described as having been prevented by 'the descendants of his enemies', οἱ τῶν ἐχθρῶν ἀπόγονοι, probably implying a lengthy period of absence: cf. for the expression Hdt. 7.153, Porph. *VP* 22 = Aristox. fr. 17 with von Fritz (1940), 19. For travelling seers cf. Grottanelli; Burkert (1983).

[3] The report by Aristotle that Empedocles died at the age of 60 is now usually considered reliable (Arist. fr. 71; Guthrie, ii. 128, Wright 5); but its lack of historical value was already seen by Bidez a century ago, and subsequent research on Greek biographical tradition either prior to or contemporary with Aristotle has done nothing to inspire any more confidence. Cf. Bidez 154–5 with A. J. Podlecki, *Phoenix*, 23 (1969), 114–37; Momigliano 23–84; Kingsley (1990), 261–4. In the absence of more solid information, the most we can say is that according to Greek biographers the age of 60 was a convenient time to die: cf. e.g. D.L. 2.44 (Socrates) and 9.3 (Heraclitus), Burnet (1892), 75–6 (Anaximenes), F. Jacoby, *Apollodors Chronik* (Berlin, 1902), 329 (Demosthenes), 350 (Moschion). For the legends about Empedocles' death see below, Chs. 16–19.

[4] Kranz 83–105, 110–12; Halleux 67; J. Longrigg, *Isis*, 67 (1976), 420–9, and *Apeiron*, 19 (1985), 93–115.

Not surprisingly, such an important figure has evoked a great deal of attention and given rise to a vast literature. The main purpose of this book is to demonstrate that—in spite of all the attention and the literature—modern scholarship has by no means come to grips with Empedocles' teaching as a whole, and to indicate that a fundamental revision of our ideas about him is in order: a revision which has major implications not just for our understanding of Empedocles himself but also for our understanding of ancient philosophy—and, indeed, of the origins of western culture.

More specifically, the aims of this study are to show that the major obstacle to a correct appreciation of Empedocles has not been (as is usually claimed) the fragmentary nature of the surviving evidence but, instead, has been a wrong approach; to clarify the reasons why, ever since the time of Aristotle in the fourth century BC, his teaching has been so misconceived; and to show where we need to turn for a more accurate representation of his work. That will involve in the first instance uncovering evidence which is more or less contemporary with Empedocles and, although so far overlooked, is directly relevant to him. In the second instance it will involve uncovering evidence for a continuous chain of tradition: an 'alternative' tradition of understanding and interpreting Empedocles which survived for centuries and in several important senses remained closer, more faithful to the authentic Empedocles than the mainstream philosophical interpretations of him in the schools of Aristotle and Plato. An inevitable consequence of bringing these strands of evidence back into the picture is the need for an extensive broadening of our perspectives, and for a reassessment of many unspoken assumptions about transmission of ideas in the ancient world.

Empedocles used poetry to communicate his teaching, and his poetry—like the writings of all early 'philosophers' prior to the time of Socrates and Plato—only survives in the form of fragments as quoted and preserved in the texts of later writers. This has naturally raised the issue of what aids and sources we can possibly use to elucidate these fragmentary remains of Empedocles' teaching. The issue is of crucial importance, because it poses very basic questions about the framework in which we need to situate and appreciate the writings of these so-called 'Presocratics'.

Although the fact is rarely admitted, contemporary study of the Presocratics has reached a crisis-point. The crisis revolves around two words: authority, and tradition. On the one hand, post-'Enlightenment' scholarship over the past two centuries has persistently viewed the history of early Greek philosophy as a progressive evolution towards some extremely vague, but numinously seductive, ideal of rationality; and in doing so it has almost unquestioningly decided to embrace Aristotle's arrogant assessment of Presocratic philosophy as no more than a stammering attempt to say what only he, at last, was able to articulate with any fluency. Aristotle, and Aristotle's successor Theophrastus, have come to be treated as the ultimate authorities for our understanding of the Presocratics—not only because they are the earliest writers to provide us with extensive information about their teachings but also because they are considered ideally qualified to appreciate the aims, and the supposed shortcomings, of Presocratic philosophy. As a result, a kind of genealogy has been established: the tradition of interpretation of the Presocratics is at its purest, its most reliable, and even its most infallible the closer we return to its dual source in these two great authority-figures.[5]

On the other hand, since the publication of Cherniss's work on Aristotle and the Presocratics in 1935 there has been a deeper awareness not only of the fact that Aristotle and his school were frequently capable of misinterpreting the Presocratics at a very fundamental level, but also of the fact that he and his followers systematically used deliberate misunderstanding and 'shameless' misrepresentation as a way of silencing their predecessors. In other words, Aristotle and Theophrastus turn out to be far from infallible guides to the interpretation of the Presocratics; and the further back we move up the streams of 'ancient tradition' the more directly we come

---

[5] Efforts at tracing the history of 'doxography'—the now-technical term for the writing up and discussing of the views of early Greek philosophers—back even further in time than Theophrastus and Aristotle are still very tentative: cf. e.g. Mansfeld (1990), 22–83. On Aristotle's predominant role in determining the nature of the doxographical tradition as known to us see Kingsley (1994a) with the refs. in n. 46.

face to face with the forces of prejudice, bias, and downright ill will.[6]

Desperate situations often give rise to desperate solutions. The most recent trend has been to turn necessity into a virtue: to glorify the subjectivity of our ancient sources by making it not just something worth studying in its own right, but the only possible object for study. To quote a recent writer,

> It is not words which are of interest but interpretations. . . . There is no way that one can cut through the layers to some 'objective truth' about the meaning of the 'fragments'. . . . When it comes to the Presocratics, scholars have no justification for asserting that what [any ancient authority] saw in the text was not there or was incorrect as a reading of that text. . . . No subsequent reading based on the words they quote can have greater validity than their own readings.[7]

In a number of ways this 'contextual' approach to the Presocratics represents the logical and ultimate conclusion to the course of interpretation pursued during the past two hundred years: so much importance is attached to the contextual statements of Aristotle, Theophrastus, and later reporters that eventually Empedocles—or any other of the Presocratics—fades from the picture altogether. Yet what this approach misses entirely is the fact that the context provided by ancient writers who quote fragments of the Presocratics is not the only context in which those fragments can, and must, be understood. On the contrary, they need also to be situated in their own historical context; and the reason why this has only been done to such a minimal extent so far is because scholars have by and large remained content to read and understand the Presocratics through the eyes of Aristotle and later writers. As soon as we start to anchor the fragments of the Presocratics historically we stop being the slaves of those writers and, instead, can begin to use their reports and interpretations by appreciating them at their real value.

As we will see, there are a number of tools that allow us to perform this task. The first is philology. Put very simply, words

---

[6] Cf. Cherniss (1935) and the further refs. in Kingsley (1994a), n. 2; (1994b), esp. nn. 17–19; (1995a), § IV. 'Shameless': Guthrie, ii. 160. As we will see later (Ch. 24), this perception of the extent to which Aristotle and his school misrepresented the Presocratics is in fact nothing new.

[7] Osborne 22–3.

often change their meaning with time. If a later writer quotes a fragment from Empedocles and understands a word or expression in it in a sense it could not possibly have had in Empedocles' time, we can dismiss the quoter's interpretation as incorrect and unhistorical. On the other hand, examination of literary evidence more or less contemporary with Empedocles can often allow us to tell what he himself will have meant by the word or expression and how he expected his audience to understand it; it will show us this thanks to, but also in spite of, the person quoting and interpreting the fragment. Matters are made even easier by the fact that Empedocles deliberately chose to write not in prose but in the tradition of epic poetry. There has been a growing sensitivity to this aspect of his work in recent years—unfortunately sometimes accompanied by an over-reliance on modernist theories of poetry and a lack of basic common sense.[8] But all in all the fact remains that Empedocles, like other Presocratics, is still treated to a great extent as a 'philosopher' living in a mental world of his own rather than as an individual concerned to communicate his ideas within the framework and limitations of the language of his time. The phenomenon of modern scholars interpreting Presocratic texts in the most anachronistic of ways by granting them what could be described as 'diplomatic immunity' from the most fundamental laws of linguistic criticism is one that we will keep encountering, and sooner rather than later.[9]

But philology is by no means the only tool at our disposal. There is a vast amount of historical, geographical, mythological, and religious evidence still surviving in one form or another which together establishes a network of points of orientation that help us to understand Empedocles, his predecessors, his contemporaries, and his successors in a way that neither Aristotle nor Theophrastus was able or willing to help us understand them. The ground covered in this book will allow several extended glimpses of such a new perspective: glimpses which should be enough to show how much there is to be learned from it not only about Empedocles himself, but

---

[8] See below, Ch. 4, and D. W. Graham, *CQ* 38 (1988), 303, 312.
[9] Below, Chs. 2, 3; and note also J. Defradas's remarks, *REG* 86 (1973), 216.

also about early Pythagorean tradition and about the background to Plato's myths. The perspective in question grants access to a strange underground world in which what from the surface appeared separate and isolated phenomena prove to be profoundly linked, where what has been assumed to be random and arbitrary turns out to be just a part of a much larger picture. Unsolved problems solve each other; unanswered questions answer themselves; and the claim that insufficient evidence survives to allow us to understand aspects of Empedocles' teachings, or of early Pythagorean tradition, or of the origins of Plato's myths, meets with the response that the evidence was always there but has simply not been looked for in the right direction.

The present work is intended as a basic introduction to certain teachings and traditions in antiquity: a laying of the ground for future work on those teachings themselves and on their wider affiliations in the ancient and post-classical world. It falls into three sections. The first involves the sometimes laborious task of breaking through the strata of misunderstandings that have accumulated in over two thousand years: misunderstandings which, in the case of Empedocles, have effectively blocked any real approach to his work by forcing us to view it in the light of later rational philosophy with its very particular concerns. The second takes us down into the strange, labyrinthine underworld of mythology and mystery mentioned earlier, while the third opens into an even stranger world of magic and the 'irrational'. The second of these sections shows how urgently the now-traditional ways of interpreting Empedocles need to be supplemented and deepened; the third shows why these traditional ways of interpretation need, ultimately, to be left behind completely.

As the starting-point for this study it was obviously appropriate to start where Empedocles started: with his four elements, or 'roots' as he called them, which he evidently introduced right at the opening of his cosmological poem by equating each of them with a divinity, and which—as his own term for them shows—were fundamental to his system. The apparently simple procedure of beginning where Empedocles began, and then allowing his concerns—rather than our own—to determine the course pursued, may seem unexceptional; but it is a procedure worth emphasizing, and one that will require further comment later. By way of contrast, it is difficult to resist the impression that the major controversies which have engulfed Empedoclean scholarship over the past fifty years or so are little more than storms in a teacup. They are controversies over questions which have tended to be imposed on the Empedoclean material rather than evolving naturally out of it; and, what is more, these questions turn out to be rather easy to answer when approached calmly and with a constant eye on what survives of Empedocles' own poetry. For instance, there is the now-famous issue of whether or not he spoke of a cosmic cycle in which the four elements are alternately blended with each other by the force of love and separated out from each other by the force of hate or strife. The issue is only peripheral to the ground covered in this book, and in no way affects the conclusions drawn. However it is worth noting that the line, already adopted for reasons of their own by certain Neoplatonists in antiquity, of denying that he subjected the universe to an alternating cycle of unification and separation is contradicted at the simplest and most fundamental level by Empedocles' own words.[10] The rules of Greek language and grammar do often allow us to judge whether an ancient interpretation of Empedocles—or an ancient interpretation of him revived in modern times—is correct or incorrect.

Then there is the question whether we have fragments from just one poem by Empedocles, or from two: one of them cosmological, addressed to his disciple Pausanias, and the other—the *Purifications*, focusing on the theme of pollution and purification—addressed to a wider public in his home town of

[10] To give just two examples: the idea—so Bollack, i. 19, 118, ii. 16, iii. 51; C. H. Kahn, *Gnomon*, 41 (1969), 443—that Emp. B17.1-2 represents mere vicissitudes of 'The One', untouched by the worldly forces of Love and Strife, is contradicted by the obvious point that ἕν cannot be the subject of ηὐξήθη and διέφυ but is simply the resultant of the first of these verbs. Only this way is the pointless statement that one becomes one 'out of one' avoided; and only this way is the clear structural parallelism between ἓν ἐκ πλεόνων and πλέον' ἐξ ἑνός preserved. As for the idea that Emp. B17.1-2 describes temporal interchange between a transcendent One and the world (Bollack and Kahn, locc. citt.), it is contradicted by Empedocles' explicit statement—immediately following—that cyclical interchange within the world is altogether without end or beginning in time (B17.6-8, 12-13). On Empedocles' cosmic cycle see now D. W. Graham, *CQ* 38 (1988), 297-312; Inwood 40, 42.

Acragas. Once again, the issue in no way affects the conclusions drawn in the main body of this book; and if anyone chooses to reinterpret the few references to two poems of Empedocles as references to different parts of one and the same poem, he or she is free to do so. However, it will become clear in due course (Ch. 23) just why we are bound to conclude on the basis of the fragments of Empedocles' own poetry that what we have are the remains of two separate poems. Here it is worth adding that to have been able to assume we only have the remains of one poem would in a sense have made my task much easier: in agreement with those who would like to abolish any distinction between Empedocles' poems, one of the main conclusions of this book will be that the insistence during the past century and more on segregating Empedocles' 'philosophical' or 'scientific' interests from his religious or mythological concerns is not only inappropriate but also completely untenable. We need now to approach him, and his peers, through and beyond these anachronistic dichotomies.

The question of modern attitudes to science and philosophy raises another issue which deserves some brief comment because it relates to a topic that will crop up more than once during the course of this book. In 1908 Clara Millerd published her study *On the Interpretation of Empedocles*. In it she described Empedocles' philosophy as belonging to the 'first bungling attempts' to understand the universe: to a stage of 'tentative blundering' in which man was 'almost ready'—but, intellectual infant that he was, still unable—to arrive at the fundamental distinction between mind and matter.[11] Millerd herself can be excused for not being aware that, already for a few years before the publication of her work, the slumbering giant of quantum physics had been starting to come alive. But there is no excusing classical scholars today for ignoring the developments in scientific theory and practice throughout the twentieth century, and continuing to pursue their own specialized interests apparently unaware of the fact that many people at the forefront of contemporary science are no longer able to accept that distinguishing between mind and matter represents

[11] Millerd 1–2, 79–81. For similar kinds of statement in more recent writers cf. e.g. Guthrie, ii. 229–30, 239.

a genuine approach to reality—let alone an achievement—or that the basic Aristotelian dictum of the 'excluded middle' (that something either is x or is not x, but cannot be both simultaneously) necessarily holds good. For anyone accustomed to the world of the Presocratics and also to the world of modern science and cosmology, it is difficult not to notice how the second of these realms appears to be moving closer and closer to the first with its increasing appeal to bold paradox, to the simple but also the enigmatic and—dare one say it—the mythological. What is more, there are modern scientists who are consciously aware of this movement 'back' to the Presocratics, and not ashamed to admit it.[12] In this situation it should be apparent that scholars concerned with understanding and interpreting the Presocratics are not just in charge of a few museum pieces which can be played around with or locked away as historical curiosities. They are entrusted with much more, and one obvious aspect of this obligation involves treating Empedocles or early Pythagoreans with a greater respect by making a more serious attempt to understand them on their own terms.

Any attempt at understanding Empedocles or Pythagoreans in such a way cannot possibly ignore 'foreign' material just because it happens to be preserved in a language other than Greek or Latin. Myths on the subject of a single, self-enclosed Greek world still abound in the scholarly imagination, in spite of the fact that the concept of 'Panhellenism', or a united Greece, was never much more than a short-lived dream in the minds of a few ambitious Athenians. To keep a grip on the realities of the situation it is important, for example, to remember that the famous Persian Wars of the early fifth century BC were as much a battle between Greeks and Greeks as between Greeks and 'barbarians'; that during those wars—as both before and after—there were many individual Greeks, not to mention entire colonies or states, who opted for what they considered greater cultural as well as political and economic gain by allying themselves with the Persians; and finally that the Greeks of Sicily—so profoundly ambiguous in their attitudes to

[12] See e.g. the popular comments by F. Capra, *The Tao of Physics*[2] (London, 1983), 24–6, on the Presocratics and on modern physics returning 'in a way to its beginnings'.

10   *Introduction*

the Hellenic world, and so closely linked with other nations in both Africa and Asia—show up probably better than anyone else the absurdity of 'the superstition that things Greek and things Latin have some special common nature by virtue of which they ought to be kept apart from all other things'.[13]

However, in this book the issue of links between Greek and oriental traditions has only been touched on occasionally, and only when immediately relevant to the question at hand. A mere accumulation of evidence would distract from more important issues and, besides, would be of little value without a detailed discussion of principles and methodology: a subject demanding another book on its own.[14] As for the transmission of Empedoclean or Pythagorean ideas to Egypt and the Arab world, some of the most significant steps in this process have been noted where appropriate. As a result it should be possible, among other things, to appreciate the full significance of the fact that the recently discovered papyrus evidence for the survival, study, and, it would seem, the religious use of Empedocles' poetry—probably dating from the late first or early second century and from Akhmīm in southern Egypt—should have originated precisely in Akhmīm.[15] Altogether, the reader will hopefully discern the outlines of a broad pattern emerging: a broadly cyclical pattern, with elements of Empedoclean and Pythagorean doctrine deriving from Near Eastern sources and, in the course of the centuries, making their way back to the East. The following story is itself only a fragment of a much larger one.

[13] For the first two points see Momigliano's succinct comments in A. Ryan (ed.), *The Idea of Freedom: Essays in Honour of Isaiah Berlin* (Oxford, 1979), 140; for the quotation, Freeman, i, p. xiv, with his further comments, i, p. viii; ii, pp. 76–7 (Acragas), 167–71.

[14] For comments on the complexities of the issues involved see Kingsley (1992), (1993*b*), (1994*c*) with n. 19. For earlier periods of contact between the Greek world and the ancient Near East, Burkert (1983) and (1992) are fundamental.

[15] My special thanks to Prof. Alain Martin for discussions about the nature and circumstances of the discovery, and for showing me his provisional restorations of the papyrus text (due to be published in 1996 in Strasbourg). See also below, Ch. 23 nn. 17–18.

I
PHILOSOPHY
*

1
Back to the Roots

*

ACADEMICALLY, doubt is a virtue. It is wise to be cautious, virtuous to allow for different points of view. The problem arises when this attitude hardens: then doubting becomes a certainty in itself, and we forget the importance of doubting our doubt. To hesitate rather than pass premature judgement is one thing; it is quite another thing when scholarship flounders in disagreement over questions that could easily be settled once and for all on the basis of the evidence available. One example of this situation is the various opinions that have been put forward regarding the nature and divine identity of Empedocles' four elements. They are an accurate indication of how little progress has been made in understanding him over the last century.

In the history of western ideas Empedocles is, as far as we know, the first person who specifically reduced all of existence to four fundamental elements. He introduces these elements in fragment 6 of Diels's collection of his poetry by describing them as 'roots' and identifying each of them with a divinity.

> Hear first the four roots of all things:
> Dazzling Zeus, life-bearing Hera, Aidoneus, and
> Nestis who moistens the springs of mortals with her tears.[1]

The obvious question is which of these divinities matches which of the four elements that Empedocles will repeatedly refer to later: earth, air, fire, and water. Nestis plainly corresponds to the element of water. Aidoneus is an alternative poetic name for the god of the underworld, Hades.[2] That is as

---
[1] DK 31 B6. Bollack places the fragment much too late in his reconstruction of the poem, in spite of the obvious implication of the first line (ii. 65; the πρῶτα of B62.4 which he compares at iii. 171 occurs in a restricted context and is irrelevant). Cf. Flashar 547.
[2] *Il.* 5.190, 20.61; Hes. *Th.* 913; *Dem.* 2; cf. Ciaceri 193 and n. 2.

far as the agreement goes. For two thousand years the issue has been disputed as to which elements Zeus, Hera, and Aidoneus correspond to. In antiquity there were commentators who claimed that for Empedocles Zeus meant fire, Hera earth, and Hades—or Aidoneus—air. This view goes back, as I have shown elsewhere, to Aristotle's pupil Theophrastus. And there were those who also assumed that Zeus represents fire, but said that Hera must be air and Hades earth. This view arose specifically through the subsequent influence of Stoicism on later interpreters.[3] Finally, in the nineteenth century another view emerged: Zeus is air, Hera earth, and Hades fire.[4]

Writers of the most recent full-scale studies of Empedocles are agreed in favouring the tradition that equates Zeus with fire, Hera with air, and Hades with earth.[5] Modern proponents of an ancient tradition, they are quick to argue that its age is an indication of its reliability—even though there was more than one tradition in antiquity and both cannot be right. As for the nineteenth-century alternative, it is dismissed as a superfluous innovation.[6] Amid such a show of confidence there is clearly some value in checking if our assumptions are correct. The results will prove far-reaching.

[3] For full refs. to, and analysis of, the ancient sources see Kingsley (1994a).
[4] F. Knatz in *Schedae philologae Hermanno Usener . . . oblatae* (Bonn, 1891), 2, followed by Burnet (1892), 242–3 and (1930), 229 n. 3; G. Thiele, *Hermes*, 32 (1897), 69; Bodrero 78–92.
[5] Guthrie, ii. 144–6; Bollack, iii. 169–74; Gallavotti 173–4; Wright 165–6. Cf. also R. K. Sprague, *CR* 22 (1972), 169 and J. Longrigg, *CR* 24 (1974), 173, but contrast Imbraguglia *et al.* i. 174. The issue is not addressed by Inwood.
[6] e.g. Bollack, iii. 170: 'Les modernes ont compliqué le problème . . .'.

2
Aither
*

THE ancient traditions about the identity of Empedocles' elements, plus the modern status of opinion, are all linked to each other by a network of interlocking assumptions. Only by unravelling these presuppositions, one by one, will we be able to get to the reality underlying them.

In later antiquity it was generally taken for granted that when Empedocles spoke of *aither* he meant the element of fire.[1] Recently it has been common to assume instead that for Empedocles *aither* was not an element at all, but a secondary combination of fire plus air.[2] Both these assumptions are tributaries of an even larger and more enduring one. This is that Empedocles' basic name for the fourth element—in addition to fire, water, and earth—was *aer*: the origin of our word 'air'.

On the one hand, this major assumption made it inevitable that sooner or later Empedocles' Hera would be equated with air. The equation of Hera with *aer* may well already have been suggested before Empedocles' time; it crops up a century after him in Plato's *Cratylus* as one possible explanation for Hera's name, and it was later to have a flurry of success among the Stoics. Greeks—like any other ancient people—loved playing with words, and with the passing of time a philosophical significance was bound to become attached to the fact that, as spelt in Greek, *Hera* is an anagram of *aer*.[3] On the other hand,

---

[1] So e.g. Achilles, *Isagoge* 3, 31.15 Maass; Stob. i. 121.16 (DK i. 289.20–1), ps.-Plut. *Placita* 1.3.10 (DK i. 289.17). On these listings of the four elements—earth, water, *aer*, *aither* (= fire)—see below, n. 9.

[2] Cf. e.g. O'Brien 287–92 (criticized by Wright 197), with refs. to earlier literature; Bollack, iii. 172, 221.

[3] Below, Ch. 3 with n. 9; Pl. *Crat.* 404c; Kingsley (1994*a*), with nn. 20, 48–9, 61. The similarity in sound between the two words was obvious grist for the poetic mill, and we already find a play on it in Homer (*Il.* 21.6, ἠέρα δ' Ἥρη), where *aer* means mist or a

equation of Empedocles' air with Hades was equally inevitable. There was an obvious precedent for identifying them in the Homeric description of Hades' realm as '*aer*-like darkness' (ζόφος ἠερόεις): an expression which originally meant 'misty darkness' but eventually—as *aer* came to mean air instead of mist—would be explained as meaning 'dark realm of air'.[4]

On either interpretation the basic assumption is that Empedocles called his fourth element *aer*. Ironically, the last person who seems to have been consulted on the matter is Empedocles himself. In what survives of his work he explicitly lists the four elements several times. Once he refers to the fourth element—apart from fire, water, and earth—as 'heaven' (*ouranos*). But every other time the fourth element is not *aer*; it is *aither*. The classic example is the statement that 'we see earth with earth, water with water, *aither* with *aither*, and fire with fire'.[5] So much for the claim—made in antiquity, still repeated today—that *aer* was Empedocles' name for the element of air; and so much for the claims that for Empedocles *aither* was either the element of fire or a secondary combination of fire plus air.

How did it come about that Empedocles has been so radically misunderstood? The answer is simple. In the earliest surviving Greek literature, and in poetic tradition down to the fourth century BC, *aither* was the basic term for what we now call air.[6] *Aer*, on the other hand, was originally only a very isolated example of air: obscure mist or cloud. Gradually its meaning broadened from mist or vapour to describing the vaporous atmosphere we breathe,[7] and so to denoting air or atmosphere

---

cloud. But conversion of the joke into an element of doctrine was another matter. On Homeric word-play cf. M. Casevitz (ed.), *Études homériques* (Lyons, 1989), 55–8.

[4] *Il.* 15.191 (cf. e.g. 8.13, 12.239–40, 23.51) with Cumont, *CRAI* (1930), 106–7. See also Millerd 30; Kahn 152.

[5] B22.2; B109.1–2. Cf. esp. B71.2, 98.1–2, 115.9–11; also B54 with Aristotle's introductory comment, *GC* 334ᵃ3–5; B37 with Aristotle's πυρὶ τὸ πῦρ and Lucr. 2.1114–15. Lucretius' uncharacteristic distinction here between fire and *aither* is a clear sign that he is following Empedocles closely: see C. Bailey's commentary ad loc. and below, Ch. 3 n. 31. Cf. also Emp. B100.5, 7, 18, 24 and, for B17.18, Ch. 3.

[6] For Homer see Kahn 141–3. *Aither* = air in 5th-c. poetry: Gilbert 33 n. 2; Kahn 146–7; West (1982), 5.

[7] This development is particularly clear in the 5th-c. Hippocratic *Airs, Waters, Places*: the *aer* of the title oscillates exclusively between the senses of mist and of a damp kind of atmosphere responsible for making people's voices deep or hoarse (6.5, 15.5; cf. 6.2, 8.5

in general. As the scope of the term *aer* increased, the scope of the word *aither* decreased; by the early fourth century BC it was only used to refer to the highest and most exalted region of air, up in the heavens. In other words instead of *aer* being a particular example of *aither*, *aither* had become a particular example of *aer*.[8] It was then, once the word had become dissociated in people's minds from air in general, that the inevitable happened: a phase of self-conscious speculation began about what substance *aither* consisted of. In terms of sheer influence on general opinion in later antiquity, the most important result of this speculation was the view of the Stoics that *aither* is a form of fire.[9]

It was bound to happen that later Greeks, lacking what we call critical insight, would read their own understanding of *aither* back into the earlier poets and philosophers; we must be careful not to repeat the same mistake. The damage has more or less been repaired in the case of the poets;[10] but the same cannot be said for the philosophers. This is partly a result of lumping together things that, even though intimately related, are not the same. There is an obviously close connection between the region of the heavens and the heat of the sun (which was generally assumed in antiquity to be made of fire); however, when in early cosmology the relationship between sun and *aither* was brought out into the open it was one of correlation, not identity.[11] The delay in getting the record

---

bis, 15.4, 19.4–5). On the early meaning of *aer* in Homer and Hesiod see P. Louis, *Revue de philologie*, 74 (1948), 63–71; Kahn 143–6.

[8] Cf. esp. Pl. *Tim.* 58d: the purest form of *aer* is *aither* (ἀέρος τὸ μὲν εὐαγέστατον αἰθήρ). Note the sharp distinction drawn here between the element of air and the element of fire, which Plato has just mentioned (58c): *aither* is a form of air, not fire.

[9] For basic refs. and comments see Lewy (1978), 430 nn. 107–8. On the equation of *aither* with fire cf. also *SVF* ii. 143.40–1, 180.10–11, 185.12–14; Philo, *De confusione linguarum* 156; M. Lapidge, *Phronesis*, 18 (1973), 254–9. A very common way of listing the four elements in later antiquity was: earth, water, *aer*, *aither* (e.g. Philo, *De praemiis et poenis* 36 γῆν καὶ ὕδωρ καὶ ἀέρα καὶ αἰθέρα). However we still see cases, even in texts heavily indebted to Stoic terminology, where the elements are presented as a quaternity with *aither* retaining its original equivalence to air (*CH* 1.17: earth, water, fire, *aither*). On the development of the idea of *aither* as a fifth element see Jaeger's comments, 143–8, 300–1; P. Moraux, *RE* xxiv. 1172, 1177–96.

[10] Above, n. 6. Gilbert's insistence (20) at the start of the century on reading the *aither* = fire equation back into Homer is now a historical, but an instructive, curiosity.

[11] So e.g. Empedocles himself, B21.3–4; also B98.2, where the traditional brightness of *aither* is again explicitly distinguished from the element of fire. On the etymology of the word *aither* see Kahn 141 and n. 2.

straight where the philosophers are concerned is also a result of the way that Greek philosophy has come to be treated academically as a separate discipline from Greek literature, and as somehow subject to different principles and rules. Desire to view western philosophy as a continuous tradition moving towards ever greater sophistication, self-consciousness, and understanding inevitably involves looking at early Greek philosophers through the eyes of later ones. Aristotle and the Stoics still determine the basic perspective, even though it is common knowledge how totally unhistorical they could be.[12] The result is chaos. For example, it continues to be assumed that Parmenides 'implied' the *aither*–fire equation in describing his principle of fire as 'aitherial'.[13] In fact, as the context shows, the point of the adjective is to define the fire as 'bright' in contrast to the darkness of night, and also to define fire as a 'celestial' principle—belonging on high in contrast to the principle of night which is weighty and belongs down below.[14] This in no way justifies the conclusion that for Parmenides *aither* consists of fire. On the contrary, when he does use the term *aither* his reference to it—alongside earth, sun, and moon—as 'common' is impossible to explain satisfactorily on the assumption that *aither* is fire.[15] However it makes immediate and obvious sense when, as in the case of all other sixth- and fifth-century writers, *aither* is accepted as referring to the 'upper and lower regions of the air ... evidently thought of as a continuum extending from the earth's surface to the stars or beyond'.[16]

---

[12] Aristotle could, for instance, not possibly have written *Cael.* 270ᵇ16–24 if he had any awareness of what *aither* and *aer* really meant in the centuries before him. On the Stoics see Burnet's succinct remarks, (1930), 32.

[13] DK 28 B8.56 (αἰθέριον πῦρ); Kahn 148. That Parmenides equated *aither* and fire is assumed without question by e.g. Diels, *Parmenides: Lehrgedicht* (Berlin, 1897), 103; Tarán, *Parmenides* (Princeton, NJ, 1965), 238.

[14] B8.58–9. Parmenides' 'aitherial gates' (B1.13, αἰθέριαι πύλαι) are gates reaching up to heaven, not gates that are made of *aither*. See Burkert's comments, (1969), 11–12, and also Eur. *Medea* 440 as well as the use of οὐράνιος in the same sense: LSJ s.v. II; West (1982), 5. On the 'brightness' of fire see also Tarán, op. cit. 231 n. 1.

[15] αἰθὴρ ξυνός, B11.2. Diels's notion (op. cit. 103) that *aither* here is the fiery substance of the heavenly bodies cannot be right: it is clearly being cited as something *in addition to* everything else named.

[16] West (1982), 5 (add DK 13 B2, 64 C4). In Stobaeus' sketch of Parmenides' cosmology, *aither* is specifically distinguished from fire (i. 195.16–17 = DK i. 224.13). On Anaxagoras' presumed equation of *aither* with fire see Kingsley (1995a), §IV. As for Heraclitus, to claim that he equated 'the purest fire' with *aither* (Kirk 316–17; Guthrie,

*

For Empedocles, as we have seen, *aither* represents the element of air. We have also seen that this is just what is to be expected of anyone writing in the mid-fifth century—let alone of someone as rooted in poetic tradition as Empedocles. But this is not all. Significantly, his commentators and interpreters often adopted Empedocles' own terminology and referred to his element of air as *aither* even though this had become an oddity in the popular usage of their own times. For instance, a passage in the *Placita* attributed to Plutarch reports that according to Empedocles

aither was the first element to be separated off. The second was fire, and then came earth. Next, out of the earth water gushed out . . .

And it goes on to add:

heaven was created out of *aither*; the sun was created out of fire.

This statement about heaven being created from *aither* agrees exactly with the point noted earlier, that Empedocles himself used the term 'heaven' as an alternative way of referring to the element of *aither*.[17]

But noting how closely the commentators follow Empedocles provides only one half of the picture. It is just as important to watch how easily they slip away from him—for example

i. 471, 480) is not only an anachronism but also gratuitous: the word *aither* does not even occur in the surviving fragments of Heraclitus.

[17] Ps.-Plut. *Placita* 2.6.3 = Emp. A49b; B22.1. This rules out changing the text of ps.-Plutarch so as to make it read 'heaven was created out of *aither* ⟨and fire⟩' (Bignone 342 n. 1), as does the fact that the alteration destroys the obvious parallelism between heaven/*aither* and sun/fire. *Placita* 2.11.2 and 2.13.2 = Stob. i. 200.17–19, 202.2–4 (A51a, A53) in no way justify the alteration: the first passage simply points out that the celestial vault consists of air (i.e. *aither*) crystallized *by* fire, while the second describes the stars as points of fire trapped in the vast expanse of heaven which are negligible as far as the substance of heaven itself is concerned. On these matters see Kingsley (1994*b*). Regarding the virtual interchangeability of the terms *aither* and 'heaven' in early epic tradition see Kahn 141–3; in tragedy and inscriptions, Dieterich 103–8, West (1982), 5. Stobaeus' report (i. 195.16–18 = DK i. 224.13–14) that Parmenides defined 'heaven' or *ouranos* as the region below *aither* and distinct from it must be wrong: Parmenides' own words plainly describe the heaven as all-encompassing (B10.5). For the similar lowering of *ouranos* in Stob. i. 196.25–197.1 = DK i. 403.19–23 (Philolaus) see Burkert 244–6; Huffman 398–400.

Chapter 2

repeating his terminology to begin with, but then sliding into using the term for air which by their time had become the normal one: *aer*.

Empedocles of Acragas had four elements: fire, water, *aither*, earth; their ruling cause is Love and Strife. He said the *aer* was separated off from the primal mixture of the elements and spread round in a circle. Next, after the *aer* fire escaped...[18]

Here in the so-called *Stromateis* the original term *aither*—faithfully retained in the passage from the *Placita*—has for very natural reasons started to erode away. We also see a similar process in Aristotle. Sometimes, when discussing a particular Empedoclean passage he conforms to Empedocles' terminology and uses the word *aither*. At other times, even when referring to the identical passage he uses the more colloquial *aer* instead.[19] It is entirely understandable that, in most later reports about Empedocles, *aither* has simply been replaced by *aer*.[20]

There is one final passage which provides valuable confirmation of Empedocles' own use of the term *aither*—although, ironically, it has repeatedly been misinterpreted[21] as showing that his *aither* was not an element but just a secondary combination of fire and air. The passage is a vivid testimony to the importance for classical scholarship of texts preserved in languages other than Greek or Latin, and an equally vivid testimony to the dangers of relying on inaccurate translations.

Philo's *On Providence* contains a passage on Empedocles' cosmology which has only come down to us in Armenian.[22]

---

[18] Ps.-Plut. *Stromateis* 10 (DK i. 288.22–4).

[19] *GC* 334ᵃ1–5, *Ph.* 196ᵇ20–3, reproduced at B53.

[20] Cf. e.g. ps.-Plut. *Placita* 1.3.10 (DK i. 289.15); Simpl. *Cael.* 528.21–4. With the *Placita* passage contrast ps.-Plut. *Stromateis* 10 (DK i. 288.22). With the Simplicius passage contrast *Placita* 2.6.3 (DK i. 292.27–8) and Empedocles himself, B38.4. Note also Simplicius' significant statement (*Ph.* 32.3–4 = DK i. 346.13–15) that Empedocles 'calls fire Hephaestus, sun, and flame; water he calls rain; and *aer* he calls *aither*'. In other words, for Simplicius *aither* is no more than a poetic alternative for the philosophically more accurate *aer*. Compare the less generous scholiast who, going one step further, takes Homer to task for mentioning *aither* when he should have used the term *aer* (οὐ γὰρ αἱ νεφέλαι ἐν τῶι αἰθέρι): Lehrs 168.

[21] Bollack, iii. 221; O'Brien 287–91.

[22] Not, as Guthrie says, in Latin (ii. 163 n. 2). For the historical background to the Armenian translation see the refs. in Kingsley (1993*a*), n. 1.

This work of Philo was first edited, and provided with a Latin translation, by Awgerean in 1822. Conybeare went back to the Armenian text at the end of the nineteenth century, and came up with a number of alternative renderings for certain expressions and phrases. It was Awgerean's Latin version, modified by Conybeare's suggestions, which formed the basis for the excerpts from the work given in Diels's editions of the Presocratics.[23] A fresh translation of the passage concerning Empedocles did not appear until 1969, when Bollack published a French version he had obtained from Charles Mercier. Work done on the text since then has been negligible in both quantity and quality.[24]

The second book of *On Providence*—the one we are concerned with—consists of a dialogue between Philo and a certain Alexander. Philo is Philo: fundamentally faithful to Judaism. Alexander is a bright spark fuelled with clever refutations of God's providence, and after a lively debate he turns to arguing that the present arrangement of the universe can be explained without any recourse to a transcendental god. The four constituent elements of the universe, he explains, are arranged concentrically, with earth at the centre surrounded by water which is surrounded by air—which in turn is surrounded by *aither*.[25] But this arrangement, Alexander claims, has nothing to do with God's providence; it is simply a physical law. Put oil, water, and sand in a bucket of water—he argues—and the sand will sink to the bottom, the oil will rise to the top and the water will stay in the middle. Armed with this analogy he then returns to his main point:

---

[23] The *editio princeps* is *Philonis Iudaei sermones tres hactenus inediti*, ed. and trans. Jean-Baptiste Aucher, i.e. Mkrtičʻ Awgerean (Venice, 1822). For Conybeare's suggestions see P. Wendland, *Philos Schrift über die Vorsehung* (Berlin, 1892), 64–5; Millerd 63 n. 7. The passages in Diels relating to Empedocles are Emp. A49a (Armenian text in Awgerean 86.15–36) and A66a (86.37–87.4). For the continuation of this second passage (87.4–16) see Ch. 3.

[24] Bollack, ii. 84–5 (text and translation); iii. 220–4 (commentary by Bollack). Regarding the so-called new edition and translation of *On Providence* by M. Hadas-Lebel (Paris, 1973), and the translation of the Empedocles passage by Abraham Terian which is published in Inwood, see Kingsley (1993a), n. 5.

[25] Armenian *arpʻi*: see ibid., n. 7. On this listing of the elements—earth, water, air, *aither*—see above, n. 9.

Well[26] the parts of the universe also seem to have been carried[27] [i.e. to their present locations] in the same way—just as Empedocles says.[28]

It is with the start of the next sentence that the real problems have begun. Awgerean's and Mercier's versions are basically identical: 'For as, separating off from the *aither*, the wind and fire flew away, . . .'. On the other hand Conybeare, followed by Diels, gave the translation 'For after the *aither* was separated off, the air and fire flew upwards'. These renderings are for a number of reasons—both grammatical and textual—impossible. As I have shown elsewhere,[29] the Armenian in fact yields quite a different sense:

For as the *aither* was separated off, it was raised upwards by the wind (*hotm*) and fire.

What are we to make of the reference here to 'wind and fire'? The answer is not hard to find. O'Brien has carefully assembled the evidence indicating that Empedocles described a massive storm of the elements which occurred when they were being separated out at the start of our world. Some of these passages refer specifically to 'fire and storms of wind'.[30] The Philo text precisely corroborates these passages, while at the same time helping to fill out the details. It is a tribute to the ease and consistency of Empedocles' vision that he evidently described these gusts of fire and wind not just as being a natural result of the chaotic initial separation, but also as helping to further that separation by blowing the purified *aither* up out of reach of the other elements. As for the winds, they are

---

[26] Bollack–Mercier translate 'Cependant', which obviously destroys the sense. Certainly *sakayn* is often simply adversative, but it can also be resumptive or emphatic in meaning. Compare Philo, *QG* 2.2.19 Paramelle, where *sakayn* together with *ew* (as here) translates καὶ μὲν δή.

[27] For Arm. *krem* = Gk. φέρω see Reynders, ii. 123, s.v. *fero*. Awgerean's translation (*affici videntur*) is wrong, as is *confici videntur* (Diels). Bollack–Mercier—'Cependant les parties aussi de ce monde-ci semblent supporter (φέρειν ou πάσχειν) quelque chose de semblable'—correctly identify the Greek verb but misunderstand it (as does Terian, 'bear the same characteristics'). The participial form *krel* can be either active or passive; here the passive is clearly required. Compare e.g. Arist. *GC* 334ª1–4 (regarding Empedocles); Philo, *De gigantibus* 22.

[28] *Prov.* 2.60, 86.15–17 Awgerean.

[29] Kingsley (1993a), 48–52.

[30] O'Brien 47–50, 268–72; Plut. *De esu carnium* 993e; Tzetz. *Alleg. Il.*, Prolegomena 291–9. For *hotm* = wind cf. Kingsley (1993a), 50 and n. 12.

clearly air as well—but air that is still agitated, still mixed up with other elements and not yet pure.[31] The agitation itself could be described as the action of something trying to become what essentially it is.

We are now in a position where we can look at the sentence of Philo as a whole, and at what follows:

As the *aither* was separated off, it was raised upwards by the wind and fire; and it was what it came to be—the broad, vast, encircling heaven. As for the fire, it remained a short distance inside the heaven; and it grew to become the rays of the sun.[32]

The precision of Philo's report is striking. So is its corroboration and elaboration of the statements by other commentators—not to mention its close relationship to the surviving fragments of Empedocles.[33] For our immediate purposes, some aspects of the report are particularly instructive. In explaining that heaven came into being out of *aither*, it agrees with the *Placita*. It also agrees with the *Placita* in presenting the creation of heaven out of *aither* and the creation of the sun out of fire as two formally parallel events. And it agrees with both the *Placita* and the *Stromateis* as well as with Empedocles himself in distinguishing sharply between *aither*—the first element to be separated out—and the element of fire.

---

[31] On wind in Empedocles as a phenomenon created by interreaction between the elements, see Olymp. *In Meteora* 102.1–3 Stüve = Emp. A64; Bollack, iii. 300–1. Cf. also Lucr. 1.759–62, 5.434–48, 6.366–78; Lucr. 5.495–505, with its contrast between the disorderly tumult of storms and the regular motion of the *aither* above, untouched by the turmoil below, is very probably a reminiscence of Empedocles—except that whereas Lucretius merges the *aither* and fire, for Empedocles the region of *aither* is above the region of fire. See Ch. 3 n. 31.

[32] *K'anzi ibrew mekneal arp'woyn t'ṙuc'eal ēr hotmoy ew hroy, ew ēr or etew: erkin layn mec i veray šurǰ pat aṙeal. isk hur sakaw mi i yerknē i nerk's mnac'eal, ew sa yaregakan čaṙagaytʻs ačec'aw* (86.18–23). For the formula *ew ēr or etew*, 'it was what it came to be', and the remainder of the first sentence see Kingsley (1993*a*), 52–4. Arm. *erkin* (used twice) translates Gk. οὐρανός: ibid. 53 n. 26.

[33] For *aither* in Empedocles as the outermost element see Ch. 3 n. 31. Philo's statement that the fire 'grew (*ačec'aw*) to become the rays of the sun' also derives from genuine Empedoclean material: cf. B41 with Macrobius ad loc., B48, and Kingsley (1994*b*).

3
Aer
*

IF *aither* was Empedocles' name for the element of air, what meaning—if any—did he give to the term *aer*? Since the time of Diels, a century ago, it has been traditional to claim that the word *aer* occurs five times in the surviving fragments. These need looking at one by one.

Line 18 of fragment 17 lists the four elements; but the manuscripts are equally divided between giving *aither* or *aer* as the name of the final element alongside fire, water, and earth. Every editor since Diels has put *aeros* in the text and rejected *aitheros*. No reason is given, and no reason could possibly be given. This is a splendid example of the way that texts of the 'philosophers' continue to be granted a kind of diplomatic immunity from the laws of editorial and linguistic criticism. We have already seen how, as a matter of course, later writers substituted *aer* for Empedocles' own use of the term *aither* when discussing or repeating what he had said. Considering the changes in terminology since Empedocles' time, the substitution was inevitable; and from the point of view of those later writers it was totally legitimate.[1] As a matter of fact this kind of substitution was not restricted to paraphrases or commentaries: it affected the transmitted texts of early writers as well. Replacement of *aither* by *aer* is a well-attested phenomenon in manuscripts.[2] It even occurs in the text of Empedocles himself.[3] In the face of this evidence there is no

---

[1] Above, Ch. 2. Note especially Simplicius' comment as quoted in n. 20: he happens to be our main authority for the text of B17.18.

[2] For refs. and comments see West (1966), 351–2.

[3] So B109.2, for which Sextus Empiricus' text gives ἤέρι δ' ἤέρα at *Math.* 1.303 although elsewhere it provides the correct αἰθέρι δ' αἰθέρα. Compare also B6.2, where alongside the significant alternative reading Ζεὺς αἰθήρ (ps.-Plutarch and Tzetzes; see Diels 132) we also have the reading Ζεὺς ἀήρ in Hipp. *Ref.* 10.7.3.13. Hippolytus' ἀήρ is more probably a consequence of an earlier reading αἰθήρ (so Marcovich, ad loc.; cf.

justification whatever for retaining the reading *aeros* in fragment 17.

We come now to an undeniable instance of Empedocles using the word *aer*. In fragment 38 he turns to describing the creation of

> everything that we now see:
> earth and wavy sea and moist *aer*,
> Titan and *aither* binding everything in its circular grip.

Here, as so often with Empedocles, the influence of the two greatest nineteenth-century specialists in ancient philosophy—Zeller and Diels—has proved decisive in determining later interpretations of the text. Instead of 'Titan and *aither*' they, along with all subsequent translators, understood the Greek as meaning 'and Titan *aither*'; and Diels, assuming not only that Titan here alludes to the sun but also that *aither* means fire, explained the words in conjunction as referring to the sun.[4] This had the advantage of reducing what seemed a list of five elements—earth, water, *aer*, fire, *aither*—down to the correct number of four. But first, as we have seen, *aither* for Empedocles was not the element of fire. And secondly, Zeller's and Diels's analysis of the Greek is grammatically impossible. 'Titan' must be an item on its own distinct from the *aither*, and there is no reason to doubt that Empedocles intended it specifically as a reference to the fiery sun.[5]

That leaves us with the apparent problem of five elements instead of four. The solution to this problem becomes clear as soon as we look more closely at the middle term: *aer*. The fact that the adjective applied to it, *hygros*, can imply not only dampness but also suppleness and fluidity[6] is irrelevant: there

---

also 10.7.2.8, where Hippolytus gives ἀέρος, Sextus αἰθέρος) than a direct corruption of ἀργής as given in the text of Sextus. This in itself is a further argument against the theory of Hippolytus' immediate dependence, in *Ref.* 10.6.2.8–10.7.6.32, on the parallel passage of Sextus. On Hippolytus' source here and its relation to Sextus' text see Osborne 89–92, 94–5. Incidentally, these variants at B6.2 raise serious doubts about claims that identification of Empedocles' Zeus with air is only a modern invention (e.g. Bollack, iii. 170; Guthrie, ii. 145).

[4] Zeller, i/2. 979 n. 5 (Titan as 'Beiname des Aethers'); Diels (1901), 123.

[5] For the grammar of the passage (the crucial point being that the conjunction ἠδέ, in Τιτὰν ἠδ' αἰθήρ, is never postponed), and for 'Titan' as referring to the sun, see Kingsley (1995a), §1.

[6] G. E. R. Lloyd, *Polarity and Analogy* (Cambridge, 1966), 25 n. 4; Guthrie, ii.185 n. 1.

can be no denying that here it has the specific sense of wetness. 'Moist *aer*' obviously refers to the same kind of phenomenon so frequently described by Homer and Hesiod in the phrase 'winds blowing moist' (ὑγρὸν ἀέντες)—moist because pregnant with rain.[7] For Hesiod before Empedocles, just as for Plato and Aristotle after him, the word *aer* was itself inevitably associated to some degree with dampness and moisture.[8] If it is true that in the sixth century BC Theagenes of Rhegium explained Homer's Hera as meaning *aer*, he too is almost certain to have understood the term not in its later sense of an 'element' but—just as Homer did—in the specific sense of mist or cloud.[9] No less instructive is the Hippocratic *Airs, Waters, Places*, written approximately a generation after Empedocles. Every single time that the *aer* mentioned in the title is referred to in this work it means either mist or fog or, failing that, an atmosphere permeated with dampness—especially with the kind of moisture carried and left by rainy winds.[10] And in much later times the notorious moistness (ὑγρὰ οὐσία) of *aer* would become a favourite topic of discussion for allegorists and scholastics.[11] Here in Empedocles the 'moist *aer*' can, as the adjective itself indicates, hardly be an element in its own right—not to mention the fact that his four elements have already been accounted for. On the contrary, this way of describing the *aer*—plus its position intermediate between sea on the one hand and sun and *aither* on the other—strongly suggests that it is some kind of modification of water under the influence of the sun: a graphic but specific reference to the mist hovering above large bodies of water as a result of evaporating moisture. This is

---

[7] *Od.* 5.478, 19.440; Hes. *Th.* 869, *Op.* 618–26.

[8] *Op.* 547–57; Kahn 145–6 and 147 n. 2.

[9] Porph., DK 8 A2, explaining the confrontation between Artemis and Hera in *Il.* 20.70–1 as a conflict between moon and *aer* respectively. The interpretation of this same passage in the *Iliad* recorded by the allegorist Heraclitus (*Alleg.* 57.5–6) is particularly relevant to the question of how Theagenes is likely to have explained it: Artemis the moon is 'weakened and obscured by the atmospheric conditions (τῶν ἀερίων παθημάτων) ... by mist (ἀχλύι) and the clouds passing by below her'. As has often been noted, Porphyry's account is not explicit as to what goes back to Theagenes himself, and it obviously contains elements of Stoic elaboration: cf. e.g. F. Wehrli, *Zur Geschichte der allegorischen Deutung Homers* (Leipzig, 1928), 89–90. In determining what is what, it is essential to bear in mind the shift in meaning of *aer*—from mist in Theagenes' time to one of the four elements for the Stoics.

[10] 6.2 and 5, 8.5 *bis*, 15.4 and 5, 19.4 and 5; above, Ch. 2 n. 7.

[11] Buffière 109–10.

confirmed by the way that Empedocles' mention of sea (πόντος) and *aer* together so pointedly evokes the familiar Homeric image of the 'misty sea'.[12] We will come back later to the question of what mist was for Empedocles.

Next we can consider fragment 100, which survives only in a quotation by Aristotle. During its description of the girl playing with a clepsydra, four times the word *aither* is used in naming the substance of air (as opposed to water, or blood); once the word *aer* occurs.[13] An attempt has been made to explain this apparent inconsistency in terminology as due to the fact that where the word *aer* is used it refers to air inside the clepsydra, surrounded and—so it is claimed—affected by water. The explanation is forced, and contradicted by the text itself.[14] On the other hand, Burnet was almost certainly correct in his account of the manuscript reading: Aristotle simply 'made a slip' at this point in his quotation of the fragment and substituted *aeros* for the more poetic *aitheros*. Our earlier examination of the way that Aristotle tended to substitute the term *aer* for Empedocles' *aither* makes Burnet's explanation even more convincing; and, in fact, when introducing fragment 100 Aristotle glosses Empedocles' *aither* by *aer* no less than three times. We have already seen examples of how this kind of substitution produced corruptions in Empedoclean texts.[15] Add to this the further fact, also noted by Burnet, that the form *aeros* as given for the one occurrence of the word in the fragment cannot be correct (Empedocles would have used the epic form *ēeros*) and there are no possible grounds for retaining the manuscript reading. Stein emended it to *aitheros* in 1852; he was clearly correct.[16]

---

[12] ἠεροειδὴς πόντος: *Il.* 23.744 and often in the *Odyssey*; Hes. *Th.* 252, 873, *Op.* 620. See Kahn 145 n. 1 and, for the interpretation of *aer* in B38.3, Bignone 326 n. 4, Bollack, iii. 262–3 ('On voit . . . les vapeurs monter de la mer').

[13] *Aither*: 5, 7, 18, 24; *aer*: 13.

[14] O'Brien 291–2. His specific idea that the *aither* 'still has fire mixed with it' which—when the clepsydra is submerged—is 'lost to the surrounding water' so that 'we have only *aer* not *aither*' is certainly wrong, based on the false explanation of *aither* as a mixture of air and fire. But even to suppose that *aither* (pure air) becomes *aer* (air mixed with water) as a result of the immersion would be implausible: the air in 13 is hermetically sealed off from the surrounding water by the girl's pretty hand.

[15] Burnet (1930), 219 n. 3; Ch. 2 with n. 19 and Arist. *De respiratione* 473ᵇ3, 4, 6; above, n. 3.

[16] Stein 69. Although the mistake is very probably Aristotle's own, it would be wrong to exclude the alternative possibility of a later corruption in the MS tradition: note that

The next presumed example of a reference to *aer* is fragment 149. Plutarch, commenting on the pithiness of Empedocles' epithets, cites the case of him calling 'the *aer* "cloud-gatherer"' (νεφεληγερέτην τὸν ἀέρα).[17] If the noun, as well as the epithet, went back to Empedocles, we would have another example of him using the word *aer* with its contemporary connotations of mist, cloud, and rain: 'cloud-gatherer' is the adjective applied in Homer to Zeus as the bringer of rain.[18] However, this is highly unlikely. The quotation only includes the adjective; the term *aer* is plainly Plutarch's own.[19] The clouds themselves might have been *aer* for Empedocles, but what gathers them together is something else. It is far more probable that Plutarch's *aer* is a substitute for an earlier *aither*, in line with the common usage of his time and all the other examples already referred to.

The last possible example is the most quickly disposed of. The word *aer* appears persistently in modern editions of Empedocles at fragment 78. However, as I have noted elsewhere,[20] not only is the mention here of *aer* in any sense at all linguistically impossible and certainly wrong: it does not even have any manuscript authority, and is no more than an unfortunate emendation of a text which was correctly restored—and explained—by the elder Scaliger in the sixteenth century.

On no less than fifteen occasions in the surviving fragments of his poetry Empedocles uses the word *aither* in referring to the element of air.[21] On the other hand, it is not even remotely probable that in the surviving fragments he has used the term

---

both cases in the fragment of *aither* in the genitive or dative have suffered extensive corruption (5: γ' ἐνθεῖναι θέρει; 24: ἕτερον).

[17] *Qu. conv.* 683e = DK i. 370.12–16.
[18] *Il.* 1.511, 517, etc.; M. P. Nilsson, *Greek Popular Religion* (New York, 1940), 5–6.
[19] So, immediately before, ἀμφιβρότην χθόνα is Empedocles' but τὸ τῆι ψυχῆι περικείμενον σῶμα is Plutarch's.
[20] Kingsley (1995a), § 11.
[21] B9.1 (reading κατὰ φῶτα μιγὲν φῶς αἰθέρι ⟨κύρσηι⟩ with Burnet), 37, 38.4, 39.1, 54, 71.2, 98.2, 100.5, 7, 18, 24, 109.2, 111.8 (see Ch. 15 n. 2), 115.9–11, 135.2. In B9.1 the play on φώς–φῶς is clearly deliberate and must be preserved (cf. B45, Parm. B14); for Burnet's κύρσηι compare B53, 59.2, 98.1, 104 and, for αἰθέρι κύρσηι, Callimachus, *Hymn to Demeter* 37, Apollonius Rhodius 2.363. For Empedocles, φώς of course is fire (B84; Parm. B8.56–9, B9).

aer more than once (in fragment 38); and on that single occasion it refers not to the element of air but to local concentrations of damp and mist. The evidence speaks for itself. It refutes the claim that Empedocles used *aer* and *aither* interchangeably;[22] and—once the veil thrown over the issue by later misunderstandings and changes in terminology is drawn back—it shows him using the two terms in clearly separate ways.

Although the surviving fragments of Empedocles are plain enough evidence in themselves for the distinction he drew between *aither* and *aer*, we also happen to have been left with some very significant information in the secondary sources. The later commentators tend, as we have seen, to make no distinction between *aer* and *aither*, but the *Placita* of pseudo-Plutarch provides one notable exception.

Empedocles says *aither* was the first element to be separated off. The second was fire, and then came earth. Next, as the earth was constricted by the force of the rotation, water gushed up out of it; and out of the water as it turned into steam, *aer* was produced. Heaven was created out of *aither*, the sun was created out of fire, and the regions around the earth were created out of the others by compression.[23]

The fact that this report apparently gives a total of five elements, not four, has naturally caused consternation.[24] It is hardly a coincidence that we have already encountered the same anomaly in a text of Empedocles (fragment 38); the parallel is illuminating, and yet that must not prevent us from looking for an explanation of the anomaly in the *Placita* in the *Placita* passage itself. The key to understanding this passage lies in noting that the emergence of the last two 'elements'—water and *aer*—is presented in terms of a *transmutation* of the previous element: water comes into being 'out of' earth, and *aer* comes into being 'out of water'. This idea of an interchange of the elements was a philosophical commonplace since the days of

---

[22] So e.g. Sachs 43; Bollack, iii. 63; Wright 23, 169.
[23] Ps.-Plut. *Placita* 2.6.3 = Emp. A49b. For 'the regions around the earth' (τὰ περί-γεια) see Kingsley (1995a), § III.
[24] O'Brien 288, with refs.

Heraclitus.[25] However, it is just as alien to Empedocles—whose four elements were strictly distinct from each other and noninterchangeable—as is the later five-element theory.[26] The question, then, is to determine how this idea of a transmutation of the elements came to be read into Empedocles. As far as the transformation of earth into water is concerned, the answer is simple: Empedocles' description of sea being 'sweated' from the earth[27] was, to mention just one example, an obvious invitation to someone predisposed to read the transmutation theory into his philosophy. In the case of water being converted into *aer*, again it is clearly a matter of too much being read into a statement by Empedocles: *aer*, as we have seen, was not a separate element. As to what that statement was, pseudo-Plutarch's language provides a useful clue. The term he uses for 'turning into steam' (θυμιαθῆναι) is an unusual one. Already in antiquity there was a tendency among copyists of the *Placita* to replace its occurrence here with the verb normally used by later philosophical writers for describing the cosmic process of evaporation or exhalation (ἀναθυμιαθῆναι).[28] Almost certainly the verb in the *Placita* goes back to Empedocles: then it will have had the specific sense of something turning into steam or smoke or mist.[29] But that, as we have seen, is precisely what for Empedocles *aer* meant. And, as pseudo-Plutarch himself tells us, this *aer* is part of 'the regions around the earth'. In other words we are dealing not with an element at all but with a climatic condition of a purely secondary order, due to modification of the last of the four elements to emerge: water. As far as this subordinate function of *aer* in Empedocles is concerned, it is also worth noting how—in its later sense of 'climate'—*aer* was to become virtually synonymous with the word *krasis*: 'mixture' or 'blending'. That, of course, is the term

---

[25] Marcovich 278–90, 352–64. Note esp. Heraclit. B36 with D.L. 9.9 (γῆν ... ἐξ ἧς τὸ ὕδωρ γίνεσθαι) and Philo, *De aeternitate mundi* 110: (γῆ) τηκομένη εἰς ὕδωρ λαμβάνει τὴν μεταβολήν, τὸ δ' ὕδωρ ἐξατμιζόμενον εἰς ἀέρα.

[26] See Lucretius' comments, 1.763–802.

[27] Cf. B55, with n. 39 below.

[28] For the history of the alternative reading ἀναθυμιαθῆναι see Daiber 382. The terms ἀναθυμίασις and ἀναθυμιᾶν in Heraclit. B12 are not genuine: Marcovich 194, 206, 213 n. 1; Kirk 368–72. On the other hand, θυμιωμένων in Stob. i. 232.22 = DK i. 147.7 should not (*pace* Marcovich 331) be emended to ⟨ἀνα⟩θυμιωμένων.

[29] Compare Plato's use of the same term with the same specific meaning, *Tim.* 66d–e. See also Lucr. 5.463–4 (... *fumare* ...).

used by Empedocles to describe the subsequent interaction and combination of his four primal elements or 'roots'.[30]

In addition to pseudo-Plutarch we must also turn, once again, to Philo. As we saw earlier, in *On Providence* he presents us with the figure of Alexander: an adversary who substantiates his godless arguments by appealing to Empedocles. The resulting format is precisely what one would expect: Alexander cites particular aspects of Empedocles' philosophy, and then sums up their relevance in terms of his debate with Philo. So, he starts by presenting Empedocles' account of the initial separation of the elements; we have already seen how valuable and accurate this description is.[31] Then we have an interruption: Alexander scores his first explicit point against Philo by leaving Empedocles aside to draw his own conclusions in line with the key issues of the debate. A great deal of trouble has arisen from the failure to realize that this is no longer a summary of Empedocles, but Alexander's own statement of his objections to the theory of providence—expressed in terms derived from contemporary Hellenistic philosophy.[32] After the interruption, however, Alexander turns back once again to Empedocles— and to outlining the salient points in Empedocles' explanation of how the earth and sea came into being.[33] Then he breaks off for a second time to launch another direct attack on Philo, using Empedocles' views as ammunition. On the one hand, he claims, Empedocles has given a perfectly satisfactory account of the origin of the oceans in terms of purely physical causes; on the other, 'it is absurd to say that such vast amounts of undrinkable

---

[30] B21.14, 22.4 and 7, 35.8 and 15. Cf. e.g. Theophr. *CP* 1.13.2, W. J. den Dulk, *Κρᾶσις* (Leiden, 1934), 55–7. On the other hand, *aither* has no *krasis* (Eur. fr. 779.2).

[31] *Prov.* 2.60, 86.15–27; above, Ch. 2. A further indication of its accuracy is the glaring discrepancy between Alexander's own ordering of the elements at the beginning of *Prov.* 2.60 (85.36–86.1), which—working from outside inwards—is *aither* (i.e. fire), *aer*, water, earth, and the arrangement ascribed immediately afterwards to Empedocles, which is *aither* (i.e. air), fire, water, earth. This second arrangement is genuinely Empedoclean (for air or *aither* on the outside see esp. B38.4). The first is the one which later came to be generally accepted, was used by Philo himself, and eventually passed into Islamic cosmology: above, Ch. 2 nn. 1, 9; Pl. *Tim.* 39e-40a; Lucr. 5.495–503; Schmidt 77–8; E. Jachimowicz in C. Blacker and M. Loewe (eds.), *Ancient Cosmologies* (London, 1975), 170 n. 31.

[32] 86.28–30; Kingsley (1993*a*), 54–6.

[33] 86.30–87.4. At 86.37–8 (2.61 *ad init.*) Alexander explicitly reiterates that he is following Empedocles here.

water have come into being as a result of providence....'. This of course is another digression in line with the major issue of the dialogue: no longer Empedocles but Alexander.[34]

Diels, followed by all subsequent editors, assumed that was the end of the Empedoclean material. He was wrong. Alexander goes on:

The humidity flowing from the earth became water. And similarly *aer* came into being, rising upwards from the water and earth as vapour[35]—just like what happens customarily in the hot baths. For the water is rarefied through heating, in accordance with the necessary transformation of the elements into each other.[36]

This last statement on the transformation of the elements is again Alexander speaking, or rather Philo.[37] Yet we have already noted Alexander's way of interspersing genuine Empedoclean material with his own comments. What is more, it is clearly significant that—as we saw earlier—pseudo-Plutarch gives precisely the same interpretation of Empedocles' 'theory' of *aer* as due to transmutation of the element of water which Philo gives here. The parallel with pseudo-Plutarch becomes even closer when we notice that both he and Philo present in direct sequence the two related processes of water being produced out of earth, and of *aer* being produced out of water.[38] There is only one plausible explanation for all this: that Philo's account of Empedocles' teaching goes back to the same source which pseudo-Plutarch has summarized more briefly.

---

[34] 87.4–10. On this passage in Philo and its relation to the Stoic debate about the usefulness of seawater cf. Gronau 94–6.

[35] 'Similarly', because—as we will soon be told—it is another example of transformation of the elements. The absence of a verb here in the Armenian is noted by *NBHL*, i. 55, col. 3, but the suggestion made there that the noun *gološi* could be an—otherwise unattested—verb is unnecessary as well as implausible: on the elliptical style of the Armenian translation see Kingsley (1993a), n. 8. Understanding *etew* ('became', 'came into being') from the previous clause helps to clarify the meaning. For *ōd = aer* cf. ibid., n. 12.

[36] *Or yerkrē coreal xonawut'iwnn ĵur etew. noynōrinak ew ōd i ĵroy ew yerkrē i ver gološi, orpēs i bałanisd sovoreal ē linel. k'anzi anōsreal ĵermut'eamb ĵroy, əst aṙ i mimeans tarerc'n harkawor p'op'oxman (Prov.* 2.61, 87.10–16).

[37] Philo uses the identical expression again in *Prov.* 2.100 (108.22–4); cf. *De aeternitate mundi* 109–11, a passage already referred to above (n. 25).

[38] Ps.-Plut. *Placita* 2.6.3 = Emp. A49b. Compare the other obvious parallelism in the two writers' formally identical descriptions of the heaven being created out of *aithēr* and the sun being created out of fire: Ch. 2 with nn. 17, 32.

A closer look at the text of Philo confirms that here Alexander has turned back again to Empedocles. In describing the first of the two processes—water emerging out of earth—he uses the term 'humidity'; but when it comes round to Philo's turn to refute him on precisely this point he answers Alexander by using the word 'sweat' instead. We know Empedocles called the sea the sweat of the earth, and from Philo's language it is plain that in using this term he has Empedocles specifically in mind; even Alexander's use of the term 'humidity' may well go back to him.[39]

If in referring to the first process Alexander has turned back to Empedocles, this strongly suggests that the second process is likely to be Empedoclean as well. But there is no need to depend on this inference alone—thanks to the analogy in the Philo text with hot baths ('just like what happens customarily in the hot baths'). Another writer, Seneca, explicitly attributes to Empedocles an analogy with hot baths: an analogy which, according to Seneca, he used as a way of describing how hot springs are produced from water flowing above the fires that burn inside the earth. The inevitable conclusion is that Seneca and Philo have recorded different aspects of one impressively coherent picture: the subterranean fires heat the water close to the earth's surface, which in turn is converted into vapour or steam.[40]

---

[39] 'Sweat': Emp. B55 and *Prov.* 2.66 (90.26–7), where Philo's language in protesting against those who compare sea with sweat (90.21–6) resumes his attack on Empedocles, Parmenides, and Xenophanes at *Prov.* 2.42 (76.26–38; Hadas-Lebel's appeal to Lucretius is irrelevant). As for 'humidity', the word *xonawut'iwn* no doubt translates ἰκμάς, as it does in *Prov.* 2.104 (114.5). There is a general tendency to restrict the early philosophical use of ἰκμάς to Diogenes of Apollonia and Ionia (so e.g. Guthrie, ii. 374 n. 4), but this is misleading: Diogenes was extremely eclectic and, as Halleux has pointed out (72), his use of the term must be viewed in a much wider context. It is worth noting that the word ἰκμάς is used specifically of sweat (Arist. *De partibus animalium* 668^{b}4) and of moisture that, like perspiration, is prone to evaporate (Halleux 72 nn. 35–6). That Empedocles used the word, or at least a derivative of it, elsewhere is indicated by ps.-Plut. *Placita* 5.26.4 (150.20 Mau; DK i. 296.21–2).

[40] Sen. *QN* 3.24.1–3 = Emp. A68. Philo's *orpēs sovoreal ē*—literally 'just as is customary'—and Seneca's 'facere *solemus*' rather obviously have one and the same point of origin. As to the identity of that source, there is little room for doubt. We have just seen that Philo used the same source of information in his account of Empedocles as did pseudo-Plutarch, and Diels has shown that both pseudo-Plutarch's *Placita* and a considerable amount of the material in Seneca's *Quaestiones naturales* can be traced back to the school of Posidonius (*Dox.* 224–32). For the Posidonian origin of some of Seneca's comments on hot springs in *QN* 3.24.1–3, and also of material occurring

Chapter 3

It is clear that Empedocles' interest in describing this picture of hot springs and subterranean fire derives from his direct familiarity with the volcanic terrain of Sicily and the surrounding islands.[41] That brings us to the question as to whether the comparison with hot baths does or does not go back to Empedocles.[42] In fact the question, when posed in this way, is no question at all. Among the best-known hot baths in antiquity were the hot springs—the *therma loutra*—provided in Sicily by nature herself. We know that Empedocles referred to these springs, which is hardly surprising. Just along the coast to the west of his home town of Acragas were the baths of Selinus; due north from Acragas were the even more famous hot baths of Himera, not to mention the other sites on the island.[43] It is not a question of whether or not Empedocles used the image of hot baths to explain the phenomenon of hot springs: the hot springs *were* the hot baths. Naturally, in the course of elaboration and reinterpretation by later writers this simple fact would come to be overlaid. One aspect of this inevitable process is the way that—along with the changing meaning of *aer*—Empedocles' words came to be interpreted in terms of transmutation of the elements by the time they reached that common source, heavily influenced by Stoicism, from which pseudo-Plutarch, Philo, and also Seneca obtained their information.[44] But Empedocles himself was not explaining the origin of air; he was explaining the origin of steam. To be more precise, he was explaining a very specific phenomenon by using very famous examples, for one of the most striking phenomena associated with the hot springs of Sicily was the vapour that rose up from them and hovered over the water as

---

elsewhere in Philo's *On Providence*, see Gronau 94–6. For Posidonius' evident interest in water sources and hot springs cf. K. Reinhardt, *Poseidonios* (Munich, 1921), 116–19, 166–74; for his interest in volcanic and related phenomena, especially in and around Sicily, Strabo 1.3.9–10, 6.1.6, 6.2.1–11 with W. Theiler, *Poseidonios* (Berlin, 1982), ii. 18, 53–5, Sen. *QN* 2.26.4–7.

[41] Cf. also Gilbert 304 n. 1; Ch. 6, with the refs. in nn. 2, 5–6.

[42] Burnet (1930), 240; W. Kroll, *RE* Suppl. vi. 431; Bollack, iii. 227–8.

[43] Pi. *Ol.* 12.19, Diod. 4.23.1, 4.78.3, Strabo 6.2.9; Freeman, i. 76–7, 166 and n. 1, 417–19. Referred to by Empedocles: ps.-Arist. *Problems* 937^{a}14–16 = DK i. 296.8–9.

[44] For the essentials regarding the so-called *Vetusta placita* see KRS 5, and in more detail *Dox.* 224–32 with n. 40 above; but regarding Diels's 'Aëtius' see Kingsley (1994a), 235 n. 3 *ad fin*. For the reference to *miliaria* in Seneca's account of the hot baths cf. Kroll, *RE* Suppl. vi. 431–2; for his reference to the baths at Baiae, Gronau 94–5.

steam. The baths of Selinus in particular were so remarkable that the design of the grotto was attributed to Daedalus, who

expelled the steam generated by the fire inside it so successfully that those who came to the place would start perspiring imperceptibly and gradually because the heat was so mild. This was a therapy for the body that was an actual pleasure, involving no unpleasant effects from the heat.[45]

Apart from serving as vivid illustrations of natural phenomena, the renowned curative properties of these vapours no doubt gave them an even greater interest in Empedocles' eyes.

Before moving on, it will be worth emphasizing one point in particular. Although Empedocles clearly distinguished between *aither* and *aer*, this was not the distinction that would later be made between *aither* as the upper and *aer* as the lower region of air.[46] *Aither* for Empedocles is heaven and the outer boundary of the universe, but it is also the air right down to the earth's surface. In this he was simply using the accepted poetic language of his time. The air around us, the air that we breathe: that is *aither*.[47] Mist and vapour: that is *aer*.

---

[45] Diod. 4.78.3. On the health-giving properties of the Sicilian vapour baths see Freeman, i. 417, 419. The various scenes of volcanic activity in and around Sicily were invariably associated with steam, fumes, and mist, although not all were so beneficial. Cf. Freeman, i. 166–7, 519–20, 525–9 (Palici); Strabo 6.2.8, *Aetna* 330–9 etc. (Etna); Strabo 6.2.10 (Thermessa).

[46] Lehrs 163, 168–70; Kahn 140; Schmidt 75–81, 87–101.

[47] B100.5, 7, 18, 24. Compare West's comments on the sense of *aither* for the tragedians, (1982), 5 ('it is not used only of the bright upper air, but also of the arena in which thunderstorms, winds and clouds disport themselves, and for the air that surrounds us, the air that we breathe in and out. Even the foul exhalations of the Cyclops Polyphemus after his dinner of human flesh can be called *aither*. . . .'). Cf. also Ch. 2 with n. 15 on Parmenides' αἰθὴρ ξυνός, B11.2.

4
The Riddle
*

EQUATING Empedocles' *aer* with the element of air was one thing for his interpreters. For Empedocles himself it was out of the question. Failure to appreciate this has meant that clues to the identity of the gods listed by him in the fragment we started with have been looked for in completely the wrong direction. As we have already seen,[1] the rationale for equating Hera with air lay in the term *aer* itself. But Empedocles called the element *aither*, and there can be no real justification at all for identifying his Hera with *aither*. On the contrary, when Hera did come to be associated in Stoic circles with the air it was precisely so that she could be equated with the lower, damp, feminine *aer*, while Zeus was the warm, masculine *aither* that lay on top of her.[2]

Then there is the old Homeric idea of Hades ruling a dark, misty realm. This eventually came to form the basis for the allegorical interpretation of Hades as representing the obscure region of *aer* around, as well as under, the earth.[3] But what relevance could it have to Empedocles? The answer has been to point to the Neoplatonic interpretation of his *Purifications*. According to that interpretation, the 'covered-over cave' which Empedocles describes fallen souls as descending into is not a literal cave but a symbol of the world we live in. And as for the traditional region of the underworld, according to the same interpretation it was transferred by Empedocles to the surface of the earth and the air immediately around us. Applied to

---

[1] Ch. 2 with n. 3; Ch. 3 and n. 9.
[2] Buffière 106–16. The later idea of equating Hera with *aither* as well (ibid. 537–8; cf. Tzetz. *Alleg. Il.*, Proleg. 272) was simply a subsequent scholastic elaboration of the Stoic allegory once Zeus had been ascribed a role unconfined to any element, as is particularly clear from Porph. frs. 355–6 Smith = Περὶ ἀγαλμάτων, frs. 4–5 Bidez; there is nothing here of any relevance to Empedocles. For Tzetzes' equation of Hera with 'fiery air' (i.e. *aither*) see below, n. 31.
[3] Ch. 2 and n. 4.

Empedocles' cosmological poem, this would seem to justify equating his element of air with Hades: 'Hades he placed in the realm of air'.[4]

This is not the place to examine the little that survives of the *Purifications* and assess just how tenable or untenable the Neoplatonic interpretation is. However, there are two points worth noting in passing. The most influential defence of it in modern times has been Rohde's; but his appeal to 'definite tradition' in favour of the interpretation—by which he means definite Neoplatonic tradition—remains a remarkable example of historical naïvety. It is well known that the Neoplatonists had their own very creative ways of interpreting the Presocratics. Apart from that, Rohde's only positive argument was that the idea of souls falling into a literal underworld rather than into the 'Hades of this world' is 'contradicted . . . by common sense (for Empedocles falls from Heaven to earth and not, please God, to Hades!)'. However, it is important to think twice before appealing to such arguments in a context such as this. To mention just one example, Milton had little regard for 'common sense' when—some two thousand years after Empedocles—he described his angels as 'fall'n . . . from Heav'n to deepest Hell; O fall!' It is a simple fact that myth is about as concerned with 'common sense' as life is.[5]

Second, it is more than a little ironic that the two scholars who have supported this interpretation most effectively in recent years did so immediately after criticizing at length the common failure among modern writers on Plato's *Gorgias* to distinguish—as sources for the myth contained in that work—between the original storyteller whom Plato refers to, and the allegorical interpreter of the story whom Plato also mentions. Even more ironical are their further conclusions that the 'Sicilian or Italian' whom Plato hintingly alludes to as the originator of the story was only the teller of the myth, not its

---

[4] Burkert 363 and n. 70. For the ancient texts presenting the Platonic interpretation of the *Purifications* see Diels (1901), 150–2, 154–5; Sturz 448–56. Apart from Burkert, Rohde, and Dodds (see following notes) the interpretation has been almost universally accepted, and the small but steady voice of opposition—Wilamowitz (1935), 489; Rathmann 100–1; Thomas 119; Zuntz 200–5, 262–3; Mansfeld, xiv. 285 n. 59—effectively drowned. The ambiguous position adopted in KRS 316 begs the question.

[5] Rohde 403 n. 75; Milton, *Paradise Lost* 5.541–2. On the historical background to Milton's ideas see G. McColley, *HTR* 31 (1938), 21–39.

interpreter; and that whereas the person interpreting the myth allegorized it by equating the underworld with our life on earth, the myth-teller himself probably described a formal *katabasis* or descent into the actual realms of the dead.[6] In the case of Empedocles' *Purifications*, the teller of the myth is obviously Empedocles; but who is the interpreter? That Empedocles himself would have defused the drama of his poem by adding his own allegorical interpretation is not the most attractive hypothesis. And to this we have to add that not one trace of such an explanation by Empedocles survives— even though nothing could have suited the Neoplatonists better than to be able to quote it in defence of their interpretation. So for example Porphyry, in the famous passage where he compares Empedocles' 'covered-over cave' to the 'subterranean cave' in Plato's *Republic*, is forced to quote from Plato alone in defence of his fundamental thesis that in both cases the cave is a symbol of this world.[7] Even more significant is a passage from Plotinus. In it he states specifically that, on the subject of the fall of souls and what exactly it meant, Empedocles

revealed neither more nor less than Pythagoras and his disciples were accustomed to reveal on this and many other matters; in his case the lack of any clear statement is understandable because he was writing poetry. So we have no choice but to turn to the divine Plato....

And yet where Empedocles was concerned, Plotinus was perfectly willing to fill in the gaps: to assume that he too equated his cave with this world of ours because, as a predecessor of the Platonists, that is what he should do. When he turns back again to the question of the 'cave' into which— according to both Empedocles and Plato—our souls descend, Plotinus' comment could hardly be more revealing: 'it refers— *I suppose* (δοκῶ μοι)—to the totality of existence'.[8] This statement bears all the hallmarks of the creative interpretation of earlier philosophers for which Neoplatonists are famous;

[6] Dodds (1959), 296–7 and Burkert 248 n. 48, referring to Pl. *Gorg.* 493a–b.

[7] *De antro nympharum* 8, citing Emp. B120 and Pl. *Rep.* 514a, 515a, 517a–b.

[8] Plot. 4.8.1.17–36. On the passage, and Plotinus' sources, see Burkert in J. Mansfeld and L. M. de Rijk (eds.), *Kephalaion: Studies in Greek Philosophy and its Continuation Offered to Prof. C. J. de Vogel* (Assen 1975), 137–46; T. Gelzer, *MH* 39 (1982), 101–31; Mansfeld, VII. 131–56 and (1992), 217 n. 33, 300–2, 305.

and, in particular, Plotinus' excessive need to qualify himself even when mentioning Plato puts a very different complexion on Rohde's reference to 'definite tradition', or Dodds's appeal to 'ancient authority'.[9] Hesitant interpretations evolve into dogmatic ones, which are then assumed to be correct just because they are stated with such certainty: this predictable chain of human events represents as great a danger for the historian of ideas as for the historian of facts.

In fact, however, even if we were to assume for argument's sake that the Neoplatonic interpretation of Empedocles' *Purifications* was correct, this still would not justify equating Hades with his element of air. The essential point which is often overlooked is that when Platonists spoke of Hades in this transposed, allegorical sense they were very clear about what they meant. Almost invariably they stated that they were referring to the sublunar world: to the dark region of air in the earth's shadow, the specifically terrestrial region of the atmosphere which is alive with the suffering and wailing of earth-bound souls.[10] That of course is the low cosmic region of 'this world', to which the fallen soul is condemned while it remains

[9] Rohde 403 n. 75; Dodds 174 n. 114. Dodds's remarks are especially surprising in view of his familiarity with the 'ingenuity of misinterpretation' that characterizes so much of Neoplatonism: cf. his *Proclus: The Elements of Theology*² (Oxford, 1963), p. xi. In particular, he correctly emphasized the Neoplatonic tendency to explain whatever was enigmatic in any other philosopher apart from Plato—as well as whatever was enigmatic in Plato himself—on the basis of Platonic principles and ideas (ibid., p. xii, quoting but also slightly misunderstanding Procl. *Th. Pl.* 1.2: ἀλλαχόθεν τὴν σαφήνειαν ἀνευρίσκοντες clearly means elucidating from other passages in Plato, not from other writers). For general remarks on the Neoplatonizing interpretation of Empedocles cf. O'Brien (1981), 86–7, 101–7 and esp. L. G. Westerink in J. Pépin and H. D. Saffrey (eds.), *Proclus: Lecteur et interprète des anciens* (Paris, 1987), 110–11.

[10] See esp. Xenocrates, fr. 15 Heinze = fr. 213 Isnardi Parente; Heraclid. frs. 95–6; Plut. *Moralia* 591b–c, 942f–943c, 944a–b, 1007e; Procl. *In Cratylum* 83.24–84.1 Pasquali, *Rep.* ii. 133.10–11; Hermias, *In Phaedrum* 161.6–7 Couvreur; Olymp. *In Gorgiam* 245.26–9, 259.15–16 Westerink; Wehrli 92–3; Burkert 366–7; Boyancé, *REG* 65 (1952), 334; Cumont (1942), 124–39. On the definition of the sublunar realm as 'the region around the earth' see Kingsley (1995a), § III; and for the intimate association between the earth and the region up to and including the moon note also the idea that the moon is populated by terrestrial *daimones* (Plut. *De gen.* 591c; Cumont, op. cit. 198–9). Iamblichus, it will be noticed, took the step of giving the region between earth and moon to Persephone (= the moon) and the space between moon and sun to her husband (= the sun: Lyd. *Mens.* 167.23–168.2); but the whole point behind these allocations was that, according to Iamblichus' extraordinarily complex analysis of reality, those two regions together represented the most earth-bound areas in the universe (below, with n. 13). For Persephone as the moon cf. Cumont, op. cit. 185 n. 2, 197–8; and for the association of Hades with the sun, Ch. 5.

there. As for Empedocles on the other hand, to identify his Hades with the element of air is—as we have seen—to identify him with *aither*. But for Empedocles *aither* is above all the realm of heaven, the outermost region of the universe.[11] The soul does not fall down into the *aither*; on the contrary, if anything it falls *from* the *aither*, from heaven, and ascends back to it again when it finally escapes incarnation.[12] There can be no equating *aither* and Hades, which—on any account—are diametrically and fundamentally opposed. It is also important to understand that when the Platonists produced their interpretations of Empedocles they had a radically different view of the world from Empedocles himself. For them even heaven, *ouranos*, was a low state, a step on the way to incarnation: the soul originated from the immaterial regions beyond.[13] But for Empedocles heaven is the place of origin and the final place of return: the region of freedom—not suffering or damnation, or even transition.

Last but not least, there is the question of the cosmological poem itself. We are told with great assurance that 'there is no place for the House of Hades in the cosmology of Empedocles: the true realm of death is this existence on earth', and that Empedocles must not be considered a shaman because he 'does not even descend to the Underworld' like any decent shaman would.[14] These assertions appear in their true light when set against a line of Empedocles which, not without embarrassment on account of its content, Diels tucked away at the end of the cosmological poem even though it almost certainly belongs at the beginning. In it Empedocles assures his disciple that, as a result of what he will learn,

> You will be able to fetch from Hades the life-force of a man who has died.[15]

---

[11] Ch. 2; Ch. 3 n. 31.

[12] Emp. B146–7; cf. the Pythagorean *Golden Verses* 70, Dieterich 89 n. 2 and 103–8, Burkert 361 nn. 55–6. Note that it was normal to refer to this celestial location as the home or 'palace' of Zeus (e.g. Eur. *Or.* 1683–90); cf. Emp. B142, with C. Gallavotti in *Le Monde grec: Hommages à Claire Préaux* (Brussels, 1975), 155–7.

[13] Burkert 365; Finamore 59–64, 130–55. On the lowering of the rank of heaven see Ch. 2 n. 17.

[14] Kahn (1971), 27 n. 50, 33.

[15] B111.9: ἄξεις δ' ἐξ Ἀίδαο καταφθιμένου μένος ἀνδρός. Cf. below, Ch. 15 with n. 1; Kingsley (1994d); and for the word μένος, Onians 51–2, 194.

There can be no possible justification for avoiding the literal meaning of this remarkable statement or trying to interpret it away allegorically.[16] On the contrary, it must be understood in the light of similar statements in Greek literature from both after and before Empedocles' time. Obviously, if a dead man's life-force has to be fetched back from Hades then Hades cannot be 'this life on earth', as the Platonists claimed. What is more, when we view Empedocles' choice of wording—as we must— against the background of established epic usage, and also when we compare the standard phraseology used throughout the history of Greek literature for expressing the notion of the dead being brought back to life, the conclusion becomes unavoidable that Empedocles was specifically intending to evoke the theme of a *katabasis*: of an actual descent into the underworld.[17] This is hardly surprising: a subterranean Hades is precisely what one would expect Empedocles within the context of the south Italian magic and mysticism of his time, were it not for the strange ideas read back into him by later centuries.[18] His Hades cannot be the air around the earth—let alone the *aither* of heaven. And according to Empedocles

---

[16] See Burkert himself, 153–4; below, Ch. 15.

[17] Compare esp. *Dem.* 335–8 (ὄφρ'... ἀγνὴν Περσεφόνειαν... ἐς φάος ἐξαγάγοι), but also Eur. *Alcestis* 359–62 (σ' ἐξ Ἅιδου λαβεῖν, κατῆλθον ἄν... ἐς φῶς σὸν καταστῆσαι βίον), Isocrates, *Busiris* 8 (ἐξ Ἅιδου τοὺς τεθνεῶτας ἀνῆγεν), Hermesianax *ap.* Athenaeus 597b = *OF* test. 61 (ἀνήγαγεν... Ἀιδόθεν), Schol. Eur. ii. 227.8 Schwartz (ἀνήγαγεν ἐξ Ἅιδου), Plut. *De sera* 566a, Luc. *Menippus* 6; Clark 102–3, 108–14. ἐξάγειν in early epic invariably meant actually to go and fetch: *Il.* 1.337, 13.379, *Od.* 14.264, Hes. *Th.* 586, etc. On the history of the broadly related notion of ψυχαγωγία or evocation, which developed a metaphorical significance of its own, cf. Burkert (1962), 47.

[18] Cf. Dieterich 107–11 and (1911), 471–3; Burkert 153–6, 364. Kahn's theory of there being 'no place for the House of Hades in the cosmology of Empedocles' refers no doubt to the extraordinary idea, inspired by Erich Frank, that a spherical earth is incompatible with a subterranean Hades; cf. Dodds (1959), 297–8, Burkert 305 n. 30, 357–8. This has no foundation whatever. The fact that, from Plato down to recent times, hell or Hades has persistently been located inside a spherical earth should speak for itself: Pl. *Phd.* 108e–114c; Stewart 94–111; Neugebauer 373; Thomas Aquinas, *Summa theologica*, Part 3, Supplement, Question 97, Art. 7; Giordano da Rivalto, *Prediche* (Florence, 1739), 22; G. Rowell, *Hell and the Victorians* (Oxford, 1974), 155. Servius' simplistic argument (on Virg. *Aen.* 6.127) that there can be no subterranean Hades because the earth is a solid sphere is inapplicable not only to Empedocles but also to at least some Stoics (cf. below, Ch. 6; *SVF* iii. 264.1–6); we will see later (Ch. 13) what the crucial factors were which led to Hades being projected outside the spherical earth. For Frank's mostly absurd strictures about what is necessary or impossible in Greek thought see de Santillana 190–201. For Empedocles' earth as spherical see Millerd 63 n. 6 and Burkert 305 with Ch. 8 n. 5 below.

himself, as we will soon see, *aither* is not an element with any significant presence inside the earth itself.

Equation of Hera, or Hades, with *aither* can be excluded. This leaves only one possible candidate, Zeus; and no one could fit the bill more perfectly. The idea that *aither*, up in heaven, is the home of Zeus is perhaps the single most enduring religious tradition from Homer down to the end of antiquity. There, for any Greek, is where he belongs.[19] Conversely, the *Iliad*'s famous portioning of the cosmos established that 'the broad heaven (*ouranos*) in the *aither* and the clouds' belongs to Zeus.[20] The tendency towards outright assimilation of *aither* and Zeus is old—much older than Empedocles—and very nearly as pervasive.[21] Empedocles himself, as we have seen, applied to the air the epithet traditionally reserved for Zeus.[22] In the crucial fragment 6, his description of Zeus as 'bright' (ἀργής) agrees exactly with his repeated emphasis elsewhere on the brightness of the element of *aither*; it also describes perfectly the dazzling brilliance of the Mediterranean sky.[23] As for the claim by recent writers that the identification of Empedocles' Zeus with air is a modern invention, this is rather obviously contradicted by the facts.[24]

That Empedocles applied to air an epithet specifically reserved for Zeus deserves a little extra comment. There is a more subtle side to him than his common image as a philosopher who just happened to write down his ideas in the form of poetry, and this subtlety lies chiefly in his relationship to Homer. Diels had already annotated his editions of Empedoclean fragments with lists of Homeric echoes and parallels, but it has been left to later writers to show that as a rule these echoes are chosen deliberately and that the parallels serve a particular function. The Homeric poems were so well known to Empedocles' audience that by borrowing one unusual word from them he could immediately evoke the word's context; by

---

[19] αἰθέρι ναίων: *Il.* 2.412, *Od.* 15.523, Hes. *Op.* 18, Theognis 757, etc. Cf. Eur. fr. 487; Cook, i. 25–6.
[20] *Il.* 15.192. Cf. Eur. frs. 839, 985; Kahn 135–7.
[21] Cook, i. 27–33; Buffière 108–9.
[22] B149 with Ch. 3 n. 18.
[23] Emp. B21.4 (ἀργέτι αὐγῆι), B98.2; Traglia 152; Ch. 2 n. 11.
[24] Ch. 3 n. 3.

using two or three words together which also occurred together in Homer he could graphically evoke a whole scene. His choice of words—even his apparent fondness for creating new words—is undoubtedly based on a keen awareness of the way that particular terms were used by Homer. The evidence shows that in the case of Empedocles, as in the case of Parmenides before him, this art of evocation and subtle transposition was precisely that: a matter of deliberate method and conscious skill.[25] Of course this art of allusion is also an art of speaking indirectly: the art of hinting, of really saying more than one is apparently saying. To Aristotle that has no place at all in the writings of a philosopher: it is just another infuriating aspect of the vague, implicit, oracular use of language which he condemned in Empedocles as pretentiousness, trickery, and sleight of hand.[26] But if we want to understand Empedocles on his own terms it is more helpful to note that this concern with communicating more than immediately meets the ear or eye links him very closely on the one hand to the figure of the Near Eastern philosopher–poet with his riddles and hinting allusions, who survived down into the Middle Ages and beyond; and on the other hand to the Greek rhapsodes and Sophists, for whom riddles, hinting, and the interpretation of Homer were intimately interrelated.[27]

Bollack has helped considerably to bring this side of Empedocles into the open. But at the same time he made the mistake of equating one kind of subtlety with another by importing

[25] See esp. Bollack, i. 277–85 and e.g. iii. 493; for adaptation of Homeric words, O'Brien 267 and Gemelli Marciano. For Parmenides see esp. Mourelatos 1–33 and *passim*.

[26] *Rh.* 1407^{a}31–7 (= Emp. A25b) with Cope and Sandys ad loc., who note that the obscurity in diction of which Aristotle accuses Empedocles 'appears especially in the symbolic terms, such as "Nestis", by which he sometimes designates the elements'. Cf. *Meteor.* 357^{a}24–8 = Emp. A25c; Karsten 60; Lloyd, *Demystifying Mentalities* (Cambridge, 1990), 23. For attribution by ancient writers of the deliberate use of riddles not only to Pythagoras but also to Empedocles—and with specific reference to Empedocles' equation of elements and divinities in fr. 6—see Mansfeld (1992), 193–5 with n. 113, 213 n. 17; also H. Munding, *Hermes*, 82 (1954), 136–7, Mansfeld, *Phronesis*, 17 (1972), 30–5. On the riddling nature of Empedocles' fr. 6 cf. also Kingsley (1994a), § VII with the refs. in n. 62; below, Ch. 23. On riddles and early Greek philosophy in general see esp. Clearchus, fr. 63 Wehrli; Huizinga 115–18, 146–52, and *passim*.

[27] Pl. *Ion* 530b–542a with Richardson (1975), 65–81; J. Grangeret de Lagrange, *Anthologie arabe* (Paris, 1828), 118, Nicholson 166–7. On Empedocles, Sophists, and rhapsodes cf. esp. Diels 159–84; Morrison 55–63; Guthrie, iii. 29–31, 42–3, 269–70; below, Ch. 16 n. 53.

French post-symbolist ideas on the nature of poetry into the study of ancient literature.[28] This has done as much to confuse matters as to clarify them. The art of allusion in Rimbaud or Mallarmé is essentially open-ended. For Empedocles it is not. The chief object of allusion—the Homeric poems—was fixed, and allusions were plainly chosen for the specific effect they produce when a term or image is transferred from one particular context to another. The original context in Homer informs and determines. In the light of these general considerations, it is difficult not to conclude that when Empedocles calls air by the name which Homer applies to Zeus, he is implying that his air and Homer's Zeus are either very intimately related or simply one and the same. To the modern reader of Empedocles this game of inference is likely to seem subjective and also rather naïve. As Aristotle did, we like to stamp our fist on the desk and insist that philosophy is all about putting our cards on the table and calling a spade a spade. But the fact is that in fragment 6 Empedocles has presented a poetic riddle;[29] if he had said what he wanted to say directly, there would never have been any trouble deciphering what he meant. He has also offered us clues, which we can either use or not. Application of Zeus' epithet 'cloud-gatherer' to air is one obvious hint which Bollack was forced to ignore because he wrongly equated Zeus with fire: 'The epithet usually associated with the name of Zeus is attributed to an element which, according to Empedocles, does not belong to him. It is impossible to see what meaning it had.'[30] On the contrary, the clue simply reinforces what is anyway an unavoidable conclusion: Empedocles' *aither* corresponds not to Hera or to Hades, but to Zeus.

This leaves Hera and Hades, earth and fire. To begin with, we can rule out any identification of Hera with fire; there is no possible justification for it. Admittedly we find her equated with the fiery aspect of *aer* in Tzetzes, 'because she is the mother of Hephaestus'.[31] But, in time and in mentality, the twelfth-century writer Tzetzes is far removed from Emped-

---

[28] Cf. Kahn's comments, *Gnomon*, 41 (1969), 439–40, 447.
[29] See above, n. 26.
[30] iii. 301.
[31] ὁ πυρώδης (ἀήρ) ... ὅνπερ καὶ Ἥραν εἴπαμεν μητέρα τοῦ Ἡφαίστου: *Alleg. Il.*, Prolegomena 295–7. Cf. ibid. 272–6; *Exegesis in Iliadem* 55.5–6 Hermann.

ocles. Closer to him in both respects is Plutarch, who does not cut quite so many corners as Tzetzes; instead, he states more carefully that 'Hera represents *aer*, and the birth of Hephaestus represents the transmutation of *aer* into fire.'[32] There should be no need to say any more about the total inapplicability to Empedocles of even this idea of transmutation.

While there is no justification for identifying Hera with fire, the grounds for identifying her with earth are almost as strong as if the equation had been stated explicitly. In fragment 6 Empedocles gives her the epithet 'life-bearing', *pheresbios*. This is an epithet specifically reserved in Hesiod and the Homeric poems for the earth and its fruitfulness.[33] As Guthrie has said, 'the epithet applied to Hera immediately suggests the fruitful earth'.[34] But there is more to the matter than that. As Guthrie (along with earlier writers) has also emphasized, Empedocles was writing in the epic tradition and so took as his point of departure the language and conventions of the Homeric and Hesiodic poems.[35] That brings us back to the issue of his deliberate use of the art of allusion. In line with what was said above, it is highly probable that his choice of adjective here is not just descriptive but also prescriptive: an indirect, but very definite and deliberate, indication that he means Hera to correspond to earth. For all its subtlety, this game of allusion is also extremely simple. Yet the usual view is that Empedocles transferred the traditional epithet 'life-bearing' from the earth to the air, and then applied it to Hera, 'in accordance with his custom of putting established phrasing in a new setting'.[36] More than anything else, however, this attempt at an explanation reflects the sophisticated chaos of twentieth-century ideas about language. The 'new setting' lies precisely in the transfer of the epithet for earth to Hera. The subtlety, the transposition, consists of that—not of some conceptual juggling that no one could hope to follow. Here, in fragment 6, it is a simple case of an obvious clue within the riddle itself.

[32] *De Isid.* 363d. Cf. also Cornutus 19, 33.14–18 Lang.
[33] Hes. *Th.* 693, *Dem.* 450–1, 469, *Homeric Hymn to Apollo* 341, *to Ge* 9. The full relevance of these passages to Empedocles' use of the epithet was noted by G. Thiele, *Hermes*, 32 (1897), 69.
[34] ii. 145.
[35] Ibid.; cf. Sturz 212, Thiele, loc. cit.
[36] Wright 165; cf. Bollack, iii. 173.

As for Empedocles equating Hera and earth, that is no more surprising than his identification of *aither* with Zeus. From the point of view of cult, Hera was intimately connected with the fertility not just of humans but also of the earth—and this was especially the case in the Greek West.[37] What is more, the evidence tends to suggest that long before the equation of Hera with *aer* gained any real currency, the equation of her with earth was being accepted in precisely those circles with which Empedocles is most likely to have been linked.[38] Last, but not least, it is healthy to bear in mind that the fertility and fruitfulness of Sicily's soil were as good as proverbial.[39]

Finally we come to Hades. Bollack claimed that Hades 'est le dieu chtonien, très exactement'.[40] But of course nothing is chthonic, very precisely: the term was an extremely broad one. It was only natural that Hades would be referred to as terrestrial when contrasting him with the celestial divinities of Olympus.[41] However, when it comes to defining him and his domain more specifically we invariably find them both being located *underneath* the earth (ὅς ὑπὸ χθονὶ δώματα ναίει); to a Greek in Empedocles' time it was in fact second nature to draw a sharp distinction between the earth (*gaia* or *chthōn*) and Hades.[42] There was a good reason for this. To the Greeks

[37] A. Klinz, Ἱερὸς γάμος (Halle, 1933), 98–104; Guthrie, *The Greeks and their Gods* (London, 1950), 67–72; A. Hus, *Greek and Roman Religion* (London, 1962), 24. The fact that the attribution of Empedocles' heaven to Zeus and of his earth to Hera provides us with a marriage between heaven and earth is not in itself a proof of the accuracy of the attribution, but it is a point of obvious significance. Cf. Klinz, loc. cit.; Dieterich 100–5; Varro, *De lingua latina* 5.65–7 (also Augustine, *De civitate Dei* 4.10 and 7.28 = Varro, *Antiq. rerum divinarum*, fr. 206 Cardauns) with J. Collart's comments, *Varron: De lingua latina, Livre V* (Paris, 1954), 184.

[38] Pap. Derv. col. 22.7 with Burkert (1968), 104–5 and n. 26. A continuing Italic tradition no doubt helps to explain the reappearance of the equation in Latin literature (see previous note).

[39] Cf. Aesch. *Pr.* 371; Pi. *Nem.* 1.14; Lucr. 1.728; Strabo 6.2.3, 4, 7, 13.4.11; Diod. 5.2.4–5, 5.69.3, 11.90.1, 23.1.1, 37.2.13; Ov. *Met.* 5.481; also Freeman, i. 67 n. 1, 91–2, 540.

[40] iii. 173.

[41] So e.g. Aesch. *Supplices* 156; cf. West (1978), 276. Note the ambivalence in the common reference to Hades as καταχθόνιος (e.g. *Il.* 9.457, Aesch. *Ag.* 1386–7): as well as meaning 'right down into the earth', the adjective also suggests 'underneath the earth' (cf. LSJ s.vv. κατά A II. 2, κατουδαῖος). Either way, there is no question of Hades' domain simply being equated with the earth.

[42] Hes. *Th.* 455, Semonides, fr. 1.14 West, Aesch. *Persae* 839, Pl. *Phdr.* 249a, Cornutus 35 (74.5–8 Lang), etc. Cf. *Il.* 15.191/3; Hes. *Th.* 847/850; Richardson 145, 159. The common habit in modern times of referring to the earth–Hades equation as

nothing was closer, more intimate, and more tangible than the earth. Hades, however, was always essentially 'other'. Even if his realm was no longer restricted to the hidden areas inside the earth but projected into the atmosphere instead, it was still understood to be hidden, unseen.[43] This helps to explain the fact that, although theoretically the opportunity for equating Hades with earth could hardly have been closer to hand,[44] the first surviving reference to the equation dates from Roman times—as an over-simplistic inference from the Greek habit of referring to Hades as the 'terrestrial Zeus'.[45] And then there is the question of the element of earth itself. In referring to it Empedocles uses the word *gaia* just as frequently as the term *chthōn*, and *gaia*—even more so than *chthōn*—is mythically as well as grammatically feminine.[46] This is just as well. As we have seen, there is only one element left which Hades can possibly represent, and that is not earth but fire.

The nineteenth-century 'innovation' was correct in its explanation of fragment 6 after all: Zeus is air, Hera earth, Hades fire, and Nestis water. On the other hand, both of the ancient traditions are wrong. It is not hard to see why. The rationale behind the tradition which goes back to Theophrastus—that Zeus is fire, Hades air, and Hera earth—is still discernible: the explanation is superficially reasonable, but no more.[47] As for the tradition that explained Empedocles' Hera as air and Hades as earth, it is an unsophisticated projection back onto Empedocles of typically Stoic ideas: a classic example of the disastrous principle of 'accommodation' (συνοικειοῦν), the principle deliberately used by Stoics to turn earlier writers into mouthpieces for their own ideas. Behind the few

---

obvious and 'natural' (Millerd 31, Bignone 543, Guthrie, ii. 144, Gallavotti 173, Wright 165) betrays a markedly post-classical perspective.

[43] The common etymology of Hades as meaning 'invisible' (Pl. *Gorg.* 493b, *Crat.* 403a, 404b, *Phd.* 80d with Burnet ad loc.; Heracl. *Alleg.* 74.6) is of course just one expression of this idea, not its cause.

[44] Cf. Pl. *Crat.* 403a; Rohde 160 and n. 13.

[45] Varro, *De lingua latina* 5.66, with the later refs. in Kingsley (1994a), n. 28.

[46] B17.18, 21.6, 27.2, 38.3, 39.1, 71.2, 85, 109.1, 115.10. The importance of the gender is especially clear in e.g. Hesiod's *Theogony*, Pherecydes, frs. 14, 68 Schibli, and Pap. Derv. col. 22.7 (above, n. 38). Cf. also Spoerri 182 n. 20; Kerschensteiner (1962), 124 n. 2; Kingsley (1994a), n. 53. For *Chthōn* as goddess cf. Aesch. *Pr.* 205, *Eum.* 6, and West (1966), 351.

[47] Kingsley (1994a), §§ ii–v. For the attribution to Theophrastus see ibid. 238–41.

assertions that Empedocles equated his Hera with air lies the simple device of citing his authority as a way of giving the equation the prestige of a respectable authority.[48]

That Diels, along with practically all subsequent writers, was so taken in by this device that he tried to ascribe the equation of Hera with air and Hades with earth both to Theophrastus and to Empedocles is one of the lesser ironies of Empedoclean scholarship. Nowadays it is no longer easy to share his assumption that what Theophrastus said must be right, and that misinterpretations of the Presocratics only crept in in later centuries—although, it must be said, the increasing willingness since Diels's time to admit that the truth is probably very different still remains strictly at the level of theory. In practice, the extent of the damage caused by Theophrastus'—and, by implication, Aristotle's—misinterpretation of the Presocratics has scarcely begun to be appreciated.[49] Admittedly in this particular case Theophrastus appears to have identified two out of Empedocles' four divinities correctly, which is more than can be said for the alternative explanation. However, that is not to say very much. The heart of the whole problem is clearly the identity of Hades; and as guides to understanding the real factors that lay behind Empedocles' attributions, both Theophrastus and the later commentators are equally useless.

With no apparent tradition left to guide us, the only way forward is one step at a time. Empedocles' equation of Hades and fire has nothing to do with the conventional, mainstream traditions of Greek philosophical thought, but it does bring us to the threshold of a very different world: one only mentioned in poetry, legend, and esoteric tradition. Before entering that world, however, there is a little tidying up to do with regard to Empedocles' views on the origin and nature of the sun.

---

[48] Ibid., § VII.
[49] See the Introduction, with nn. 5–6.

5
The Sun

*

THE conclusion that Empedocles equated Hades with the element of fire is, we have already seen, inevitable. But one objection, and an apparently forceful one, has been placed in the way of accepting it. Simply stated, according to our ancient sources Empedocles described one half of the entire heavens as now filled with fire: a picture which rather obviously contradicts the notion that he confined the element of fire, or at least the greater part of it, to a subterranean Hades inside the bowels of the earth.[1]

The objection is in fact null and void. As I have shown elsewhere,[2] this theory of a celestial fiery hemisphere does not go back to Empedocles but is a mistaken interpretation of him by Theophrastus. The problem arose from Empedocles' account of what happened at the start of our world. He himself had described how *aither*—the first element to separate out from the primal, chaotic mass—rose up until it reached the fixed boundary of the cosmos. Then came fire: heavier than *aither* yet more volatile, it ran up to the *aither* and spread through it, creating that blazing fiery hemisphere which Theophrastus mistook for a phenomenon still existing now. But that was not the end of the matter. Next the fire reacted with the *aither*, glazing or crystallizing it and creating the brilliant, crystalline heaven which lights up so spectacularly in Mediterranean climates when the sun rises every day. Small pockets of fire became trapped in this crystallized *aither*, becoming what we now perceive as the fixed stars; and then the most crucial event of all occurred. So much fire at the upper boundary of the cosmos made the universe top-heavy. The fire started to weigh downwards, tilting the universe and the

---

[1] Cf. esp. Bignone 543.
[2] Kingsley (1994*b*).

celestial pole in the process; and as it slid down the circular boundary of the universe it accumulated into one concentrated body of fire—our familiar sun. The creation of the planets and moon was also described as part of this cosmic transformation. What had been a hemisphere of fire was a hemisphere of fire no longer: instead, it was embodied above all in the orbiting sun.

The elegance and consistency of Empedocles' description should need no emphasizing: in one sweep of the imagination he explained the origin of the stars, the brilliant crystalline appearance of the heavens, the tilting of the celestial pole, and the creation and rotation of our sun. But this was evidently lost on Theophrastus—who after all, where the Presocratics are concerned, was more interested in criticizing than in understanding. Instead, he made the mistake of failing to realize that Empedocles' hemisphere of blazing fire was no more than a past stage in the history of the universe. His error was easy enough to make: there is no lack of examples of writers in antiquity misunderstanding accounts, by earlier authors, of past events in the cosmos as referring to phenomena that still exist.[3] Yet with Theophrastus it is difficult to avoid concluding that if he had not made this mistake he would have made some other—as he did when identifying the divinities in Empedocles' fragment 6.

The objection which we started with, to accepting that Empedocles equated Hades and fire, is in fact no objection at all. But, on closer examination, the underlying problem turns out not so much to have been resolved as simply to have been shifted elsewhere. For how are we to reconcile Empedocles' evident equation of fire and Hades with the fact that he repeatedly uses the visible sun as his chief example of elemental fire in the world around us?[4] Sun and underworld have, to us, such radically separate connotations that the one idea would appear to exclude the other. And as far as the world-view of the ancient Greeks is concerned, the fundamental Aristotelian distinction between a celestial or supralunar realm—including of course the sun—and the sublunar world—including the earth and

---

[3] Cf. e.g. Kerschensteiner (1962), 136; West 214–15.
[4] Refs. in Kingsley (1994b), 319 n. 16; above, Ch. 2 with nn. 17, 32–3.

The Sun

everything in it—has come to seem not only representative but also definitive.[5]

And yet the facts are not quite so simple. First we need to look at the surviving evidence, fragmentary as it is, of Empedocles' own poetry. In one place he describes the creation process that gave rise to the ancestors of mankind. Fire, pushing up from the depths of the earth, 'raised up the benighted shoots' of humanity (ἐννυχίους ὄρπηκας ἀνήγαγε) and then left them—still sexually undifferentiated—on the earth's surface as it rushed to the heavens to join the fire that was accumulating there.[6] Modern commentators have noted that the reference to these 'shoots' of humanity as 'benighted' has a very pointed significance: these prototypes of mankind were raised up out of the earth at a time when our sun did not yet exist and there was no such thing as daylight.[7] It has also been noted that Empedocles' choice of vocabulary at this point appears to hint that these ancestors of humanity arose out of a place in the depths of the earth equivalent to what is now known as the underworld—the world presided over by Hades.[8] In other words Empedocles is not only stating that the fire which eventually rose up to become the sun had its origin in the bowels of the earth, but also implying that the source of daylight and illumination derives ultimately from the dark depths of the underworld. The element of paradox here is striking, but not unexpected: it is entirely consistent with the general trend in both Parmenides' and Empedocles' accounts of the creation of man and of the universe as a whole.[9]

There is another fragment of Empedocles which needs mentioning here, because it appears to paint a very similar picture in an even more delicate way. One fragmentary line, preserved for us by Proclus, states that 'there are many fires burning beneath the surface of the earth' (B52). In the next section of this book it will become clear how much there is to be learned about

---

[5] Cf. e.g. F. Rochberg-Halton, *JNES* 43 (1984), 116–17.
[6] B62; below, Ch. 6 with n. 29.
[7] Bollack, iii. 429 and n. 4, Wright 216, who note that this implication of the adjective—especially as applied to the noun ὄρπηκες, 'shoots' or 'saplings'—is confirmed by ps.-Plut. *Placita* 5.26.4 *ad init.* (DK i. 296.16–17).
[8] Wright, loc. cit., on the words ἀνήγαγε and ἐννυχίους (to her mention of Soph. *OC* 1558 ἐννυχίων ἄναξ Ἀϊδωνεῦ add *Tr.* 501).
[9] Cf. Bollack's comments, iii. 429–30.

Empedocles' doctrine of subterranean fire by placing this brief statement in its historical, geographical, and mythological context. Yet there is another implication in this line which deserves noting here, not only because of what it tells us about Empedocles' ideas but also because of what it tells us about the way he seems to have chosen to convey them.

In the last chapter we saw how Empedocles deliberately used the art of verbal allusion to convey more than he was communicating at the surface level: how by borrowing a single word from the Homeric poems he could immediately evoke the word's context, and by using two or three words together which also occurred together in Homer he could graphically evoke a whole scene. In doing so he took his place in a tradition which extended both back in time before him and forward into the future; but what is most important to appreciate is the precision with which he used this technique to convey his meaning implicitly, by hinting, suggesting, saying more than he is saying apparently. As far as the fragment cited by Proclus is concerned, Empedocles seems just to be asserting the existence of vast fires underneath the earth's surface—an assertion which of course is highly significant in its own right, considering his evident identification of the god of the underworld with the element of fire. Yet there is almost certainly another dimension of significance to the line as well. As Bollack has noted,[10] the words 'there are many fires burning' ($\pi o \lambda \lambda \grave{a}$ $\pi v \rho \grave{a}$ $\kappa a \acute{\iota} \epsilon \tau a \iota$) are an obvious echo of the almost identical combination of words at the end of *Iliad* book 8, where Hector tells his army to make preparations

> so that all night long until the early birth of dawn
> we can burn many fires ($\kappa a \acute{\iota} \omega \mu \epsilon \nu$ $\pi v \rho \grave{a}$ $\pi o \lambda \lambda \acute{a}$), and the blaze
> will reach to heaven.

Empedocles' careful use of the art of Homeric allusion makes it impossible to ignore the probability that, through his verbal reminiscence of this passage, he was implying a link between his own 'many fires' and a blaze which reaches up to heaven— especially as the identical link is already made in the passage about the ancestors of humanity. This is not even to mention the further fact that the image of subterranean fires emitting a

[10] iii. 229, referring to *Il.* 8.508–9, 554–63.

blaze that 'shoots up to heaven', that 'reaches into the heavens' and 'licks the stars', was a standard way of describing volcanic activity on Empedocles' Sicily.[11] Just as significantly, the very last words of *Iliad* book 8 draw a detailed analogy between the 'many fires' which 'were burning' (πυρὰ καίετο πολλά) throughout the night and the brilliance of the stars in heaven. What makes this web of allusions all the more meaningful is the fact that for Empedocles the blaze which rose up to heaven from his 'many fires' actually created the celestial bodies. Analogy becomes identity, metaphor fact. And yet there is nothing in these hints that can be proved, or methodically demonstrated, just as there is no reason to suppose that in the portions of Empedocles' poetry which have been lost he was any clearer or more specific; as we have already seen, he liked to teach and communicate by using the method of hinting. Even so, however, the central message of the passages we have been considering comes through with clarity: the fires of heaven—including the fire that rose up to become the sun— had their point of origin in the depths of the earth.

There seems to have been a time early in his career when Aristotle appreciated the 'Homeric' qualities of Empedocles' poetry; later, however, he had nothing but criticisms for the 'laughable' way in which Empedocles tried to combine the use of poetic imagery with the role of philosopher.[12] The fragmentary pieces of evidence we have just examined are enough in themselves to show how inappropriate those criticisms are if one wants to understand what Empedocles himself had to say. And they are also enough to defuse any doubts about how Empedocles could possibly equate the element of fire with the god of the underworld while citing the visible sun as an example of fire in the world around us. There may be a contradiction here for us, but the contradiction plainly did not exist for him.

And yet, however clear in its implications, this Empedoclean evidence deserves to be placed in its wider context. In fact the idea of a profound affinity between sun and underworld—and,

[11] Ch. 6 with n. 29; Ch. 17 with n. 18.
[12] Arist. fr. 70 = D.L. 8.57; *Poetics* 1447^{b}17–20 = Emp. A22; *Meteor.* 357^{a}24–8 and *Rh.* 1407^{a}31–9 = Emp. A25; above, Ch. 4. Cf. Gemelli Marciano 24, 209.

more specifically, the idea that in some sense the sun belongs in and is at home in the underworld—lies at the roots of classical Greek and Roman mythology. It appears in many forms, and is especially prominent in the mythology of the Greek West. Close, and obviously significant, analogies exist with themes and motifs in oriental mythology, both Near Eastern and Indo-European.[13] Certainly this notion of correspondences between the underworld and the sun occupies a central place in Babylonian myth—where it forms part of a whole complex of paradoxical associations between sun and darkness, light and blackness, the visible and the invisible—and in Egyptian myth as well.[14] As far as the classical world is concerned, this association between sun and underworld did not remain confined to the oldest strata of Greek myth. Empedocles' immediate predecessor Parmenides, who was clearly his teacher in several respects as well, merely revived and gave new expression to this same paradox when he described in the introduction to his poem how he was taken by the daughters of the Sun down into the House of Night—which by implication is the home both of the daughters of the Sun and of their father.[15] There are a number of features in this introductory

[13] Dieterich 19–34, Burkert (1979), 93–4. Cf. also E. Wüst, *RE* xxi. 1003; A. Ballabriga, *Le Soleil et le Tartare* (Paris, 1986); and the further refs. below.

[14] For the Egyptian evidence cf. Griffiths 303–9. For the Babylonian material see now Heimpel, and e.g. Livingstone (1989), 101 (sun–Marduk–underworld). A similar taste for the paradoxical is displayed in the Babylonian designation of Saturn as both 'black star' and 'star of the sun', which is due at least in part to a pun on the word *ṣalmu*—meaning 'dark' or 'black' but also evidently used as a name for the winged solar disk. Cf. S. Parpola, *Letters from Assyrian Scholars to the Kings Esarhaddon and Assurbanipal*, ii (Neukirchen–Vluyn, 1983), 342–3, to be supplemented on the Greek side by *Epinomis* 987c3–5—see Taylor 194 n. 1, with the further refs. in Tarán (1975), 309 n. 738 (unjustifiably overruled by Tarán himself)—and R. Beck, *Planetary Gods and Planetary Orders in the Mysteries of Mithras* (Leiden, 1988), 86–90, and on the Akkadian side by S. Dalley's comments on the word *ṣalmu* in its application to the sun, *Iraq*, 48 (1986), 85–101 (where, however, the Sun–Saturn correspondence and the *deliberate* appeal to the ambiguity of the word are not mentioned). Regarding the principle of 'philological association' as used in Babylonian esoteric exegesis, cf. Livingstone (1986), 2–3 and *passim*. For the possibility of an etymological link between the Babylonian deity Adêsu and the Greek Hades, see Dalley, op. cit. 101.

[15] Parm. B1.8–10, with the punctuation corrected as in Mansfeld, *Die Offenbarung des Parmenides und die menschliche Welt* (Assen, 1964), 234–9, Burkert (1969), 7–9, and with Pellikaan-Engel's comments, 36–7, 51–3. Compare esp. the very similar journey of the Sun back home to his mother, wife, and children in 'the depths of the night' (βένθεα νυκτός) in Stesichorus, fr. S17 Davies: Dieterich 21, J. S. Morrison, *JHS* 75 (1955), 59–60, Burkert (1969), 9. For Parmenides' own journey as a descent into the underworld cf.

section of his poem which point to oriental, and specifically Mesopotamian, origins; and there can be little doubt that this particular paradoxical motif derives from a Babylonian mythological prototype.[16]

The later philosophical trends which enforced the—to us still familiar—distinction between the supralunar and the sublunar, the world above and the world below, never succeeded in eradicating this more paradoxical, ambiguous world-view. For example it recurs, prominently, in the Graeco-Egyptian magical papyri; while the close association, and at times identification, of sun and underworld, Helios and Hades, continued to find expression in the Greek world both in religious cult and at the literary level.[17] This helps to explain how, in late antiquity, the sun seen in a vision could meaningfully be described as 'Tartarus-like' ($\tau\alpha\rho\tau\alpha\rho\text{o}\epsilon\iota\delta\acute{\eta}s$); and it was entirely in the spirit of this 'underground' tradition about the sun and its affinities that early Islamic cosmology presented the sun as 'created from the fire of the earth', related the heat of the sun to the heat of hell-fire, and could speak in one breath of 'the fire of which the sun and the devils are made'.[18]

But above all it was alchemical tradition that was responsible for preserving and maintaining this basic association between sun, earth, and underworld. Alchemists, from the end of antiquity through to and beyond the Middle Ages and Renaissance, were so concerned with the paradoxical discovery of light in the depths of darkness that they undercut all the familiar distinctions between upper and lower, celestial and terrestrial. For them fire was only secondarily a celestial phenomenon: in origin it came from, and belonged at, the centre of the earth. This 'central fire', as they sometimes called

Burkert, op. cit., and note as well the implication of B1.26–7 that the road Parmenides has travelled—while still alive—is the same as the route normally taken by people at death: see below, Ch. 17 n. 6.

[16] Appendix I.

[17] *PGM* I.33–4, IV.437–49, 1596–1715, 1958–69, V.246–51, VIII.75–84, XXIII.3–5 (24–6), and *passim*. $\Pi\lambda\text{o}\acute{\upsilon}\tau\omega\nu$ ὁ ὑπὸ γῆν ἰὼν ἥλιος: Porph. fr. 358.5 Smith = Περὶ ἀγαλμάτων, fr. 7 Bidez, with E. Wüst, *RE* xxi. 1003. Cf. also Iamblichus *ap.* Lyd. *Mens.* 167.23–168.2, with Ch. 4 n. 10 above; and, in general, Eitrem, *Beiträge zur griechischen Religionsgeschichte*, iii (Christiania, 1920), 131–41.

[18] P. J. Alexander, *The Oracle of Baalbek: The Tiburtine Sybil in Greek Dress* (Washington, DC, 1967), 11.24–30; A. M. Heinen, *Islamic Cosmology* (Beirut, 1982), 146 (IV.6–7), 210.

it, was considered by them the key to the alchemical transformation process. According to them it was the real source of light—so much so that they referred to it as the 'sun in the earth', the 'subterranean' sun. On the one hand this 'earthly' or 'invisible sun' was the 'fire of hell', the 'black sun', the 'darkness of purgatory'. On the other, as the sun that 'rises out of the darkness of the earth', it was the origin not only of the visible sun but also of the light of the stars. And, significantly, they indicated that the nature of this hidden, generative fire was volcanic.[19]

Remarkably, considering what we have seen of Empedocles' own views on the subject, the earliest of the surviving references in alchemical literature to this paradoxical idea occurs in an Arabic text where the idea is attributed, quite specifically, to Empedocles. This apparent coincidence deserves a little careful attention here, because it has a significance which is not immediately self-evident—both for our understanding of Empedocles and for our appreciation of how he came to be understood in later generations by different people with different interests.

The Arabic text in question was known by the name of *Muṣḥaf al-jamāʿa*, 'Tome of the Gathering'. The original survives only in a few small fragments; however, part of it was translated into Latin under the title *Turba philosophorum*, the 'Gathering' or 'Assembly' of the philosophers. In both cases the title refers to a gathering of ancient philosophers under the presidency of Pythagoras.[20]

The relevant passage—which so far is only known from the Latin—has Empedocles deliver a speech that opens with the words, 'I signify to posterity that air is an attenuated form of water' (*significo posteris quod aer est tenue aquae*). The Empedocles

---

[19] For documentation and discussion of the sources see Jung (1968b), 148–52, (1970), 28 with n. 144, 54, 94–5 with n. 24, 98–100 with nn. 40, 42, 441 and n. 278, 512, (1968a), 249–50; and cf. also J. Ruska, *Tabula smaragdina* (Heidelberg, 1926), 226. For the particular relationship between sun and central fire outlined in the passage quoted from Mennens by Jung (1970), 441 n. 278, see Vodraska 39, 196–9; M.-T. d'Alverny and F. Hudry, *Archives d'histoire doctrinale et littéraire du moyen âge*, 49 (1975), 139–260; Marsilio Ficino, *De vita* 3.16.1–8 Kaske–Clark.

[20] For further details of the Latin work, and of its Arabic prototype, see Kingsley (1994c), § III.

The Sun

of the *Turba* then goes on to compare the earth to an egg, with the earth itself corresponding to the shell, 'the water which is under the earth' (*aqua quae sub terra est*) to the egg-white, and the fire underneath the water to the yolk. Finally 'there is the point of the sun at the middle of the yolk, which is the chick' (*solis autem punctus in medio rubei, qui est pullus*).[21]

This idea of a 'point of the sun' at the very centre of the earth is at first sight both paradoxical and bizarre, so much so that modern scholarship has cheerfully followed Julius Ruska in altering the text to give it a more acceptable meaning. But Ruska's 'emendation' is in itself far from plausible;[22] and, above all, it is unnecessary. The one person who, since Ruska's edition in 1931, has seen the need to retain the manuscript reading has been Carl Jung. Owing to his familiarity with alchemical literature he was quick to realize that the strange idea of the sun growing out of the middle of the earth was fundamental to alchemical doctrine. More specifically, he noted not only that even the earliest of Latin commentators on the *Turba philosophorum* found the reading 'point of the sun' in their texts—and interpreted it without any hesitation—but also that this same basic idea of a fiery, generative point at the heart of matter can be traced back to Gnostic writings of the very first

---

[21] *Turba* 52.1–53.22 Plessner. The Latin form of Empedocles' name, Pandolfus, is due to the common confusion in Arabic of *qāf* (ق) and *fā'* (ف): Ruska 24. More obvious corruptions of the name as a result of transmission via the Arabic are Bendaklis (Kaufmann 4, Schlanger 58) or Pantocles, as in Hermann of Carinthia, *De essentiis*, ed. C. Burnett (Leiden, 1982), 174.16. For Hermann's coupling in this passage of Empedocles with Plato as a prime authority on the origin and reascent of the soul see esp. Pl. *Phdr.* 248b–249b = Emp. C1 (DK i. 374–5); Plut. *De exilio* 607c–e (DK i. 359.13–18) with Mansfeld, VII. 135–7; Tert. *An.* 54.1 (*Dox.* 205); Plot. 4.8.1.17–46, 4.8.4.21–5.6; Porph. *De antro nympharum* 8; Hierocles, *In carmen aureum* 24.2–3; *Theol. Arist.* 1.30–42 (23.6–25.6 Badawī, Lewis 227–9); *Picatrix* 306.11–13 with n. 6. For Hermann's ascription to Empedocles of the idea that the soul is an entity which after its ordeal in this world eventually 'returns to its original rank and dignity', *ad originalem dignitatem rediturum* (174.17–18), cf. Emp. B119 and 146–7; Plut. *De Isid.* 361c αὖθις τὴν κατὰ φύσιν χώραν καὶ τάξιν ἀπολάβωσι; Hierocles, loc. cit. τὴν ἀρχαίαν ἕξιν ἀπολαμβάνει; and al-Ḥimṣī's addition to the text of Plotinus at *Theol. Arist.* 1.31 (23.10 Badawī, Lewis 227) about Empedocles exhorting souls 'to return to their own original, sublime, and noble world', *an... yaṣīrū ilā 'ālam al-awwal al-'alīy al-šarīf*.

[22] Ruska 51, 112, 178 with n. 2, followed by Plessner 54 ('evidente Verbesserung') and Rudolph 106 ('sicher richtig'). Ruska's emendation of *solis punctus* to *saliens punctus* on the basis of Arist. *Historia animalium* 561ᵃ9–13 requires that the author of the *Muṣḥaf* had a familiarity with Aristotle equal to Ruska's, but of this there is little trace; and the sense of the Aristotle passage does not even correspond to what we find in the *Turba* (in *Hist. anim.* the 'pulsating point' is said to be in the white of the egg, not the yolk).

centuries AD.[23] To this we can add that in Greek alchemical tradition the image of a point at the centre of a circle had for centuries been a common symbol used to denote the sun, the alchemical egg, and the generative, fiery principle of red sulphur.[24] The manuscript reading is plainly correct, and Empedocles' teaching in the *Turba* is clear: the fire at the centre of the earth gives birth to the visible sun.

The agreement between this teaching and the teaching of the historical Empedocles is striking; but it is important to appreciate that—and why—this agreement is not altogether a coincidence. The *Turba* is a difficult work. However, as Martin Plessner was the first to point out in detail, the process by which certain doctrines are put into the mouths of certain Presocratic philosophers in the opening section of the work is far from arbitrary. To be more precise, he claimed that this initial section of the *Turba* was based on a detailed knowledge of Greek 'doxographic' tradition—that is, the tradition followed by authors in later antiquity of systematically 'writing up the opinions' expressed by earlier philosophers on subjects such as the origins of man and the universe. He managed to show that the views ascribed in the *Turba* to the various Presocratics often correspond both to ideas attributed to them by the Greek 'doxographers' and to the statements of the Presocratics themselves as preserved in the surviving fragments of their works; he noted a particularly close affinity between the text of the *Turba* and the summary of Greek philosophical ideas provided by Hippolytus of Rome in the third century AD, and came to the conclusion that the author of the Arabic *Muṣḥaf* had succeeded 'in producing a text which adds some genuinely new material to the doxography of the Pre-Socratics and represents the oldest evidence hitherto known of the penetration of the doxographic tradition into Islamic literature'.[25]

Viewed both historically and geographically, there is little in this conclusion to be surprised at. As Plessner argued, the

[23] Jung (1968a), 218–21 with n. 149. Cf. Jung (1970), 42–56 with nn. 32–3, 35; (1968b), 152 with nn. 86–7. For the later history of the idea see Duhem's comments, v. 120–2.

[24] BR i. 108.13, 26, 122 n. 1, 244; C. O. Zuretti, *CMAG* viii. 2–3, 5, 10, 16–17, 24, 26, 34, 37–8; R. Turcan, *RHR* 160 (1961), 22.

[25] (1954), 337; see also Plessner 101–5, 131–2, and *passim*. For the history of Plessner's various publications on the *Turba* cf. ibid., pp. ix–xii and 135.

Arabic prototype of the *Turba philosophorum* was almost certainly written either just before or just after the year 900 by the alchemist 'Uthmān Ibn Suwaid from Akhmīm—the Greek Panopolis—in Upper Egypt.[26] Akhmīm has quaintly been described as a town which outwardly 'has no history';[27] but by the time of Ibn Suwaid it had already had a long history as a centre of alchemical theory and practice. He was by no means the only alchemist living in, or originating from, the town during the ninth century—any more than Zosimus of Panopolis was an alchemist living in isolation there some six hundred years earlier. There is evidence from both periods alike for the proliferation of alchemical, or alchemically minded, circles either in Akhmīm itself or in other associated centres with which people from Akhmīm were in contact; and this evidence, together with the notorious conservatism of alchemical lore and the basic convention of handing the knowledge down from generation to generation, very strongly implies the existence of a continuing and unbroken tradition in the place from the third and fourth centuries AD down into the early Islamic period.[28]

In fact there can be no doubting the need to trace the origins of the Arabic *Muṣḥaf* and Latin *Turba* tradition back to the Greek world. As a whole, the Arabic *Muṣḥaf* was clearly an original work; parts of it are only understandable in the context of Islamic religion and culture. But at the same time many of its terms and ideas, and sometimes even entire passages, can be traced back to Greek literature. Regarding the title of the work, and the form in which the text is cast—a gathering of and discussion between alchemists, or 'philosophers' as they used to call themselves—the very fragmentary remains still survive of a Greek alchemical work called, precisely, 'Gathering of the philosophers' (Περὶ συνάξεως τῶν φιλοσόφων). Here, clearly, we have the Greek prototype both of the Latin title *Turba*

---

[26] Kingsley (1994c), § III.
[27] G. Wiet, *EI²* i. 330a.
[28] For alchemical circles in Zosimus' time cf. Lindsay 338–42, Fowden 125–6, 167, 173–5, 188. For the 9th and 10th centuries see Ibn al-Nadīm 355.22–3, 359.1–6 = Dodge, ii. 855, 865, with Kraus (1942), pp. lxii–lxiii (Ibn Suwaid); 312.6–7 = Dodge, ii. 732 with Kraus, op. cit., p. lxiii n. 1 (Salāma ibn Sulaimān); 358.3–5, 359.4–5 = Dodge, ii. 862, 865 with Ch. 24 below (Dhū 'l-Nūn). On the conservatism and continuity of alchemical tradition see Ch. 15 with nn. 12–13.

philosophorum and of its Arabic equivalent. Although this line of affiliation has long been recognized, the tendency has been to restrict the connection to the name of the works alone and deny any similarity of content. As it happens, the parallels go much further than that: the Arabic *Muṣḥaf* may have been an original work in its own right, but it was also an extensive reproduction and reworking of older, Greek materials.[29]

To these points we have to add one other. The first nine speakers in the Latin *Turba*, who provide the cosmological introduction to the rest of the work, are given the names—corrupted and distorted during the double transmission from Greek into Arabic and from Arabic into Latin—of Presocratic philosophers.[30] Here again there are broad parallels with Greek alchemical literature: this time with a work, by an Olympiodorus of Alexandria, which cites the views of Presocratics on cosmology within a specifically alchemical context. The work in question happens to be a commentary on an alchemical text by Zosimus of Panopolis; as such, it provides one further testimony to the close interchange of alchemical and philosophical traditions, up and down the Nile, between Alexandria in the north and Panopolis or Akhmīm in the south.[31] There is no real chance that the author of the work is identical with the Alexandrian Neoplatonist called Olympiodorus; and yet the kind of reasons often put forward by those reluctant to attribute the work to the Neoplatonist are significantly wide of the mark, revealing more about modern attitudes to alchemical literature than anything else. It is only too easy, for instance, to dismiss the subject of alchemy as 'nebulous', 'eccentric', and an unsuitable interest for a 'philosopher'—in spite of the plain evidence for 'eccentric' interests on the part of Neoplatonists, and in spite of the fact that there is much 'about late antique Alexandria to suggest that its intellectual atmosphere was more

---

[29] For refs. to the points in this paragraph see Kingsley (1994*c*), nn. 77–81.

[30] Ruska 23–5, with the corrections by Plessner (1954), 334 and nn. 19–21; Rudolph 104; above, n. 21.

[31] For the Olympiodorus text cf. BR ii. 79.11–85.5 = iii. 86–91, with Berthelot's comments, *La Chimie au moyen âge* (Paris, 1893), i. 253 and Plessner's, (1954), 337; the text is re-edited in Viano. On contacts between Alexandria and Panopolis cf. e.g. H. M. Jackson, *Zosimus of Panopolis on the Letter Omega* (Missoula, Mont., 1978), 3–5 with Fowden's correction, 120 and n. 15. Note also the type of interaction, attested on papyri, between Alexandria and Oxyrhynchus: E. G. Turner, *Journal of Egyptian Archaeology*, 38 (1952), 84–93.

catholic than is generally realized'.[32] Again, just as significantly, certain scholars have dismissed the author of the alchemical Olympiodorus text as 'some muddle-headed charlatan' because he supposedly took his information about the Presocratics from two brief passages in Aristotle and then 'completely distorted' and 'corrupted beyond recognition' what Aristotle had said.[33] The verdict is vindictive enough to be impressive, but is itself a distortion of the facts. Closer analysis shows that the main source of information used in the Olympiodorus text was not Aristotle at all; instead, it was the doxographic tradition which had its roots in Theophrastus. The author's basic fidelity to, and acquaintance with, this tradition is unmistakable.[34] As Reitzenstein already saw and stated over seventy years ago, this work of Olympiodorus is just one of the

[32] That the author of the alchemical text was not Olympiodorus the Neoplatonist has been adequately demonstrated by J. Letrouit, *Emerita*, 58 (1990), 289–92 and further in Letrouit (1995), § 7; see also L. G. Westerink's comments in R. Sorabji (ed.), *Aristotle Transformed* (London, 1990), 331–6. For alchemy as 'nebulous' or 'eccentric' see Westerink, i. 22–3. The quotation is from the more balanced appraisal by Fowden, 178–9; compare Berthelot, *Les Origines de l'alchimie* (Paris, 1885), 144–5. Obviously relevant in this context is Proclus' *On the Hieratic Art*, with its close affinities to alchemy: cf. Bidez, *CMAG* vi. 139–51, plus Kraus (1943), 34 n. 3, and Lindsay 60–1. Note also Brisson's remarks in Σοφίης μαιήτορες: *Hommage à Jean Pépin* (Paris, 1992), 481–99, on the alchemical allusion in the Neoplatonist Olympiodorus' commentary on Plato's *Phaedo* (1.3.8–9 Westerink), where ἐκ τῆς αἰθάλης τῶν ἀτμῶν must mean 'from the sublimation of the vapours'—exactly as we find the word αἰθάλη used both by Zosimus of Panopolis (BR ii. 84.18, 113.5) and in the Olympiodorus commentary on Zosimus (BR ii. 84.12–85.5). On the other hand, Zosimus apparently wrote a life of Plato (*Suda*, s.v. Ζώσιμος Ἀλεξανδρεύς, Fowden 120 with n. 16); but there is no justification for describing either Zosimus or his alchemical writings, with Kraus (1943), 34, as 'Neoplatonic'. See Kingsley (1993b), 19–20; and note how pointedly Zosimus refers to Plato as merely 'thrice-great', in contrast to Hermes who is 'ten thousand times great': BR ii. 230.17–18 = Letrouit (1995), § 3.3.

[33] Westerink, i. 23 (citing Arist. *Ph.* 184ᵇ15–22 and *Met.* 983ᵇ20–984ᵃ8), followed and amplified by Rudolph 103.

[34] For example, with BR ii. 81.3 (Melissus) cf. esp. D.L. 9.24 (DK i. 258.18); with ii. 81.9–11 (Parmenides), Simpl. *Ph.* 28.7–8 = Theophr. FHSG § 229.4–5, Hipp. *Ref.* 1.11, and Stob. i. 35.13–14; with 82.5–6 (Thales), Simpl. *Ph.* 23.21–22 = Theophr. FHSG § 225.1–2. For BR ii. 82.21–83.1 τὴν γὰρ γῆν οὐδεὶς ἐδόξασεν εἶναι ἀρχὴν εἰ μὴ Ξενοφάνης ὁ Κολοφώνιος, cf. esp. Stob. i. 123.9–11 = DK i. 124.16–17—contradicted by Arist. *Met.* 989ᵃ5–6—and Mansfeld (1990), 151–5 (Viano, § 4.1 with n. 72, misses the specific origin of this tradition in Xenophanes B27). With ii. 83.7–8 (Anaximenes) compare D.L. 2.3, Simpl. *Ph.* 24.26–31 = Theophr. FHSG § 226A.15–21, ps.-Plut. *Stromateis* 3, Hipp. *Ref.* 1.7.1–2, Cic. *ND* 1.10.26, etc.; and for ii. 84.12–14 cf. Norden 247–8. The doctrine of τὸ μεταξύ attributed to Anaximander at BR ii. 83.11–14 and 84.16–17 is not attributed to him explicitly by Aristotle, but only by the later doxographers: see Kahn 43–6; Plessner 98. The correct reading at ii. 81.4 is Μέλισσος, not Μιλήσιος: cf. BR on ii. 81.17, with Letrouit (1995), § 7.1.

surviving Greek alchemical texts that appear to demonstrate direct or indirect familiarity with the doxographic tradition which took its formal point of departure from Theophrastus.[35]

The situation is much the same with the *Turba philosophorum* itself. Plessner well noted the close analogies between the initial sections of the work and the details of Presocratic 'doxography' preserved for us by Hippolytus of Rome;[36] but he also noted respects in which the *Turba* relied on doxographic sources independent of what we find in Hippolytus. Plessner has recently been criticized for naming more than one doxographic source for the work, and the unfortunate attempt has been made to impose an orderliness on the material which is entirely artificial by trying to derive all the *Turba*'s information about the Presocratics from Hippolytus alone.[37] This criticism of Plessner might, possibly, have had some force if we possessed all the doxographic texts ever written or circulated in antiquity; but that is far from being the case. The chance recovery of information about Empedocles from a work by Philo which is only preserved in Armenian is a striking reminder of how extremely fragmentary our knowledge is about the literature and information that once used to be available—especially in Alexandria.[38] And, what is more, we encounter exactly the same phenomenon in the doxography preserved by the alchemist Olympiodorus: a text containing summary accounts of Presocratic doctrines which markedly fail to correspond to any one surviving doxographic source alone.[39] Here, too, the desire for neatness must yield in the face of the evidence available.

In fact, the *Turba* contains too many signs to be ignored of

---

[35] R. Reitzenstein, *NGG* (1919.1), 5, 28–37. G. Goldschmidt, *Heliodori carmina quattuor* (Giessen, 1923), 11–19, adds little.

[36] Above, n. 25. See further Rudolph 112–21.

[37] Rudolph 110–11. Compare the analogous criticisms of Rudolph by Genequand (1991), 946 for positing Hippolytus as sole and direct source of information about the Presocratics in the case of the so-called 'Ammonius doxography'. Rudolph himself draws an explicit parallel between the supposed indebtedness of both the Ammonius text and the *Turba* to Hippolytus, (1989), 211; in both cases his over-simplification of the evidence is manifest.

[38] For the Philo material see Chs. 2–3; and cf. P. Tannery's general comments on the Olympiodorus text, *Mémoires scientifiques*, vii (Toulouse and Paris, 1925), 126–7.

[39] See above, n. 34. In her treatment of the Olympiodorus doxography Viano (§ 4.1–2) oversimplifies its relationship to Simpl. *Ph.* 22.22–25.13 and minimizes the significance of those points where the two sources differ.

dependence on other sources apart from Hippolytus.[40] In this context one particular point is worth noting. Two of the *Turba*'s opening speeches have no features at all in common with the text of Hippolytus. One is the ninth, by Xenophanes; the other the fourth, by Empedocles.[41] It is obviously relevant to our understanding of the Empedocles speech that the one by Xenophanes does contain unmistakable echoes from other doxographic sources.[42] When the material known to us through Hippolytus was not used, the void was evidently filled from elsewhere.

As already mentioned, Empedocles' cosmological speech in the *Turba* has a clear basic structure. After the opening statement that air is a rarefied form of water we are presented with a scheme of water under the earth and then, at the earth's centre, of fire underneath the water. Plessner was at his least successful in dealing with this particular speech; Ruska had already pointed out that the doctrine of water inside the earth was a genuine item of Empedocles' teaching, but he overlooked even that.[43] In fact however here we have not just one, or two,

[40] To mention just one example: the second speech, by Anaximenes, is based on a correlation between the distance and position of the sun relative to the earth, and the changing seasons (*Turba* 45.1–46.3 Plessner). For the comments at the start of the speech about warmth, coldness, and the rarefying and thickening of air Plessner (46–7) was able to cite parallels from Hippolytus and Plutarch; but he was unable to cite any parallel for the correlation between seasons and sun. Rudolph (105, referring to *Ref.* 1.7.6) noted that the same Hippolytus passage mentions the distance of the stars in connection with changes of temperature on earth; but as Rudolph admitted, it only mentions the link between stars and weather so as to deny it—and of a positive link between seasons and sun there is no mention at all. In fact the correspondence between sun and seasons which the *Turba* attributes to Anaximenes is explicitly ascribed to him in the *Placita* (2.19.1–2, cf. Stob. i. 203.16–17; DK i. 93.31–3). Regarding the historical background to what for us has become a self-evident correlation cf. Pfeiffer 22; West (1971), 102, 108–9; Kingsley (1990), 253 and n. 50.

[41] Cf. Rudolph's table, 120–1.

[42] The priority given to earth and water (82.22–83.26 Plessner) is, in particular, hardly a coincidence: cf. Xenophanes, DK 21 B29 and B33, and e.g. Sext. *Math.* 10.314, ps.-Probus, *In Bucolica* 6.31 (343.21–4 Hagen) and ps.-Plut. *Vita Homeri* 2.93.2 (963–6 Kindstrand) = *Dox.* 92–3, Porphyry *ap.* Philoponus, *In Physica* 125.27–32 Vitelli (DK i. 121.8–12), plus the further refs. listed by P. Steinmetz, *RhM* 109 (1966), 41 n. 83. Also, for *Turba* 82.5–6 Plessner cf. D.L. 9.19 (DK i. 113.23); Rudolph 110; Mansfeld (1990), 153–4. The passage in *Turba* 82.16–19 on the corruption of the elements may well be an alchemical application of what we find in D.L. 9.19 (DK i. 113.27).

[43] Ruska 177 n. 3; Plessner 57 and n. 118 ('the specific teaching of subterranean water is, admittedly, not attested for Empedocles in the doxographic literature'), repeated by Rudolph 106.

but three pieces of Empedoclean doctrine mentioned in succession: first the idea that *aer* is a rarefied form of water, second the idea that there are large amounts of water under the earth, and third the idea that underneath this water there is a central fire. We find these ideas attributed to Empedocles in the *Placita*, in the Aristotelian *Problems*, in Plutarch, Seneca, and Philo of Alexandria.[44] This is significant enough. But what is even more remarkable is that according to the Empedoclean doxography which—as we know from Philo—was available in Alexandria, these three ideas were not just isolated pieces of doctrine: they were all different aspects of one underlying theory. The fire at the centre of the earth heats the underground water lying above it; and as this water rises to the earth's surface the process of evaporation begins which ends in the water being transformed into *aer*—a word that to Empedocles himself meant 'mist', but was soon interpreted as meaning the element of air.[45]

That is only half of the story, however. The introductory section of the *Turba* certainly contains doxographic material, yet it is not just a doxographic text—any more than the section on the Presocratics in Olympiodorus is just a doxographic text. On the contrary, the Olympiodorus piece plainly testifies to a self-conscious principle of composition: a principle that involved taking items of doctrine attributed to individual Presocratics in the doxographic literature, using those items of doctrine as starting-points for brief meditations on alchemical themes, and attributing these meditations to the Presocratic himself. As for the alchemists' motives in adopting such a procedure, it is clear that their purpose was to trace an affiliation between Presocratic ideas and their own alchemical tradition.

Many of the misunderstandings about the opening section of the *Turba philosophorum* stem from failure to appreciate this method of composition. There is no point in criticizing the work because of the way that it continually, and unpredictably, alternates between ascribing more or less genuine items of doctrine to each of the Presocratics and putting into their

---

[44] Ps.-Plut. *Placita* 2.6.3 (... ἀναβλύσαι τὸ ὕδωρ, ἐξ οὗ θυμιαθῆναι τὸν ἀέρα); ps.-Arist. *Problems* 937ª11–16; Plut. *Frig.* 953e; Sen. *QN* 3.24.1–3; Philo, *Prov.* 2.61 (87.10–16); above, Ch. 3, below, Ch. 6.

[45] Ch. 3.

mouths straightforward alchemical teaching. What is essential in approaching this, or any other, ancient text is that we learn to take it and understand it on its own terms. In the first nine speeches of the *Turba* we find exactly the same kind of procedure adopted as in the text of Olympiodorus: the dividing line between the doxographic and the alchemical is just as fluid—often more so than scholars have noticed.[46]

Ironically, the resulting difficulty in determining precisely what is what only serves to validate the approach adopted by the writers of these texts, because it highlights the underlying similarities between Presocratic and alchemical traditions. As we have seen, the basic structure of the speech by Empedocles in the *Turba* no doubt derives from doxographic information originating in Alexandria. But, even here, matters are not so simple. It is striking that all the surviving doxographic reports which mention the existence, according to Empedocles, of fire underneath the water inside the earth are concerned with this subterranean fire only indirectly: only because of its effect on certain natural phenomena and processes, but never as a cosmological fact in its own right.[47] In other words the fundamental existence, according to Empedocles, of fire at the centre of the earth is either ignored or simply taken for granted. On the other hand, what in the surviving Greek evidence is at best implicit becomes explicit in the *Turba*: there the fire is clearly located and portrayed at the centre of the earth. Regardless of whether or not there might once have been some Greek doxographic material now lost to us which also happened to be more explicit, the central reason for this change of emphasis is plain. The idea of fire in the depths of the earth was an essential

---

[46] So e.g. with regard to the first speech, by Anaximander, both Plessner (42–3) and Rudolph (104) claimed that while the first and the last part contain doxographic material corresponding to what we find attributed to Anaximander in Hippolytus, the middle part of the speech is purely alchemical in content. And yet the statement in the very middle of Anaximander's speech, about moist vapour being drawn up from the earth by the sun to help the body of water maintain itself, plainly corresponds to the theory of the origin of rain attributed to Anaximander in the same passage of Hippolytus (*Turba* 39.12–13 Plessner; Hipp. *Ref.* 1.6.7 = DK i. 84.19; Kahn 63–4). This is just one instance of how even the most overtly alchemical passages can remain intrinsically faithful to Presocratic ways of thinking.

[47] Ps.-Arist. *Problems* 937^{a}11–16 and Plut. *Frig.* 953e = Emp. A69; Sen. *QN* 3.24.1–3 = A68, with Philo, *Prov.* 2.61 (87.10–16); ps.-Plut. *Placita* 5.26.4 (τοῦ ἐν τῆι γῆι θερμοῦ) = A70a; Procl. *Tim.* ii. 8.26–8 = B52.

item of alchemical doctrine. Technically speaking, it was bound up with the shape—and associated symbolism—of the alchemical apparatus; in more general terms, it can be traced back to ancient Near Eastern metallurgical and mystery traditions which reflect a period when metal-working had strong 'underworld' connotations and was still associated with underground hollows and caves. In short, the prominence given by the Empedocles of the *Turba* to the existence of fire at the centre of the earth is due less to any emphasis on its existence in the doxographic literature than to the role of this idea of a central fire in alchemy, where it was considered the key to the transformation process. As it happens, descriptions of the fire inside the earth broadly comparable to the one attributed to Empedocles in the *Turba* still survive in Greek alchemical literature; and it is also worth noting that Ibn Suwaid of Akhmīm—the person almost certainly responsible for the Arabic original of the *Turba*—wrote another work called 'The Book of the Great Blazing Fire' (*Kitāb al-jaḥīm al-aʿẓam*).[48] The word here for 'blazing fire', *jaḥīm*, is unusual and striking: the ordinary word for fire in Arabic is *nār*. On the other hand, *jaḥīm* is the standard term in Koranic and religious literature for 'hell-fire'; and whatever technical sense it might have had for Ibn Suwaid, there can be little doubt that he used it with an eschatological implication.

This aptitude on the part of alchemical writers for not only reproducing Empedoclean doxography, but also developing and supplementing it in the right direction, becomes even clearer in the case of the 'point of the sun' at the centre of the earth. As noted earlier, the idea of a sun-point at the heart of matter is a classic instance of alchemical symbolism; and even the terminology of a central 'point' of light indicates that here we have one of the many themes absorbed into the *Turba* literature from Gnostic circles. Historically, this process of absorption could hardly be easier to understand: apart from

---

[48] Ibn al-Nadīm 359.6. Note also, ibid. 359.3–4, the other work attributed to him, 'Book of the Red Sulphur': for red sulphur cf. above, with n. 24, and for its importance—both terminologically and as a book title—in later Sufism, Addas 112 with n. 10, also Massignon 155. Regarding the shape and symbolism of alchemical apparatus see BR i. 127–73, ii. 123 n. 4. For the idea of fire inside the earth in Greek alchemy cf. e.g. ibid. ii. 21.3–6, 409.11–415.9 with Kraus (1943), 37 (earth-egg); and the refs. below, n. 52.

Zosimus of Panopolis' well-known adherence to Gnostic traditions in his alchemical writings, there are equally significant—if less well-known—indications of links between Panopolis or Akhmīm and the type of Gnostic literature discovered some fifty miles further up the Nile near Nag Hammadi.[49] And yet, however Gnostic and alchemical this idea of a 'point of the sun', in attributing it to Empedocles the author of the *Muṣḥaf al-jamāʿa* attributed it correctly.

Ultimately these correspondences and overlaps between alchemical, Gnostic, and Empedoclean teaching can only be explained in one way: as a result of underlying similarities of outlook and world-view. It was not just a matter of Greek and then Islamic alchemists receiving doxographic information but more a matter, as it were, of their riding on the back of the doxography: using it, but also supplementing and even recreating it. We must not forget that the Graeco-Egyptian alchemists had access to Presocratic ideas by various routes and channels which were strictly distinct from doxographic tradition. On the one hand, direct familiarity with Empedocles' work cannot be excluded: the recent discovery of fragments of papyrus from Akhmīm—probably written some two centuries before the time of Zosimus, and containing otherwise unknown portions of Empedocles' poetry—is one of the more obvious cases in point.[50] On the other hand, there are important lessons to be learned from the way that mythological ideas about the underworld which are of ultimately Pythagorean and Orphic origin made their way down, in Christian dress, to Akhmīm; while there are even greater complexities involved in the process that led to Pythagorean and Orphic theories about the cosmic egg

---

[49] On Zosimus and Gnosticism cf. Festugière (1967), 211 and n. 30; Fowden 120 with n. 17; Letrouit (1995), §§ 3.3.7, 3.3.13. Equally indicative of the links between Akhmīm and the Nag Hammadi finds is the 5th-c. Berlin Coptic Papyrus 8502, which partially duplicates Gnostic material from Nag Hammadi and was apparently discovered in the cemetery at Akhmīm: H.-C. Puech, *Coptic Studies in Honor of Walter Ewing Crum* (Boston, 1950), 94 n. 1, 152 and n. 2. It will also be noted that, in their battle against heretical Gnostics, the Pachomian monasteries of Upper Egypt evidently fought a concerted feud with 'philosophers' from Akhmīm: cf. L. T. Lefort, *Les Vies coptes de Saint Pachôme* (Louvain, 1943), 117.4–118.31—the preceding lacuna is restored in A. Veilleux, *Pachomian Koinonia* (Kalamazoo, 1980), i. 73–6—and J. Doresse, *The Secret Books of the Egyptian Gnostics* (London, 1960), 135–6 with Puech's comments, op. cit. 93 n. 1. On the term 'philosophers' in this context see Kingsley (1994c), § III with n. 79.

[50] See the Introduction, with n. 15.

being adopted by the Egyptian alchemists.[51] As will emerge in the final section of this book, the blending of Empedoclean, Pythagorean, and alchemical traditions which occurred in Hellenistic and post-Hellenistic Egypt formed part of a highly intricate picture based not just on one-directional influence but also on underlying affinities. And, as we will also see, cosmological ideas about fire inside the earth or about the origin of the sun were far from being the only points of affinity between Empedocles and the alchemists.

One final detail deserves mentioning here. Behind the *Turba*'s images of a central fire and the point of the sun lies, as noted earlier, the basic alchemical notion that fire—all fire— has a hellish origin. Where Empedocles is concerned, it would be natural to assume he was the only person writing in Greek who ever chose to use the word 'Hades' in the way that he evidently did: as a semi-technical term of reference for the element of fire. In fact however another text shows other people, centuries later, doing just the same thing. The text is alchemical; in our manuscripts it is preserved along with Greek texts of Zosimus of Panopolis, and on this very point its affinity with ideas found in Zosimus is clear.[52] Once again, the apparent coincidence may seem strange or just accidental; but it will start to make better sense as we proceed. That will mean, in the first instance, getting back to Empedocles himself—and to the background to his own identification of Hades with fire. Before we can start to trace the branches of Empedoclean tradition, we need first to get back to its roots.

---

[51] For the cosmic egg see R. Turcan, *RHR* 160 (1961), 11–23 with Boyancé's comments, *MEFRA* 52 (1935), 95–112, P. Dronke, *Fabula* (Leiden, 1974), 79–99 with 81 n. 2; and for the ultimately Egyptian background to the Orphic motif of the cosmic egg, Kingsley (1994*c*), § 1 with the refs. in n. 25. Pythagorean eschatology and Akhmīm: below, Ch. 10 n. 26.

[52] BR ii. 295.17, 20 (cf. 293.1, 10, 23–4, 296.17, 297.19–20); iii. 123 with n. 4. The passage in question derives from the *Dialogue Between Philosophers and Cleopatra*: Reitzenstein, *NGG* (1919.1), 1–22; Letrouit (1995). The association of Hades and fire is (as Jean Letrouit has pointed out to me) already implicit in Zosimus' *On Virtue* (BR ii. 108.2–110.5).

II
MYSTERY
*

6
An Introduction to Sicily
*

There are many fires burning beneath the surface of the earth.

IN view of the unavoidable conclusion we have already arrived at, that Empedocles associated the underworld with the element of fire, these words of his could hardly be more suggestive. There should be no need to point out that the reference to fire 'beneath the surface of the earth' (ἔνερθ' οὔδεος) is itself a way of evoking the region of Hades.[1]

Incidental reports about Empedocles' cosmology by later writers help to place this brief fragment in a wider context. They add the information that the subterranean fires which he refers to were viewed by him not only as the cause of hot springs but also as responsible for the formation of stones, rocks, crags, and cliffs.[2] In the first instance this emphasis on the significance of underground fire needs, of course, to be understood in the context of the remarkable volcanic phenomena both on and in the neighbourhood of Empedocles' home island of Sicily. We have already seen (Ch. 3) how interested in these phenomena Empedocles was, and how naturally he incorporated them in his cosmology. Burnet was using little more than practical good sense when he pointed to these local phenomena and emphasized that they hold the obvious key to understanding Empedocles' identification of Hades with fire: 'nothing could have been more natural for a Sicilian poet, with the volcanoes

---

[1] Emp. B52; cf. *Il.* 14.274, *Od.* 11.302, Aesch. *Persae* 629–30, Soph. fr. 686, etc. For Empedocles' ἔνερθ' οὔδεος compare also the term ὑπουδαῖος, used by Plutarch of Kronos (*Quaestiones Romanae* 266e) and by Oppian (*Halieutica* 3.487 Lehrs) of the nymph Cocytis, 'who they say once lived beneath the surface of the earth (ὑπουδαίην ἔμεναι) and lay in Hades' bed; but when he seized the maiden Persephone from the crag of Etna . . .'.

[2] Sen. *QN* 3.24 = Emp. A68; ps.-Arist. *Problems* 937ª11–16 and Plut. *Frig.* 953e = A69.

and hot springs of his native island in mind, than this identification'. His observation was in vain. Bignone retorted by claiming that according to Empedocles 'only a little proportion of fire is underground', and, ever since, Burnet's case has been dismissed as 'adequately refuted by Bignone'.[3] However, Bignone has refuted nothing. Empedocles emphasized how much fire ($\pi o\lambda\lambda\acute{a}$) there is underneath the earth, Bignone emphasized how little ('piccola'). It is simply a question of whom we prefer to follow.

In a similar vein to Bignone, Bollack tried to minimize the extent and the significance of Empedocles' subterranean fire by claiming that the reference was not to volcanic phenomena but only to the origin of hot springs. It should hardly need saying that this distinction is meaningless and totally artificial. The various hot springs on Sicily are themselves just one aspect of the region's volcanic activity, and there can be no understanding Empedocles' reference to fire in the earth without attempting first to appreciate the geographical and geological conditions on the island.[4] Even the claim that Empedocles never spoke specifically of volcanic phenomena is untrue. We have Plutarch's word that he described how rocks and crags were created as they were thrust upwards by the fire 'seething' inside the earth: clearly the description is modelled on volcanic eruptions.[5] It is also revealing that Proclus cites the line about the 'many fires burning beneath the surface of the earth' as an allusion by Empedocles to the 'streams of fire under the earth' ($\dot{v}\pi\dot{o}$ $\gamma\hat{\eta}s$ $\dot{\rho}\acute{v}a\kappa\epsilon s$ $\pi v\rho\acute{o}s$). These 'streams' can only be volcanic; almost certainly they refer to the molten lava associated above all with Etna.[6] Proclus' interpretation of the fragment could

---

[3] Burnet (1930), 229 n. 3, cf. (1892), 242–3; Bignone 543 (see above, Ch. 2 n. 17; Ch. 5 with n. 1); Guthrie, ii. 144–5.

[4] Bollack, iii. 227. For a synoptic account of the hot springs, fountains, lakes, mud volcanoes, fire volcanoes, and other related phenomena in Sicily see Freeman, i. 74–8; cf. also ibid. 164–9, 519–30, and Pace, iii. 519–27.

[5] Plut. loc. cit. ($\dot{v}\pi\dot{o}$ $\tau o\hat{v}$ $\pi v\rho\dot{o}s$ ... $\tau o\hat{v}$ $\dot{\epsilon}v$ $\beta\acute{a}\theta\epsilon\iota$ $\tau\hat{\eta}s$ $\gamma\hat{\eta}s$ ... $\phi\lambda\epsilon\gamma\mu a\acute{\iota}\nu o\nu\tau o s$). Cf. Gilbert 304 n. 1.

[6] Procl. Tim. ii. 8.26–8. It so happens that in Greek literature $\dot{\rho}\acute{v}a\kappa\epsilon s$ $\pi v\rho\acute{o}s$ are almost invariably associated with Etna: cf. Thucydides 3.116 $\dot{o}$ $\dot{\rho}\acute{v}a\xi$ $\tau o\hat{v}$ $\pi v\rho\dot{o}s$ $\dot{\epsilon}\kappa$ $\tau\hat{\eta}s$ $A\check{\iota}\tau\nu\eta s$; Pl. Phd. 111e–113b $\ddot{\omega}\sigma\pi\epsilon\rho$ $\dot{\epsilon}v$ $\Sigma\iota\kappa\epsilon\lambda\acute{\iota}a\iota$...; ps.-Arist. Mirabilia 833^{a}17–21; D.L. 5.49 $\Pi\epsilon\rho\grave{\iota}$ $\dot{\rho}\acute{v}a\kappa o s$ $\tau o\hat{v}$ $\dot{\epsilon}v$ $\Sigma\iota\kappa\epsilon\lambda\acute{\iota}a\iota$, with P. Steinmetz, Die Physik des Theophrast (Bad Homburg, 1964), 214–15; Diod. 5.6.3 = Timaeus, FGrH 566 F164; Appian, Bella civilia 5.114, 117.

possibly just be a guess; if so, it was an enlightened one. There can be no doubting that with his 'many fires' Empedocles had in mind a far more extensive phenomenon than little, isolated pockets of fire.

To obtain some idea of how that extensive phenomenon used to be viewed in antiquity, we can turn to begin with to Strabo's description of Sicily:

the entire island is hollow beneath the earth, filled with rivers and fire.[7]

We will come to the rivers later; as far as the fire is concerned, Strabo's statement is virtually a commentary on the fragment of Empedocles quoted at the start of this chapter. However, it is primarily a geographer's commentary, and even more important than simple geography is the dimension of myth. For a counterpart in mythology to Empedocles' description of fires burning 'beneath the surface of the earth' (ἔνερθ' οὔδεος) we have only to look to Callimachus, with his portrayal of the fiery 'giant underneath the surface of the earth' (κατουδαίοιο γίγαντος) shifting the weight of Etna from one shoulder to the other.[8] Pindar also refers to a giant pinned down by Etna and causing its eruptions as he lies prostrate under all of Sicily. Regarding his mythical location, Pindar is quite specific: the giant is in Tartarus.[9]

This already brings us more or less to the heart of the matter as far as Empedocles' equation of fire with Hades is concerned. Subterranean fire, as we have seen, was of fundamental importance in his cosmology; in the first instance that points us to the volcanic fire under Sicily, and of course under Etna. As for Pindar, by placing his giant simultaneously in Tartarus and underneath Etna he was simply giving expression to the common and fundamental idea that the volcano, with its craters and caverns, is an opening into the underworld. This idea is only one instance of the tendency—reflected in literature as well as in local cults and myths—to associate

---

[7] ἅπασα ἡ νῆσος κοίλη κατὰ γῆς ἐστι, ποταμῶν καὶ πυρὸς μεστή, 6.2.9.

[8] *Hymn to Delos* 141–3. For the shifting from shoulder to shoulder cf. Frazer (1919), 197; also West (1966), 314.

[9] *Pyth.* 1.15–28. For the relation between Pindar's and Callimachus' giants cf. L. Malten, *Hermes*, 45 (1910), 552.

volcanoes and volcanic regions with entrances to Hades.[10] It of course explains the versions of Persephone's abduction which have her dragged down to the underworld via the caverns and craters of Sicily.[11] It also explains why, when in the *Phaedo* Plato compares fire in the underworld with the action of volcanoes and describes streams of lava as branches of the mythical river Pyriphlegethon, he refers specifically to Sicily; in other words he is simply reaffirming Pindar's association of Tartarus with the region of fire under Etna.[12] Finally it is worth noting, as a healthy reminder of the strength of a tradition which modern scholarship has virtually forgotten, that by the time of Milton this very same association of underworld and Etna was still a standard poetic analogy.[13] Whether the fire is related to Tartarus, as in Pindar, or to Hades, as in Empedocles, or to both, as in Plato, is of no real consequence: from at least the sixth century BC onwards, any mythical distinction between Hades and Tartarus was purely optional.[14]

Then there is the question of Hephaestus. Twice in the surviving fragments Empedocles uses Hephaestus' name when referring to the element of fire. That is not a particularly significant fact in itself: substitution of this god's name for the element he was so closely associated with is part of the poetic tradition Empedocles inherited.[15] In one of these two cases, the

---

[10] For Etna and the surrounding region see Ganschinietz 2381–2 and the next two notes below, plus J. K. Wright, *The Geographical Lore of the Time of the Crusades*² (New York, 1965), 225. For volcanic regions in general cf. esp. Ganschinietz 2379–82; S. Eitrem, *RE* xx. 259; E. Wüst, ibid. xxi. 995; Burkert (1969), 24; Milton, *Paradise Lost* 1.670–751 with P. Brockbank's comments in Brockbank and C. Patrides (eds.), *John Milton, Paradise Lost Books I–II* (London, 1972), 53–4. Similarly, hot springs are understood as tributaries of the rivers of the dead: Strabo 1.2.18, 5.4.5–6, 13.4.14; Ganschinietz 2379; Croon 67–87; Kedar 997. Apart from being untrue, Hardie's statement (284) that 'the Greeks did not associate entrances to the Underworld with volcanic phenomena, but rather with lakes and caves' is based on an entirely artificial distinction: as if lakes and caves, like hot springs, could not themselves be volcanically related phenomena. See above, with n. 4.

[11] R. Foerster, *Der Raub und die Rückkehr der Persephone* (Stuttgart, 1874), 63–98; Freeman, i. 530–42; Malten, *Hermes*, 45 (1910), 506–52; J. K. Wright, op. cit. 311. Cf. also Pace, iii. 465, 468, 480, and esp. 498–9 on the importance of natural and artificial caves in the Sicilian cults of Demeter and Persephone.

[12] *Phd.* 111d–112a.

[13] *Paradise Lost* 1.233–8.

[14] Cf. K. Scherling, *RE* ivA. 2442–3.

[15] *Il.* 2.426, 9.468 etc.; cf. Malten 328–30; L. Graz, *Le Feu dans l'Iliade et l'Odyssée* (Paris, 1965), 200–7, 349–50.

reference to 'Hephaestus' as an element alongside earth, rain, and *aither* is an obvious example of this weakened sense. However, in the other case the mention of his name alongside Nestis and Harmonia suggests a more specific allusion to the god.[16] This presents an interesting situation. As we have seen, for Empedocles the definitive god of fire—as named in fragment 6—was Hades; and yet he seems also to have allowed a subordinate association with Hephaestus. It is easy to dismiss this as a simple inconsistency; or to be more precise, it *was* easy while it was assumed that fire corresponds to Zeus instead of Hades. Characteristic of this attitude is the assertion that

> In Empedocles' naturalistic theology the names [of gods given to the elements] have only a secondary significance, and they exhibit a great variety in the terms used; the same element is capable of being referred to by the names of different gods, such as fire which is called Zeus as well as Hephaestus. All that is important is the fact that they are considered to be gods.[17]

But in fact there is clearly more to the matter than that. The crucial fragment 6 nominates Hades, embodiment of death and destruction, as the god of fire; the almost equally important listing of the elements in fragment 109 defines fire as 'destructive'.[18] From these two decisive passages it is fair to conclude that Empedocles saw the power of fire as fundamentally destructive. And yet that is only half of the story, because he also gave fire a central role as Aphrodite's chief accomplice in the creation of our world; it is plainly significant that he refers to fire in this creative role not as Hades, which would be absurd, but as Hephaestus, the craftsman.[19] To draw the obvious conclusion that he associated fire in its fundamentally destructive aspect with Hades and in its creative aspect with Hephaestus is hardly to ascribe to him a subtlety or sophistication which, from our position of superiority, we might like to believe is beyond his means. On the contrary, it implies no

[16] B98.1–2 with Malten 329; B96.2–4.
[17] Olerud 83.
[18] This is plainly the chief sense of ἀίδηλον here; cf. *Il.* 2.455 etc., and Parm. B10.3 with Mourelatos 237–40. That the word also contains a pun on Hades is perfectly plausible: cf. e.g. Soph. *Ajax* 608, Gemelli Marciano 48.
[19] For the creative role of fire in Aphrodite's cosmology cf. esp. B73, Guthrie, ii. 188–90, Wright 24–5. Use in this context of the name Hephaestus: B96 with Wright 210, B98.

more than an elementary awareness of those two apparently contradictory roles of fire which—needless to say—are so characteristic of its action and which were also to attract so much attention in later Greek philosophy.[20]

Hades and Hephaestus: on the surface they are two very different gods with nothing to connect them. But the apparent gulf between them rapidly disappears as soon as we look a little more closely at what the second of these gods will have meant, not for us but for Empedocles. The mythology and cult of Hephaestus spread to the rest of the Greek world from the north-east Mediterranean, where he appears to have had associations with subterranean—and specifically volcanic—fire.[21] His transference westwards to Sicily was evidently not via the Greek mainland, but direct. There, in Sicily, he took over the cult and attributes of an indigenous non-Greek god, Adranus: Adranus had his temple on the edge of Mount Etna, with a sacred grove and a fire 'that was never extinguished and never died down'.[22] It was here in the West—in Sicily and the surrounding islands—that Hephaestus' connections with fire expressed themselves most overtly in the form of a direct connection with volcanic fire. On Lipara he was the chief god of the island, and personification of the volcano; Thermessa, between Lipara and Sicily, was known as Hiera because—at least in historical times—it was considered 'sacred' to Hephaestus.[23] On Sicily itself, and in the immediate vicinity of Empedocles' town of Acragas, there was the 'hill of Hephaestus': a local cult centre where the god was believed to make his presence under the hill known by extraordinary feats of spontaneous combustion.[24] But above all Hephaestus was connected with

---

[20] Cf. esp. Cic. *ND* 2.15.40–1 with A. S. Pease's notes ad loc. For comments on the paradoxical role of fire—both destructive and creative—which still holds an important place in Greek ritual and folklore, see Danforth 130–1. For the 'theological departmentalizing' of fire in antiquity see also M. Detienne and J.-P. Vernant, *Cunning Intelligence in Greek Culture and Society* (Hassocks, 1978), 281–2 = *Les Ruses de l'intelligence* (Paris, 1974), 263–4.

[21] A. Hermary, *LIMC* iv/1. 628–9; Cook, iii. 228–35, F. Brommer, *Hephaistos* (Mainz am Rhein, 1978), 3. Hence the tendency to explain hot springs, also, as caused by Hephaestus: Malten, *RE* viii. 321, 324, Frazer (1919), 209.

[22] Malten 326; Aelian, *De natura animalium* 11.3, 20, with the refs. in Kingsley (1995*b*), n. 39.

[23] Freeman, i. 87–91; Ciaceri 153; Malten 322, 326; Pace, iii. 547; Delcourt 190.

[24] Solinus 5.23–4; Freeman, i. 76, Ciaceri 152, Pace, iii. 526, 598.

Mount Etna, not just in Sicilian cult and myth but in classical tradition right down to the end of antiquity. There, underneath the earth, was his home—and especially his workplace.[25] The common reluctance to give this fact its due significance is a result of failing to appreciate that, in spite of Hephaestus' formal inclusion in the Olympian pantheon, he essentially never lost his role as a god of the inner depths of the earth.[26] In short, any seeming inconsistency in Empedocles' referring to fire now as Hades, now as Hephaestus is more apparent than real. On the contrary, the name of Hephaestus is itself just one more pointer in the direction of the underworld.

Hades and Hephaestus, the destructive and creative: to place these two aspects of Empedocles' fire in their true perspective we need finally to set them against the background of the idea, so common in antiquity, that the underworld is a place of paradox and inversion. In particular it is the place where polar opposites coexist and merge, and especially the place where the paradox of destructive force being converted into creative power is realized at its greatest intensity. Two thousand years of classical tradition relating to volcanoes—primarily Etna—and the underworld are summed up by Milton's Lucifer: 'Here in the heart of Hell to work in Fire'.[27] This is not to quote Milton as direct evidence for ideas held by Empedocles but, once again, simply to emphasize the resilience and endurance of a tradition which classicists who treat Empedocles as a 'philosopher' ignore at their peril. Milton himself equated Lucifer with Hephaestus, and there is no shortage of descriptions in ancient literature of the extraordinary effects produced by Hephaestus as he works with the volcanic fire inside the earth.[28]

[25] Aesch. *Pr.* 365–74, Eur. *Cyclops* 599, Callimachus, *Hymn to Delos* 141–6, Virg. *Aen.* 8.414–54, *Aetna* 29–32, Solinus 5.9, Dante, *Inferno* 14.56–7, etc. Cf. Simonides, fr. 552 Page; Ciaceri 12–13, 151–3; Malten 323; and for the religious background, Kingsley (1995*b*) with Ch. 18 below.

[26] Cf. Malten 318; E. O. von Lippmann, *Entstehung und Ausbreitung der Alchemie*, i (Berlin, 1919), 608–9.

[27] *Paradise Lost* 1.151. The most typical example of the merging of opposites in the underworld is the mythical commonplace of Hades as a place of both fire and deepest darkness: J. Kroll, *Gott und Hölle* (Leipzig, 1932), 396–7. For other opposites cf. Dieterich 201–2; Eitrem, *RE* xx. 259–60; also Emp. B122–3. The idea of the world of the dead as a place of paradox and inversion was widespread in antiquity: cf. Burkert 347–8, Zandee 73–8, Harva 347–9.

[28] *Paradise Lost* 1.738–47. Cf. e.g. Aesch. *Pr.* 365–74; Thucydides 3.88.3; Callimachus, *Hymn to Delos* 141–6; *Hymn to Artemis* 46–63.

That this particular association of ideas was known to Empedocles is undeniable: to gain some impression of how important it was in shaping his cosmology, we only have to look at how he described the creation of the ancestors of men and women. On the one hand, the way in which he words his account of their formation inside the earth is plainly meant to invoke Hesiod's famous description of Hephaestus creating Pandora. On the other hand, his image of these first examples of humanity being spewed up by fire shooting into the sky is an obvious example of volcanic imagery.[29] In Sicily, Hades and Hephaestus are two sides of one and the same coin: two aspects of the volcanic fire just inside the earth.

[29] With Emp. B62, 96 and 98 compare Hes. *Op.* 60–3, *Th.* 570–2; cf. Wright 210, 217. Note esp. the allusion in B62's ἐρατὸν δέμας and ἐνοπὴν ἀνδράσι to Hesiod's εἶδος ἐπήρατον and ἀνθρώπου αὐδήν; *Op.* 56 and 82 help to understand what Empedocles meant by B62.1's πολυκλαύτων (a good example of 'Hesiodic pessimism': Wright 215). For Pandora herself conceived of as rising up out of the earth, and understood as the ultimate ancestor of mankind, cf. West (1978), 164–6. B62 and volcanic imagery: cf. Bidez 110, Gilbert 304 n. 1; also Guthrie, i. 291–2, Wright 199. Note the similarity between the description in B62 of the fire shooting up out of the earth to join the fire already in the heavens and the routine poetic descriptions of Etna's flames reaching up to the heavens: Lucr. 1.722–5, 6.644–5, 669–70; Virg. *Aen.* 3.571–4; *Aetna* 608–10; Sen. *Herc. Oet.* 285–6; Claud. *Rapt.* 1.164–5; and cf. Pi. *Pyth.* 1.19.

7
The *Phaedo* Myth: The Geography
*

> Nowhere do we better learn than in Sicily the folly of those arbitrary divisions which have made the study of history vain and meaningless.
>
> Freeman, i, p. viii

IN the last chapter we saw Plato bring Tartarus into association with Etna; we also saw that in doing so he was following a precedent already established, for example, in Pindar. But in his *Phaedo* he goes into considerably more detail about the geography of the underworld than Pindar does, and it will be worth turning to look at what exactly he has to say. This may seem to involve something of a detour as far as Empedocles is concerned; however, detours in the underworld are invariably instructive.

To begin with, Plato describes how

Underneath the earth there are ever-flowing rivers of an immense size, some with hot waters (θερμῶν ὑδάτων), others with cold. And there is much fire (πολὺ δὲ πῦρ), and great rivers of fire, and many rivers of liquid mud—some purer, others more turbid—just like in Sicily with its rivers of mud flowing ahead of the lava and then the stream of lava itself. These fill each of the regions [in the underworld] depending on where they happen to be flowing at any one time; and they all move up and down according to a kind of see-saw motion inside the earth. This see-saw movement is a result of the fact that one of the chasms in the earth happens to be vaster than any of the others, and is bored right through the entire earth. It is this chasm which Homer had in mind when he described it as

'far, far away, in the deepest abyss underneath the earth';

elsewhere both he and many other poets refer to it by the name of Tartarus (111d-112a).

80   *Chapter 7*

Later he adds more detail when he comes to describe the underground river called Pyriphlegethon:

> Near to its source it falls into a vast region burning with much fire (πυρὶ πολλῶι καιόμενον); there it creates a lake larger than the Mediterranean, boiling with water and mud. From there, turbid and muddy, it flows off in a circle, and after passing through various other regions in its winding course inside the earth it arrives at the far end of the Acherusian Lake. But instead of mingling with the waters of the lake it continues to circle several times underneath the earth before discharging into a lower level of Tartarus. This is the river they call Pyriphlegethon; its streams of lava cause fragments of it to erupt at various points on the surface of the earth (113a–b).

Considering that these passages set the scene for one of Plato's most famous myths, they have been given remarkably little attention. The few references to them in modern literature consist almost entirely of futile arguments as to whether the descriptions are meant to be mythical or scientific in nature—futile, because the modern distinction between myth and science is hardly if at all applicable to Plato's time.[1] And as for Plato himself, he is the one writer to whom one definitely should not look for any such clear-cut distinction. Certainly he was interested at a theoretical level in distinguishing between *mythos* and *logos*; between 'myth' as the conveyer of ideas that are beyond logical demonstration, and the positive certainty arrived at by reasoned argument. Even here, however, the matter is complicated by Plato's refusal to draw a clear line between *mythos* and *logos*, and by his repeated insistence that what to a superficial person is just a 'myth' may have all the decisive attributes of a *logos* for someone whose perception runs deeper.[2] But for our purposes even more important is the fact that the *logos* side of this apparent distinction is by no means equivalent to our category of 'science'. The reason for this is, very simply, that for Plato the study of the physical world was

---

[1] It would appear that this is gradually, if belatedly, coming to be realized. Cf. e.g. Guthrie, iv. 362, 'in Plato's time no firm line could be drawn between myth (or religious belief) and what was taken for scientific fact'. For examples of the pointless debate over myth versus science in the *Phaedo* myth cf. Friedländer, i. 261–9; Frutiger 61–6.

[2] *Gorg.* 523a1–2, *Phdr.* 245c1–2; G. E. R. Lloyd, *The Revolutions of Wisdom* (Berkeley, Calif., 1987), 10 and nn. 27–8, 136 and n. 116. See also Frutiger's comments, 5, 144, Zaslavsky 12, 15, and repeatedly ambiguous comments by Plato such as 'mythologizing in a *logos*' (*Rep.* 376d9, 501e4).

by definition merely provisional, just a matter of probability, and as such basically excluded from the domain of certainty. The *Timaeus* as a whole is a perfect example of the interchangeability and virtual identity in Plato's philosophy not of science and *logos* but of science and myth.[3] And to anticipate slightly, later[4] we will encounter a particularly vivid demonstration of just to what extent so-called 'scientific' endeavour in the fifth and early fourth centuries BC was governed by mythological considerations—and, above all, by the reading of Homer. As we will also see, that vivid example is extremely relevant to the question of Plato's own interests and sources of inspiration in the *Phaedo* myth itself. To return to the repeated attempts that have been made at interpreting Plato—and the *Phaedo* myth in particular—in terms of our own familiar distinction between myth and science, these are rather typical of the danger of importing anachronistic issues into the study of the ancient world as a substitute for attempting to understand it on its own terms.

In line with this guiding assumption of a myth–science dichotomy, the consensus of modern opinion in interpreting the *Phaedo* myth has been that Plato put all the details together from various sources, imaginatively harmonized them, and then added the real-life allusion to Sicilian volcanoes as a touch of 'scientific colouring'—based no doubt on personal observations he made during his earlier visit to Sicily—so as to give a practical, tangible dimension to the story as a whole.[5] One writer even cites the myth's obvious reference to Etna as a perfect example of Plato's style of composition, because it shows the 'reckless freedom' with which he converted his observation of the volcano into grist for the mythical mill.[6]

---

[3] For further discussion of these issues cf. e.g. Lloyd, op. cit. 9–11, 135–40, 181–3.

[4] Ch. 13.

[5] The term 'scientific colouring' is Guthrie's, (1952), 168; he adds that 'no one but Plato has made it [the *Phaedo* myth] into the wonderful imaginative whole which we find it to be'—a claim we will better be able to judge the accuracy of later on. Frutiger (63–4) tried to distinguish between what in the *Phaedo* myth may derive from scientific observations made by Plato during his journeys and what simply belongs to the category of 'myth'; but, all other factors apart, the ancient reports of Plato conducting 'scientific observations' on his travels have virtually no historical value whatever. See Wilamowitz's comments, i. 252 and, for this and other matters relating to Plato's first visit to Sicily, Ch. 8 below.

[6] Thomas 94–5.

This of course will not do, and not only because the distinction between myth and 'science' here is hardly meaningful or workable. One point of crucial significance has been overlooked when considering the story at the end of the *Phaedo*. This is the fact noted earlier: that the association of Etna with Tartarus and the underworld was an essential feature of Sicily's mythical geography. Plato did not invent the association; it existed long before him. By the time he was writing, the physical dimension of Sicily's volcanoes and the mythology of the other world had already been inextricably interwoven. In short, it is not nearly as simple as it might seem to detach Plato's reference to Sicily from the *Phaedo* myth as a whole.

But this is not all there is to be said on the matter; far from it. In fact we possess all the information we could possibly need to show that Plato's reference to the geography of Sicily is not just some incidental flourish in the way of a passing analogy but, on the contrary, explains the very fabric of the myth itself.

The best place to start is with the rivers flowing underneath the surface of the earth. These subterranean rivers, so elaborately portrayed by Plato, happen to have been a typical feature of Sicily's limestone terrain. He describes how

the regions in the hollows of the earth are ... all joined to each other by channels—some narrower, some broader—bored under the earth in numerous places, with interconnecting passages through which much water flows from one place to another as if into mixing bowls or 'craters' ($\kappa\rho\alpha\tau\hat{\eta}\rho\alpha\varsigma$).

He then goes on to describe how when these subterranean streams are filled up

they flow through the passages and through the earth and arrive at their various destinations, some forming seas, some lakes, others rivers and springs ...

In fact these details provide a precise account of the island's geography, with its rivers disappearing into underground channels only to emerge somewhere else; it should be sufficient to quote as an example Strabo's description of one particularly striking cavern (σπήλαιον) containing a large pipe-like channel and a river that flows invisibly through it for a considerable distance before surfacing (ἀνακύπτοντα) and becoming visible again.[7]

As for Plato's reference to the channels forming seas, lakes, rivers, and springs (κρῆναι), nothing needs saying about the seas; Sicily was famed for its extraordinary lakes;[8] its unusual rivers—visible and invisible, large and small—were objects of special wonder and interest throughout antiquity;[9] and exactly the same was the case with Sicily's remarkable springs.[10]

Then there is the mention of waters flowing into 'craters', *kratēres* (111d). Burnet, struck by the fact that this allusion to craters occurs just before Plato refers explicitly to Sicily, suggested a connection with the volcanic craters of Etna. What he failed to note is the fact that 'craters' *of water* were a virtually unique and particularly famous feature of Sicily—not just from a geographical but also from a religious point of view.[11]

Plato frequently uses the term *chasma* or 'chasm' in describing his mythical topography; it is also a normal word in descriptions of Sicily.[12] Another key term in his myth is the idea

---

[7] *Phd.* 111c5–d5, 112c5–d6; Strabo 6.2.9. On the Strabo passage see F. Lasserre's comments, *Strabon: Géographie*, iii (Paris, 1967), 262. Springs, water volcanoes, and the passage of watercourses through rocks and gorges (cf. Freeman, i. 82–3) are naturally all part and parcel of this phenomenon of underground rivers. Compare also Strabo's summary description of Sicily, 'the entire island is hollow underground, filled with rivers...' (loc. cit.), plus his account elsewhere of the river Arethusa (6.2.4; cf. Timaeus, *FGrH* 566 F41). With Strabo's ἀνακύπτοντα compare Plato's ἀνακύψας, *Phd.* 109d2, e3–5.

[8] Cf. Freeman, i. 75–6, 521–30, 536–42.

[9] Ibid. 74–84, 540–1. Cf. Solinus 5.16 *fluminum miracula abunde varia sunt* and esp. the name of Polemon's work 'On Sicily's Astonishing Rivers', Περὶ τῶν ἐν Σικελίαι θαυμαζομένων ποταμῶν: Macr. *Sat.* 5.19.26; L. Preller, *Polemonis Periegetae fragmenta* (Leipzig, 1838), 125.

[10] Freeman, i. 74–7, 364–7, 519–20; above, Ch. 3.

[11] Burnet 134. For Sicily's water craters cf. Diod. 11.89.1 (τοὺς ὀνομαζομένους κρατῆρας), Strabo 6.2.9, Macr. *Sat.* 5.19.19–21, 5.19.25 = Callias, *FGrH* 564 F1, 5.19.26–8 = Polemon, fr. 83 Preller (Palici); Freeman, i. 75, 161 (Hybla Geleatis); Ov. *Met.* 5.424 (Kyane); below, Ch. 11.

[12] *Phd.* 111c8, e6, 112a5; so e.g. Diod. 5.3.3 (referring to Persephone's descent into the underworld), 5.7.3, 11.89.4, cf. Strabo 6.2.9. It is, incidentally, worth noting that the idea of a *chasma* stretching an enormous distance into the earth (*Phd.* 111e6–112a6) is naturally suggestive of volcanic phenomena, and this for the Greeks was almost invariably a pointer to Sicily. Compare the volcanically inspired description in the 3rd-c. AD *Shih I Chi*: 'In Tai-Yü there is an abyss a thousand miles deep, in which water is always boiling. Metal or stones thrown into it are attacked and reduced to mud...' (Needham, iii. 610).

of the earth's 'hollows'.¹³ This has given rise to the notorious question, on which so many 'philosophical' issues have been made to hang, as to where and how Plato came by the notion of 'hollows' in the earth. The general answer we are given, on the basis of some perilously superficial reasoning, is that the idea is 'distinctively Ionian' as opposed to 'Western' or Pythagorean. In fact, however, the question answers itself as soon as we note that the very same word 'hollows' (κοῖλα or κοιλώματα) was an entirely routine way of referring to Sicily's craters and to the numerous crevices in its extraordinary terrain.¹⁴

Again, Plato emphasizes in the myth that the rivers he is describing take on the quality of the earth they pass through; this happens to have been a particularly noticeable feature of the rivers in Sicily.¹⁵ As for his dramatic account of the periodic see-saw motion of the waters, rushing up to the earth's surface and then withdrawing completely into the bowels of the earth, it has been understood both in antiquity and in modern times as an attempt to explain the phenomenon of tides.¹⁶ This may be correct; but while Plato's precise description of the movements of the water sounds rather too radical to have been inspired by tidal phenomena alone (which, incidentally, are

---

¹³ τὰ κοῖλα or ἔγκοιλα: 109b5, c2–3, d6, 110c7, 111c5.

¹⁴ Cf. e.g. Schol. Soph. *OC* 1593 κοίλου πέλας κρατῆρος: τοῦ μυχοῦ· τὰ γὰρ κοῖλα οὕτως ἐκάλουν ἐκ μεταφορᾶς, ὅθεν καὶ τὰ ἐν τῆι Αἴτνηι κοιλώματα κρατῆρες καλοῦνται. The scholiast goes on to discuss the general belief that Persephone was snatched down to Hades through κοιλώματα of this kind. References to Sicily's hollow recesses and hollowed-out depths could hardly be more common: see Diod. 11.89.2 and Strabo 6.2.9 on its μυχοί and ἀμύθητοι βυθοί, Lucr. 6.683, 696, *Aetna* 1, 491/2, etc., and Strabo 5.4.9 for the vast system of 'hollows', κοιλία, stretching underground from Sicily through to Cumae. For Plato's hollows as 'Ionian' cf. e.g. Burnet 127; Friedländer, i. 269–72; Hackforth 173–4. However, even on a purely theoretical level, it is essential when trying to distinguish between 'Ionian' and 'Pythagorean' to remember that Pythagoras himself came from Ionia. On the danger of categorizing supposedly 'philosophical' concepts as Ionian see also above, Ch. 3 n. 39. Friedländer (i. 265) further confuses the issue by claiming that in the *Phaedo* myth 'the theory of the hollows belongs to geophysics'—as if such an impressive title has any meaning whatsoever in relation to the 4th century BC.

¹⁵ *Phd.* 112a6–7: cf. Strabo 6.2.4 and 9, Solinus 5.17 *Himeraeum caelestes mutant plagae; amarus denique est dum in aquilonem fluit, dulcis ubi ad meridiem flectitur.* Aristotle in his commentary on the *Phaedo* passage makes what is implicit there explicit by explaining that Plato must be referring primarily to the taste and colour of the water (τοὺς δὲ χυμοὺς καὶ τὰς χρόας, *Meteor.* 356ª13–14). For change in the colour of Sicily's waters cf. also Macr. *Sat.* 5.19.27 = Polemon, fr. 83 Preller.

¹⁶ *Phd.* 111d2–112e2; cf. ps.-Plut. *Placita* 3.17.3 = Stob. i. 253.4–7 (*Dox.* 383), Burnet 136.

unknown in the Mediterranean), it is virtually tailor-made as an explanation of the Sicilian *fiumare* which oscillate in time between huge torrents and completely dry, rocky channels.[17] Finally, nothing could be more typical of Sicily than the prolific combination of hot waters and cold referred to by Plato. His term for 'hot waters', *therma hydata*, is also the term normally used for describing Sicily's hot springs—and about them enough has already been said.[18] What is more, as a result of inevitable processes that were mentioned earlier, these Sicilian springs became essential components in local mythology about the underworld.[19] Once again this is directly relevant to the story in the *Phaedo*, and particularly effective in undermining the anachronistic distinctions between myth and science or between eschatology and physics that have been superimposed upon it.

Then there is the mud. Plato refers to mud a number of times. Once, in describing the 'many rivers of liquid mud, some purer, others more turbid', he states explicitly that he has the geography of Sicily in mind (111d8–e2). But when he returns later on to the topic of water and mud in discussing the Pyriphlegethon (113a7–b1), it is just as clear that here again his description is inspired by Sicily. The island was in fact notorious for its water and mud volcanoes.[20] To be specific, it would hardly be possible to find a more exact verbal parallel than the one between Plato's account here of the Pyriphlegethon creating an enormous lake, 'boiling with water and mud, from where—turbid and muddy—it flows off in a circle' (ζέουσαν ὕδατος καὶ πηλοῦ, ἐντεῦθεν δὲ χωρεῖ κύκλωι θολερὸς καὶ πηλώδης) and a typical geographer's account of the famous double lake of the Sicilian Palici:

> Their water is turbid (θολερὸν) ... and is carried along, bellying into swirls and seething, just like the circling eddies of boiling waters (ζεόντων ὑδάτων) as they bubble up to the surface.[21]

---

[17] Cf. Freeman, i. 79 ('wide, stony beds, at one time empty or with the scantiest supply of water, but growing at other times into wide and rushing torrents').
[18] *Phd.* 111d6–7. Cf. esp. Strabo 6.2.9 θερμῶν γοῦν ὑδάτων ἀναβολὰς κατὰ πολλοὺς ἔχει τόπους ἡ νῆσος; above, Ch. 3.
[19] Freeman, i. 364–7, 525–8; Richardson 18, 163, 181; Ch. 6 with n. 10. For cold springs see below, Ch. 9.
[20] Cf. esp. Solinus 5.24; Freeman, i. 74–5, 161.
[21] Polemon, fr. 83 Preller *ap.* Macr. *Sat.* 5.19.27 (330.14–331.2 Willis).

At last, with these boiling waters we come to the fire. Plato's reference to 'much fire' (πολὺ πῦρ) exactly echoes Empedocles' 'many fires' (πολλὰ πυρά) which, as we have seen, obviously refers to Sicily; his combined reference to underground rivers and underground fire echoes exactly Strabo's description of the island.[22] Burnet was considerably more alert than other commentators to the possibility of a Sicilian colouring to Plato's myth, and yet when he came to comment on the passage describing a 'vast region burning with much fire' he suggested that Plato was basing his account on voyagers' reports about the Senegal.[23] First, however, the reference to 'much fire' is a repetition of the identical phrase earlier on, where Plato specifically refers to Sicily (111d–e). Second, whereas any stories about Senegal which Plato might possibly have heard will just have mentioned fires, he himself defines the lake in question as boiling with water and mud—and that, as we have seen, is a precise description not only of eruptions at Etna but also of other famous sites of volcanic activity in Sicily. In fact, here too, the verbal similarities to geographical descriptions of Sicily could hardly be more exact. On the one hand, Plato speaks of a huge subterranean lake with small but violent outlets and 'burning with much fire' (πυρὶ πολλῶι καιόμενον). On the other, Diodorus Siculus in his own account of the double lake of the Palici describes them as craters which are small at the surface but of an immense depth, 'very similar in nature to cauldrons burning with much fire (ὑπὸ πυρὸς πολλοῦ καομένοις); and they throw up boiling water'.[24] The similarities do not stop there. Plato goes on to speak of his boiling subterranean waters as being thrown up in 'jets'; Strabo, in his own description of the Palici, uses the identical term; while both he and Diodorus use it elsewhere in describing the volcanic phenomena at Etna as well as on the Liparan islands just off the north coast of Sicily.[25]

Enough—more than enough—has already been said to

[22] Phd. 111d7, 113a7; Emp. B52; Strabo 6.2.9.
[23] 113a6–7 with Burnet ad loc., referring to Hanno's Periplous 13–15.
[24] Phd. 113a6–7; Diod. 11.89.2.
[25] Phd. 113b6 ἀναφυσῶσιν; Strabo 6.2.9 κρατῆρας ἔχουσιν ἀναβάλλοντας ὕδωρ εἰς θολοειδὲς ἀναφύσημα καὶ πάλιν εἰς τὸν αὐτὸν δεχομένους μυχόν; Diod. 5.6.3 and 5.7.3 = Timaeus, FGrH 566 F164; Strabo 5.4.9 (ἀναφυσᾶσθαι). For Strabo 6.2.9 see above, nn. 7, 11, 14; for Plato's mention of ῥύακες in 113b5, Ch. 6 n. 6.

show that even the smallest details of Plato's mythical landscape, with its rivers above and below the ground, were inspired and put together as a result of intimate acquaintance with the geography and natural phenomena of Sicily. They are not the figments of Plato's own imagination, and they have not been pieced together from various disparate sources; they are thoroughly consistent, homogeneous, and plainly derive from one specific area of the ancient world. The significance of this conclusion begins to emerge when we combine it with the fact that, as pointed out by Thomas, the *Phaedo*'s description of these mythical and physical rivers provides the structural foundation for the myth as a whole: a fact which no amount of arbitrary distinctions between 'mythical' and 'scientific' elements or 'eschatological' and 'cosmological' sections in the myth can possibly obscure.[26] All that remains is to see what this conclusion means.

---

[26] Cf. Thomas 94–5. It should hardly need emphasizing that for the Greeks of Plato's time eschatology was still inevitably and essentially bound up with matters of cosmology—as indeed it always remained in the West down to the end of eschatology in the 19th century.

8
The *Phaedo* Myth: The Sources
*

As we have seen, at its most basic level the myth at the end of the *Phaedo* is Sicilian in origin. The next step is to isolate the means by which this mythical material reached Plato—or he reached it.

That proves to be no problem. The *Phaedo* is the most overtly Pythagorean of all his dialogues, in both its ideas and its setting.[1] But nowhere is the debt to Pythagoreanism more evident than in the closing myth. As Plato makes perfectly clear, the ultimate point and purpose of the myth is its eschatology: its message about what happens to the soul after death.[2] On the one hand, there can be no doubting that for the main source of Plato's eschatological myths in general, and particularly of the eschatology in the *Phaedo*, we must look to the Pythagoreanism of southern Italy and Sicily.[3] On the other hand, we have already seen that there are no grounds whatever for cutting off the specifically eschatological or other-worldly passages in the *Phaedo* myth from their 'physical' or topographical foundations—while there are sound reasons for refusing to do so. Stating the matter in this way only serves to highlight what should already be becoming obvious: that the coincidence of a Sicilian topography combined with an eschatology of evidently Sicilian or Italian origin is no mere accident.

A closer look at the structure of the myth allows us to fill in these broad outlines and give them greater definition. The

---

[1] For some basic comments see Bostock 4, 11. For the ideas in the dialogue cf. also Burnet 17, 22–4, Burkert 272; for the setting, Burnet 1, 9–10, 19–20, Burkert 92, 212, Philip 41 n. 8.

[2] *Phd.* 107a–d, 114c–117c. Cf. Friedländer, i. 388 n. 16; Guthrie, iv. 363–4.

[3] See in general Dieterich 112–24 (and, for Dieterich's term 'Orphic–Pythagorean', Burkert 133 and Chs. 10–12 below); Thomas 60–1, 72, 108–57. For the eschatology in the *Phaedo* cf. e.g. ibid. 120–2; Kerschensteiner 132–3; Hackforth 172.

overtly eschatological parts of the myth are its introduction (107d–108c) and its conclusion (113d–114c). When we turn to the passage immediately following the introductory section (starting from 108c), we are turning to what has been referred to as 'the most striking and original part' of the myth—that is, it is a part in which Plato supposedly departs from his Pythagorean sources.[4] In fact, however, the first section of this second part (108c–109a) gives an account of the sphericity of the earth which is undoubtedly western in origin—deriving from either Italy or Sicily—and happens to be representative of contemporary Pythagorean cosmology.[5] In other words, just where Plato is claimed to be departing most radically from Pythagoreanism he is doing precisely the opposite. But there is also another factor here which needs taking into account. At this particular point in the myth, Plato has Socrates carefully qualify his exposition by explaining that the picture of the earth he is now describing is not the result of his own personal deductions but represents what 'someone has succeeded in persuading me'.[6] A few modern writers have insisted on making the 'someone' purely fictional, explaining the cosmological theories which Plato specifically attributes to this unnamed person as being the creation either of Socrates or of Plato himself.[7] But historically this is nonsensical;[8] otherwise, it is just another sad indication of the lengths to which some scholars have been prepared to go to preserve Plato's originality even at the cost of ignoring his own obvious hints. As it happens, Plato's pointed reference here to an anonymous source is a plain acknowledgement on his part that the views he is putting forward are not

[4] Hackforth 172.
[5] Cf. Furley 23. The idea of a spherical earth as presented in this section of the *Phaedo* undoubtedly goes back to Parmenides (Kahn 115–18, Guthrie, ii. 64–5, Burkert 303–6, Furley 24–6); but—regardless of whether Parmenides originated the idea or adopted it from earlier Pythagoreanism—by Plato's time it was taken for granted in Pythagorean circles. Cf. esp. Zeller, i/1. 531 and n. 2; Guthrie, i. 293–5. For the relevance of this Pythagorean background to our passage in the *Phaedo* see Frank 186, Friedländer, i. 273, Thomas 85–8, Kerschensteiner 134, Hackforth 173, Guthrie, loc. cit. Regarding Parmenides and Pythagoreanism cf. Burkert 280 and n. 13 with Kingsley (1990), 250 n. 34, 255–6; Ch. 20 with n. 45.
[6] ὡς ἐγὼ ὑπό τινος πέπεισμαι, 108c7–8; cf. 108d2–3, 108d9–e2, 108e4–109a7, and below, n. 11.
[7] So Burnet and Hackforth ad loc.; W. M. Calder III, *Phronesis*, 3 (1958), 124.
[8] Calder's view (loc. cit.) that the sphericity of the earth was unknown to the Presocratics is untenable: above, n. 5.

original either to Socrates or to himself: comparison of appeals to the authority of an unnamed 'someone' in Plato's other dialogues makes it virtually certain that the person in question was a Pythagorean.[9] As far as the *Phaedo* itself is concerned, this is further confirmed by a passage earlier on in the dialogue where the mention of teaching received orally from 'certain people' refers without any doubt to Pythagoreans.[10] A highly consistent picture is beginning to emerge.

So we come to the remaining section of the myth (109a–113c). This is the part taken up with the idea of the 'true earth' and its vast geography of rivers and hollows: a geography which, as we have seen, derives from Sicily. The first point to note here is that Plato cites as authority for this section the same 'someone' whom he has already appealed to for the theory that the earth is a sphere. This is perfectly clear in the Greek, although translators and commentators have gone a long way to obscure the fact—in line, once again, with the spoken and unspoken dogma that Plato distinguished radically between the categories of science (in this case the sphericity of the earth) and myth (the strange mythical geography). In fact, when this passage as a whole is understood as it should be it has the interesting implication that Plato placed the idea of a spherical earth firmly in the category of myth.[11] The implication is

---

[9] Cf. esp. *Gorg.* 493a–c, *Rep.* 583b (ὡς ἐγὼ δοκῶ μοι τῶν σοφῶν τινὸς ἀκηκοέναι); Adam, ii. 378–80, Dodds 225 n. 5 and (1959), 26–7, 297, 303, Burkert (1968), 100. That the 'someone' in the *Phaedo* myth is a Pythagorean was seen by Frank, Kerschensteiner, Friedländer, and Furley, locc. citt. (n. 5). Cf. also Frank 364 nn. 219–20; Rohde 468–9; and below, Ch. 9 with n. 36, Ch. 12 with n. 52.

[10] *Phd.* 61d6–e9; the 'certain others' in 61e8 are clearly other Pythagoreans apart from Philolaus. For the intimate connection between ideas presented in this early part of the *Phaedo* and the ideas contained in the final myth see below, Ch. 10 with n. 20.

[11] The essential point which has been overlooked is that the words ὡς ἐγὼ ὑπό τινος πέπεισμαι, 'as someone has persuaded me' (108c7–8) are picked up by the statement πρῶτον μὲν τοίνυν τοῦτο πέπεισμαι . . . ἔτι τοίνυν . . . : '*first* I am persuaded that' the earth is a sphere, '*and secondly*' of the earth's fantastic geography (109a6–9). Equally decisive is the fact that when Socrates alludes for the first time to the someone who has 'persuaded' him, he makes it clear that he is referring not only to the doctrine of the earth's sphericity but also—and even more plainly (108c5–8)—to the teaching about the 'many wonderful regions of the earth': that, of course, is a reference forward to the later passage with its description of the 'true earth' and its hollows (cf. esp. 109b4–6, 110c3, 111a2–3, 112e4–5; also 108d9–e2). This disposes altogether of Friedländer's influential attempt (i. 265) to draw a basic distinction between the 'convincing', scientific nature of the first section (sphericity of the earth) and the purely 'mythical' nature of the second (geography of the true earth). Friedländer has also been misled

significant enough in itself, because of what it tells us about Plato's own attitude to the relation between science and myth. However, for our immediate purposes there is an even more important conclusion to be drawn from Plato's appeal here to the same 'someone' whom he has already cited as authority for the theory of the earth as a sphere. That conclusion is that for the source of this Sicilian geography we must look once again in the direction of Italian and Sicilian Pythagoreanism.

Then there is the very idea here of a 'true earth': the idea that the world we appear to live on is only a dismal replica of another earth of cosmic dimensions. Almost without exception scholars have slid over this idea of a true earth on the assumption that it is a concept invented by Plato himself. But even if we were not obliged—as we plainly are—to view the details of the *Phaedo* myth against the historical backdrop of Pythagoreanism, things would still not be so simple. We can hardly draw any dividing line between Plato's notion here of another, aitherial earth—purer and more beautiful than our own, to which purified souls go to live when they die (109b–e, 114b–c)—and the various Pythagorean ideas of 'another' or 'aitherial' earth, a 'celestial' or 'Olympian' earth, inhabited as well. We know that in antiquity there were a number of different ways of identifying this other earth: as an invisible planet, or the moon, or the stars, or heaven itself.[12] But what is

here, along with many other writers (so e.g. Dicks 95–6), by his further mistake of giving the crucial verb πείθειν the rationalizing sense of 'convince by argument'. For the correct meaning of the word cf. e.g. *Laws* 903a-b, where scientific discussion and argument are explicitly contrasted with the πείθειν, the seductive persuasiveness, of myths; *Gorg.* 493c5–d3, where 'persuading' and 'myth-telling' are virtually synonymous (πεῖσαι . . . πείθω . . . μυθολογῶ); Laín Entralgo 112–13, 119–20, 124–6. For the role of persuasion in the myth of the *Phaedo* see also Boyancé 155–61; and for general comments on the mysterious and quite irrational power of πειθώ or persuasion, Detienne (1973), 62–8.

[12] 'Other earth' as an invisible planet: Arist. *Cael.* 293ª23–4; as the moon: Procl. *Tim.* ii. 48.17–21, iii. 142.12–16. 'Heavenly earth' as the moon: Porph. fr. 360.25–9 Smith, Procl. *Tim.* iii. 172.20–1; as sphere of fixed stars: ibid. ii. 48.24–5. 'Olympian earth' as moon: Plut. *De fac.* 935b–c; unspecified in Dam. *Phd.* 2.131. 'Aitherial earth' = moon: Macr. *Somn.* 1.11.7, 1.19.10; Procl. *Tim.* ii. 48.15–17 (cf. i. 147.1–9); Schol. Arist. 505ª1–2 and 41–2 Brandis = Simpl. *Cael.* 512.18. Counter-earth as invisible planet: Arist. *Cael.* 293ª24, ᵇ20, *Met.* 986ª11–12, Plut. *De animae procreatione* 1028b, Stob. i. 196.22–5, 221.5–7 (cf. ps.-Plut. *Placita* 2.29.4), ps.-Plut. *Placita* 3.9.2, 3.11.3, Alex. *Met.* 40.32, Calcid. 122 (166.10), Ascl. *Met.* 35.22–4, Simpl. *Cael.* 511.19–512.8; as moon: Plut. *De fac.* 944c (part of moon only), Porph. *VP* 31, Ascl. *Met.* 35.24–7, Schol. Arist. 505ª1 and

significant about all these explanations, and makes them comprehensible, is their common denominator. That, although it may not be immediately apparent, is eschatological: behind the various identifications lies the one fundamental idea of a place where the souls of the dead go to dwell.[13] Any attempt to organize these explanations into a chronological sequence—some earlier than Plato, others later—is tempting, but doomed: all of them would seem already to have been in existence by Plato's time. So, for example, the Pythagorean Philolaus is held to have originated the idea of a celestial 'counter-earth' in the form of an invisible although inhabited planet, but he is also said to have propounded the mythical idea of the moon as an inhabited earth; and according to Heraclides Ponticus, one of Plato's companions, Pythagoreans spoke of the stars as heavenly 'earths', doubtless inhabited as well.[14] This proliferation of different but related theories is almost certainly to be explained not in terms of some linear development in Pythagorean dogma, but as one more sign of the creative freedom with which individual Pythagoreans were

----

41–4 = Simpl. *Cael.* 512.17–20; as heaven: Clem. *Str.* 5.139.1. Cf. Lobeck, i. 499–501, Cumont (1942), 184 n. 2, 187–8, Burkert 231, 338, 344, 347–8. Later, the term 'counter-earth' was applied to the southern hemisphere and its mysterious inhabitants: Cic. *Tusc.* 1.28.68; Plin. *HN* 6.24.81; Apul. *Met.* 1.8.6; Solinus 53.1; Achilles, *Isagoge* 30, 65.16–66.25 Maass; Neugebauer 373 and n. 1, 378. Here it will be remembered that, beneath the scientific rationalizations, there was an intimate association between the underside of the earth and the realm of the dead: cf. e.g. Virg. *G.* 1.242–3, Burkert 347–8 with n. 57, Neugebauer 372–8. Clement's statement (loc. cit.; as usual in his time, 'heaven' here plainly means heaven of the fixed stars) that the Pythagoreans equated counter-earth with heaven has a close affinity with Gnostic exegesis, but is also related to the idea of the fixed stars as worlds with earths or as one single earth (Heraclid. fr. 113 = *OF* fr. 22; Procl. *Tim.* ii. 48.24–5).

[13] Cf. Zeller i/1. 533–4, Frank 197 and, for the origins of the Pythagorean idea of a 'counter-earth', Burkert 347–8 with Ch. 13 below.

[14] For Philolaus' inhabited counter-earth see ps.-Plut. *Placita* 3.11.3 = Philolaus A17a (Huffman 238), Burkert 348 and n. 58; for his inhabited moon, Stob. i. 222.3–8 (cf. ps.-Plut. *Placita* 2.30.1; Philolaus A20 = Huffman 270) with Burkert 346. Stars as earths with their own atmospheres: Heraclid. fr. 113 = *OF* fr. 22. Wehrli's attempt (99, ad loc.) to restrict the meaning of *asteres* here to planets as opposed to stars is not only forced but clearly wrong—cf. W. Kroll, *De oraculis Chaldaicis* (Breslau, 1894), 20, Guthrie (1952), 185—and that these earth-like stars are to be understood as inhabited can be safely assumed (ibid.). Anaxagoras' name is invariably cited when discussing the idea of an inhabited moon before Plato's time, but Guthrie's warnings are almost certainly sound (ibid. 247–8, with Guthrie, ii. 308 n. 4). For the idea of an inhabited moon in Pythagoreanism see also Iam. *VP* 30 and 82. The idea occurs as well in Epimenides (B2); cf. Burkert 150–3, 346–7.

allowed to develop and elaborate certain ideas as the spirit moved them.[15] With regard to the basic idea of another earth, Burkert has pointed out that instead of focusing on any one of the various astronomical interpretations of it to the exclusion of any other we need in the first instance to view them all together against the background of the mythical, and above all eschatological, matrix out of which they arose. Where the *Phaedo* myth is concerned, this background of eschatology points us back once again to Pythagoreanism.[16]

Here too, there is no lack of additional details to corroborate Plato's indebtedness at this particular point in the myth to Pythagorean ideas. When he comes to describe the all-important shape of his 'true earth', he presents it as an enormous twelve-sided ball: evidently a reference to the dodecahedron. The special place which he reserves for this figure—both here and in the *Timaeus*—is one of the most obvious signs in either dialogue of his debt to Pythagoreanism.[17]

And so we come back to the hollows and the rivers of water and fire—themselves essential features in Plato's description of the 'true earth'. When we stand back from this part of the myth, in whatever direction we look we see it surrounded on every side by elements that have plainly been inherited from Italian or Sicilian Pythagoreanism: the eschatology, the cosmology, the symbols. Viewed in this context, nothing could be less surprising than the fact that the mythical landscape of rivers and hollows itself has its origin in the geography of Sicily. The evidence is mutually confirmatory, and allows only one conclusion: Plato's mythical geography—in particular his intricate picture of Tartarus and the great fires inside the earth—came to him from western Pythagoreans.

As to how exactly he inherited these ideas, all the probabilities point in the same direction. It is a matter of common agreement that Plato's own visits to southern Italy and Sicily

---

[15] See on this point Zhmud' 280, 288–9; Kingsley (1990), 261 with n. 99.

[16] Burkert 345–8, who also (347 n. 55) saw the need to situate the *Phaedo*'s 'true earth' in the same context; and cf. Detienne (1963), 148–9 with further refs., 160–2.

[17] *Phd.* 110b6–7 (cf. 108d9–e1); *Tim.* 55c. For the debt to Pythagoreanism see Burnet 131 and (1930), 292–5; Sachs 76–84; M. Cantor, *Vorlesungen über Geschichte der Mathematik*, i³ (Leipzig, 1907), 174–8, A. Mieli, *I Prearistotelici*, i (Florence, 1916), 262–4, T. L. Heath, *A History of Greek Mathematics*, i (Oxford, 1921), 160–2 and *Euclid's Elements*, ii² (Cambridge, 1926), 97–9; Burkert 458–60. Cf. also de Santillana 108–14.

were a decisive factor in his acquaintance with Pythagorean circles—and, in particular, in his acquisition of the Pythagorean ideas he incorporated in his myths. Theoretically he could have been given the details contained in the *Phaedo* myth by someone in Athens. And yet there were infinitely greater possibilities for picking up these ideas during his first visit to Sicily, which he appears to have made not long before sitting down to write the *Phaedo*.[18]

We can, if we want, be more specific. Among Plato's ancient biographers there was a tendency to explain the first Sicilian visit as due to a philosopher's commendable interest in studying the remarkable craters at Etna. However, this way of casting Plato in the role of natural philosopher is a rather blatant travesty of his real interests: a naïve explanation almost certainly concocted by some well-intentioned but simple-minded writer, and very possibly based on nothing more than the reference to Sicily's volcanoes in the *Phaedo*.[19] This is not to say that a visit by Plato to Etna is unlikely, but that any such visit needs to be placed in its proper context—and above all be understood against the background of Plato's own interests and concerns during the period in question. He appears to have made the crossing from southern Italy to Sicily for the first time after staying for a while with Archytas at his Pythagorean

---

[18] See in general Wilamowitz, i. 249; Thomas 66–7, 72, 79, 120, 156; Long 69–70. For the relevance of the first Sicilian journey to the writing of the *Phaedo* see Wilamowitz, i. 252; Thomas 67–8, 85–8; Kerschensteiner 132–3, 135; also Bostock 11–12; Furley 23. The weight of the evidence favours supposing that the *Gorgias*, too, was written after rather than before Plato's first visit to the West: cf. J. Geffcken, *Hermes*, 65 (1930), 14–30; Dodds (1959), 19–27, esp. 26–7; Guthrie, iv. 236, 284–5; Vlastos 115 n. 39, 128–9.

[19] The artificiality of the explanation was well emphasized by Wilamowitz, i. 252; the apologetic tone in which it is couched is itself enough to raise suspicions as to its accuracy. See esp. Olympiodorus' life of Plato: 'Because a philosopher must necessarily be interested in observing natural phenomena, Plato set off for Sicily to view the craters of fire in Etna; with all respect he did not—as you claim, Aristides—go there for the Sicilian food' (2.94–7 Westerink and more briefly in the anonymous *Prolegomena*, 4.11–13 W.; further refs. in Riginos 73 n. 16). For the humorous idea that Plato went to Sicily because of the local cooking (Σικελικῆς τραπέζης χάριν) see Adam's note to *Rep.* 404d Συρακοσίαν τράπεζαν; L. Edelstein, *Plato's Seventh Letter* (Leiden, 1966), 64 and n. 148; Riginos 71. On the need felt by later writers to provide a specific motive for each stage of Plato's journeys cf. Kerschensteiner 48–9; for the literary battles fought between supporters and opponents of Plato over the 'causes', αἰτίαι, of his voyages to Sicily in particular, Riginos 70–4; and on ancient biographers' zest for reconstructing the life-history of famous authors from incidental details in their writings, Lefkowitz.

school in Tarentum. Undoubtedly a major motive behind his crossing over to Sicily was an interest in observing variations in political practice. However, Wilamowitz warned—sensibly as so often—against projecting the straightforward political motives behind Plato's subsequent visits to Sicily back on to the first one;[20] and there is a strong ring of truth to the suggestion in the *Seventh Letter* that, at the time of the crossing, Plato had no conscious sense of direction and was only feeling his way around.[21] Archytas will as a matter of course have provided his guest with introductions to pupils of his on the island; this is confirmed indirectly by evidence to the effect that on a subsequent visit Plato was on terms of the closest friendship with Sicilian Pythagoreans, and stayed at the house of one of them in Syracuse during the ten-day festival of Persephone.[22] Whatever his exact state of mind or intentions during his first spell on the island, there can be no doubt that he will have been keen to hear whatever his hosts were able to tell him both about Sicilian traditions and about Pythagoreanism. If he did not already pick up the essentials of the myth which he reproduces in the *Phaedo* from Archytas in Tarentum,[23] there will have been plenty of opportunities for him to hear them in the most perfect of surroundings: on the island itself.

---

[20] Wilamowitz, i. 252 n. 1. For the stay with Archytas cf. ibid. i. 246–52, ii. 82–4; Harward 16. Crossing from Italy to Sicily: *Letters* 7.326b–327b.

[21] The letter is informative about Plato's general preoccupations with politics and philosophy at the time (324b–326d), but poignantly adds that he arrived there 'perhaps by chance, or rather as the result of design on the part of one of the higher powers' (326d–e). Edelstein's rejection of this comment on higher powers because it 'really tells nothing about Plato's purpose in coming to Syracuse' and because of its 'reticence concerning essential data one wishes and is entitled [!] to know and about which other sources provide information' (op. cit. 14) is as shoddy as the rest of his arguments that the letter is a forgery. He ignores the earlier statements in the letter regarding Plato's political preoccupations, overlooks the obvious element of gross fabrication in those 'other sources' to which he refers (above, n. 19), and above all fails to allow that even great philosophers may sometimes do things without having a clear aim or conscious purpose at the time of actually doing them. For another authentic touch in the *Seventh Letter* see Ch. 21 with n. 24.

[22] *Letters* 7.339a–b, 349d (Archedemus); Burkert 92 and n. 41.

[23] Cf. Frank 186 (also 299–300), Kerschensteiner 134, Friedländer, i. 96, 273.

9
The *Phaedo* Myth: The Structure
*

It is worth digging a little deeper and taking a brief look at two of the subterranean rivers mentioned in the *Phaedo* myth: the Pyriphlegethon and its opposite number, the Cocytus. This will help us to understand the structure of the myth as well as its origins and, ultimately, see what its broader relevance is to Empedocles.

To begin with the Cocytus: Plato describes it and the region where it forms a lake as being dark or a deep dark blue in colour: the colour of *kyanos*.[1] That could seem a meaningless detail, and commentators pass it over very quickly. But the darkness of the colour is very appropriate for a subterranean river; and indeed, darkness is a quality of the Cocytus specifically emphasized in later literature.[2] However, there is more to the matter than that. Etymologically, 'Cocytus' is the river of mourning and tears, and this is how the name was understood by Greeks down to the end of antiquity; but at the same time we happen to know that in Greek literature and religion the colour of *kyanos* was itself, *par excellence*, a colour of mourning.[3] The precision and aptness in Plato's choice of

---

[1] χρῶμα δ' ἔχοντα ὅλον οἷον ὁ κυανός: 113b8–c1, c8. For the identification and colour of *kyanos* cf. Halleux (1969), J. Irwin, *ZDMG* 133 (1983), 327, 335 n. 34, 340; below, n. 3.

[2] So Virg. *G.* 4.478–9 *limus niger . . . Cocyti*.

[3] For mourning and the Cocytus cf. Pl. *Rep.* 387b–c; Heracl. *Alleg.* 74.2; Macr. *Somn.* 1.10.11 *in luctum lacrimasque compellit*; Eust. *Od.* 10.514 διὰ τὸν ἐπ' αὐτοῖς [i.e. the dead] ἔνδακρυν θρῆνον; LSJ s.v. Κωκυτός I and II. For *kyaneos*—the colour of *kyanos*—and mourning cf. *Il.* 24.94–5, *Dem.* 42 with Richardson ad loc., 182–3, 319, 360, 374; Halleux (1969), 57. As a broad rule it seems that in early literature *kyaneos* was associated especially with divine mourning, for and by gods; the colour of ordinary human mourning was *melas* or dull black: West (1966), 280. The special lustre associated with *kyaneos*—E. Irwin, *Colour Terms in Greek Poetry* (Toronto, 1974), 79–80; cf. Pl. *Tim.* 68c—plus the fact that in Mediterranean countries and the Middle East dark blue was, and still is, a colour of mourning in addition to black—Halleux, op. cit. 57 n. 52; Corbin (1978), 157 n. 121—suggest that in early Greek poetry the word may

adjective are impressive, and rather too great to be no more than a coincidence. Once again it is difficult not to be struck by how little of the detail in the myth is arbitrary, out of place, or meaningless.

But this too is not all. The colour of *kyanos* appears to have had a special connection with the mysteries of Demeter and Persephone and—through the cult of the two goddesses—with water and the underworld. Here, yet again, we are brought back in the first instance to Sicily: to the island 'consecrated' to Demeter and Persephone, and which Zeus was supposed to have given to Persephone as a wedding-gift after her rape by Hades.[4] The spring at Syracuse where Hades is said by tradition to have snatched Persephone down into the underworld was called, precisely, *Kyane*; and here too, exactly the same ambiguity or spread of meanings applies as the one just noted above. Plainly there is a reference in the name of the spring to the colour of its water.[5] But at the same time, in relation to the mysteries of Persephone and Demeter the colour of *kyanos* has very definite associations with mourning, and the choice of name for the spring is undoubtedly related to the annual rites of mourning performed beside it in honour of Demeter's lamentation and tears over the loss of her daughter Persephone.[6] This precise overlapping of double meanings is, again, hardly an accident, especially considering what we have already seen of the intimate relation between the details of the *Phaedo* myth and Sicilian geography—not to mention the occasions where beliefs and ideas associated with the Sicilian cult of Persephone have helped to throw light on the half-mythical, half-physical geography presented in the myth. And to put things in their right perspective, it helps to remember that Syracuse was a major centre for the mysteries of Persephone and Demeter, which exerted an immense influence throughout and beyond Sicily in classical times. As for the

---

indeed have been interchangeable on certain occasions with *melas* but was not for that reason identical in meaning. For the distinctive blue tint almost certainly already associated with *kyanos* and *kyaneos* in Homer, cf. Halleux 47–66, esp. 54, 58, 64–5; D. C. Innes, *JHS* 96 (1976), 196.

[4] Pi. *Nem.* 1.13–20 with schol. ad loc., Diod. 5.2.3 = Timaeus, *FGrH* 566 F164, Cic. *In Verrem* 2.4.48.106; Zuntz 71, Griffith 171.

[5] K. Ziegler, *RE* xi. 2234, 'die mit dunklem Wasser'.

[6] H. Herter, *RhM* 90 (1941), 247; Richardson 163.

Kyane itself, it seems in Plato's time to have been the single most important spot on the whole island for the cult associated with the mysteries.[7] Finally, it will also be noted that there appear to have been very close links between early Pythagoreanism and the cult of Demeter and Persephone. One particular point to bear in mind here is that according to the Sicilian cult of the two goddesses the natural and artificial hollows which dotted the surface of the island had a religious as well as geographical significance, and—just as in the more mystical side of Pythagoreanism—were viewed either as entrances to the underworld or as symbolic of the underworld itself.[8]

This may seem all there is to say on the matter; and yet it is not. Mention of the Kyane has helped to bring Plato's description of the Cocytus into the orbit of beliefs associated with Persephone, but there is another place in the Greek West which is even more immediately relevant to the picture he conveys. Right next to Cumae and at the tip of the Bay of Naples, some three hundred miles north of Syracuse, was Lake Avernus: a strange spot which matches down to the smallest detail the description of the Cocytus in the *Phaedo*. According to Plato, 'opposite' the Pyriphlegethon and before sinking underground to meet it at the Acherusian Lake, the Cocytus

is first said to emerge at a spot which is awe-inspiring and wild (δεινόν τε καὶ ἄγριον), the colour of *kyanos* all over; this place is referred to as Stygian, and the lake that the river forms as it flows into it, the Styx (113b6–c8).

According to Italian tradition Lake Avernus was situated in the vicinity of a region called the Pyriphlegethon, and not far from the Acherusian Lake. This Acherusian Lake stood near the Pyriphlegethon and the opposite side of Cumae from Avernus

---

[7] For Syracuse as centre of the Sicilian mysteries of Persephone, and for the antiquity of the rites at Kyane, cf. esp. Diod. 4.23.4, 5.4.1–2; L. Malten, *Hermes*, 45 (1910), 525 and n. 3; Zuntz 70–4; Griffith 172. Regarding the extent of influence of the Sicilian cult and mythology of Persephone cf.—apart from Zuntz and Griffith—Guthrie, ii. 130; F. di Bello in *OMG* 191–3.

[8] Cf. Bidez 114–16 (with Chs. 10–12 below on 'Pythagoreanism' and 'Orphism'); Pace, iii. 498–9; Thomas 66–7, 72, 79, 142–3, 148–53; Burkert 112–13, 143 and n. 128, 155–6, 159–60, 178 and n. 89, 182–5 and (1969), 21–7; above, Ch. 6 with nn. 10–11. For Pythagoreanism, Persephone, and Demeter see also D.L. 8.11 = Timaeus, *FGrH* 566 F17 and Iam. *VP* 56.

and, as in the *Phaedo*, was distinct from it but also supposed to link up both with it and with the Pyriphlegethon underground.⁹ On the other hand, Lake Avernus was supposed to be connected even more directly with the waters of the Cocytus which were right next to it, lying between it and the ocean.¹⁰ The area immediately surrounding Avernus was described as 'wild' (ἄγριος), and it inspired such 'awe' in the Greeks (δεισιδαιμονία) that they left it untouched by human hand.¹¹ But in addition to its proximity to the Cocytus, the water of Avernus was believed to be 'Stygian' and the lake itself the Styx.¹² And just like the Kyane at Syracuse, it was considered sacred to Persephone; while the lake itself was so uniformly pure and unruffled in appearance that it was described as the colour of *kyanos* all over (φαίνεται τῆι χρόαι κυανοῦν).¹³

What are we to make of all this? Has Plato—or rather his Pythagorean source for the myth—produced some kind of arbitrary synthesis by conflating two entirely separate traditions associated with two completely different regions? The answer, very simply, is no. We can begin by noting that close links definitely existed between the cult of Persephone at Syracuse and at Cumae. Syracuse exerted a strong cultural influence on Cumae and the surrounding region of Campania in a number of different ways; and, in particular during the sixth and fifth centuries BC, a high level of interaction between Italian and Sicilian cults of Persephone—especially between Syracuse and the Bay of Naples—was the rule rather than an

---

⁹ Lycophron, *Alexandra* 695–704, ps.-Arist. *Mirabilia* 839ᵃ12–13, 21–3, Strabo 1.2.18 and 5.4.5, Sil. 12.121–7, Eust. *Od.* 10.514; Beloch 188–9, McKay 1–6. Apart from the fact that some of our sources for details of this topography antedate Plato (below, nn. 11–12), any possibility of Italian geographical and mythological tradition founding itself on a myth at the end of the *Phaedo* is of course out of the question.

¹⁰ Lycophron 704–5, Sil. 12.116–121; J. Geffcken, *Timaios' Geographie des Westens* (Berlin, 1892), 31, 143, Nissen, ii/2. 735–6.

¹¹ Strabo 5.4.5 (followed by Eust. *Od.* 10.514); cf. also Lycophron 704–5 (χεῦμα Κωκυτοῖο λαβρωθὲν σκότωι). The area was finally cleared by Agrippa. For references to Avernus in earlier Greek literature cf. *TrGF* iii. 371–3 (Aeschylus), with J. S. Rusten, *ZPE* 45 (1982), 33–8; Soph. fr. 748; Heraclid. fr. 128; Timaeus, *FGrH* 566 F57 with Jacoby ad loc.

¹² Aesch. fr. 273a *ad fin.*; Strabo 5.4.5 (τούτου δ' ἀπείχοντο πάντες τὸ τῆς Στυγὸς ὕδωρ νομίσαντες); Sil. 12.120–1; and see the discussion and further evidence cited by Rusten, loc. cit. Cf. also Lycophron 704–7; Beloch 169.

¹³ Diod. 4.22.1; cf. Lycophron 698, Beloch 169 and n. 2, Geffcken, op. cit. 142.24–143.4, McKay 8–15.

exception.[14] But behind these generalities lies one, very specific, mythological and geological reality. According to local traditions, the whole region between the Bay of Naples and Sicily was linked by a single network of underground passages and channels of fire. Pindar does not just portray Typhon as trapped under Etna, but describes him as pinned down at one end by Sicily and by the area around Cumae at the other. Strabo goes into more detail:

> Starting from Cumae and reaching down to Sicily, the entire area in between is full of fire. It contains hollows ($\kappa o\iota\lambda i\alpha s$) deep down, which are connected with each other as well as with the land-mass to form one single whole. This explains why Etna has the characteristics everyone describes it as possessing, and the same applies to the Liparan Islands, to the region around Dicaearchia, Naples, and Baiae, and to the Pithecusan islands [off shore from Cumae].

In other words this Sicilian–Italian network was used not only to explain why the volcanic phenomena in Sicily and Campania were so strikingly similar, but also to account for the existence of the volcanic Liparan islands off the north coast of Sicily and—as well—for the fierce volcanic activity which supposedly broke Sicily off from Italy in the first place.[15]

What this means is that, although so far we have focused exclusively on Sicily, it is time to widen our horizons: even Strabo's famous description of the 'whole island' as 'hollow beneath the earth, filled with rivers and fire' only tells a part of the story. The *Phaedo* myth itself provides unmistakable hints pointing us beyond the boundaries of Sicily alone when it describes the Pyriphlegethon as creating a lake larger than the Mediterranean ($\lambda i\mu\nu\eta\nu\ \pi o\iota\epsilon\hat{\iota}\ \mu\epsilon i\zeta\omega\ \tau\hat{\eta}s\ \pi\alpha\rho'\ \hat{\eta}\mu\hat{\iota}\nu\ \theta\alpha\lambda\acute{\alpha}\tau\tau\eta s$) and as producing volcanic eruptions at various different places on the earth's surface ($\ddot{o}\pi\eta\iota\ \ddot{\alpha}\nu\ \tau\acute{\upsilon}\chi\omega\sigma\iota\ \tau\hat{\eta}s\ \gamma\hat{\eta}s$).[16] It is as though behind Sicily as an island there lies a greater Sicily, a mythological as well as geological region stretching far beyond the visible confines of the island and, in particular, up to Cumae

---

[14] Cf. e.g. Loicq-Berger 194–202; Zuntz 114, 150, 162, 175–7.
[15] Pi. *Pyth.* 1.18–20; Strabo 5.4.9; Diod. 5.7.4; Aesch. fr. 402, Diod. 4.85.4, Strabo 6.1.6. From the geological point of view the Greeks were perfectly correct: for comments on the extent and nature of the volcanic seam in question see Nissen, i. 253–4.
[16] 113a7–8, b5–6.

and the Bay of Naples. In practical terms this means that ideally—although not necessarily—one would expect the author responsible for the geographical details which lie behind the *Phaedo* myth to have had some familiarity with the region around Cumae as well as with Sicily itself. This is a point we will return to later.

After the Cocytus we can turn back again to the Pyriphlegethon. So far we have only considered this river of fire from the point of view of its physical aspects and characteristics. However, towards the end of the *Phaedo* myth Plato introduces another element into his description of the rivers of the underworld: the fate of the soul. In the context of the eschatological scheme that he outlines, the Pyriphlegethon plays a very specific role. Of the souls that go down to the underworld after death

> those who are found to have committed major but redeemable crimes—such as doing violence to their father or mother through anger (ὑπ' ὀργῆς βίαιόν τι πράξαντες) and then living the rest of their lives repenting their deed, or who have committed murder in a similar but different way—are invariably thrown into Tartarus. However, after staying there for a year the swell ejects them: the murderers into the river Cocytus, those who have laid hands on their father or mother into the Pyriphlegethon.

When we look at what Plato is saying here,[17] we see that of all the four underworld rivers—Ocean, Acheron, Cocytus, and Pyriphlegethon—he associates the river of fire specifically with anger. This is no coincidence: in Greek as well as many other languages, fire and anger are intimately related.[18] But what is

---

[17] 113e6–114a7. The construction is chiastic, as so often in Greek (cf. e.g. *Rep.* 518a7–b3): those who have acted against their father or mother in anger pass into the Pyriphlegethon, while those who have 'committed murder in a similar but different way' pass into the Cocytus. The 'similarity' must, as the context shows, lie in the repentance which is a prerequisite for redemption, while the 'different way' (ἄλλωι τρόπωι) points to a condition or motivation different from the one of anger (ὑπ' ὀργῆς)—for example cold calculation. Cf. Burnet's brief note on 114a2 and, for the connection between anger and parental murder, Eur. *Electra* 1183–4 with J. D. Denniston ad loc. Violence towards one's parents, regardless of whether or not it ended in murder, was considered a crime of the first degree: cf. e.g. Arist. *Rh.* 1386ᵇ28–9.

[18] Cf. e.g. Eur. *Andromache* 488–90, *Electra* 1183–4 (murder of mother through the 'fire' of anger and hate: see previous note), Plut. *De gen.* 577a ἀνθρώπους διαπύρους πρὸς ὀργὴν καὶ θυμοειδεῖς, Hesychius, s.v. διαπυρισθέντα· διοργισθέντα and, for

most significant about the relationship in this particular case is that the correspondence between the Pyriphlegethon and the fires of anger is mentioned explicitly in the literature of Neoplatonism, centuries after Plato. This interpretation of the river of fire, this idea of a metaphysical correspondence between the geography of the underworld and the microcosm of human passions, is assumed nowadays to be a uniquely Neoplatonic creation—a prime example of the way in which Neoplatonists read their own extravagant theories and subtleties back into the simple myths of Plato. The suspicion has also been voiced that the correspondence is quite possibly to be traced back in time as far as the Neopythagorean movement in Alexandria, still centuries later than Plato.[19] That is closer to the truth but, as we have just seen, it too is not correct: the allegorical association of ideas goes back to Plato himself. And yet even this is not to say all there is to be said on the matter. A reading of the passage in the *Phaedo* shows that Plato's own attitude to the association is no more than at the very most one of casual and passing interest. The detail is thrown into the background, simply implied rather than formulated explicitly—which would not be the case if he had invented it or even if he saw it as particularly significant. In other words, the allegorical interpretation which is usually viewed as either Neopythagorean or Neoplatonic not only goes back to Plato himself, but by Plato's time had already receded into the background.

This has a number of implications, and two in particular. First, there is a definite warning here about the way in which we tend to approach the history of western ideas. In spite of the supposedly great advances in scholarship made since the time

general comments on fire, anger, and passion, Onians 147–60. In Plato, cf. *Rep.* 440c; *Tim.* 87c-88a; *Laws* 664e, 671b–c, 716a, 783a. The significance of the Pyriphlegethon/anger association in the *Phaedo* myth has not gone completely unnoticed: cf. Dorter 172; Burger 199–200, 270 n. 43.

[19] For the Neoplatonic interpretation see Macr. *Somn.* 1.10.11: the ancients 'identified the Phlegethon with the fires of anger and desire' (*Phlegethontem ardores irarum et cupiditatum putarunt*). Cf. also Philo, *QG* 4.234, 'in Acheronte et Puriflegetonte concupiscentiarum', ed. F. Petit, *L'Ancienne Version latine des Questions sur la Genèse de Philon d'Alexandrie* (Berlin, 1973), 93 (the Armenian version is unusable); Cumont, *Revue de philologie*, 44 (1920), 229–40; P. Courcelle in *EH* iii. 95 (presenting the standard view that these are Neoplatonic fantasies foisted back onto Plato), 102–4; J. Flamant, *Macrobe et le néo-platonisme latin à la fin du IVe siècle* (Leiden, 1977), 575–81.

of the Renaissance, the fact is that our view of the so-called development of western philosophy is dictated by a Darwinian idea of history which is extremely simplistic. Histories of ancient philosophy are still written in accordance with the fundamental belief that, before the golden age of Athenian achievement during the fourth century BC, the globe was populated by naïve children groping around in the dark for truths that only Plato and Aristotle would be able to discover and articulate. Any sophistication which we, or ancient philosophers who lived later than Plato and Aristotle, or for that matter Plato himself, might be inclined to ascribe to those primitive beings is almost invariably dismissed as a historical error: as the projection back in time of subtleties and refinements only possible in a later age. But here we have an example of something very different. A subtle allegory of underworld geography has already slipped into virtual non-existence by Plato's time, and when Neopythagoreans or Neoplatonists reformulated the same idea they were not creating anything new but were simply continuing a long tradition—which is exactly what in this case they claimed they were doing.[20] In the course of this tradition Plato is neither a beginning nor an end; he is simply a point along the way. To talk of a Neoplatonism before Plato is strictly speaking just to juggle with words, but it may have the effect of emphasizing that some at least of what is usually considered to be 'new' appears to be far older than has been supposed. What is particularly worrying about this one case is that it is a question of an oral tradition received by Plato, not a written one. That inevitably raises the possibility that there may have been many ideas around in comparatively early times of which no trace remains, not only because of the loss of written texts but also because some of the most important ideas were never put down in writing to begin with.

Second, we are now in a position to start being a little more precise about the question of Plato's sources for the *Phaedo* myth. Earlier we saw that the intricate topographical details in the myth were not in any sense Plato's own invention, but are a faithful reproduction of the mythology and geography relating to the volcanic 'network' stretching up from Sicily to Campania. Now

[20] Macr. *Somn.* 1.10.8–11 (43.8–26 Willis).

we have begun to see that the myth contains not only geographical and mythological details but also an allegorical or metaphorical elaboration of those details, and that this elaboration is no more original to Plato than the details themselves. In short, he has plainly inherited not just a mythical topography that has its roots in Sicily and southern Italy, but also the elements of a psychological explanation which relates this topography to the nature and fate of the soul.

Here another of Plato's dialogues helps us understand the main factors involved. As mentioned earlier, the *Gorgias* preserves in very brief form a tradition of myth and mythological interpretation which consists of two distinct levels. At the first level we have the literal description of a mythical underworld. At the second, we have an allegorical explanation of the underworld scene which interprets the sufferings it portrays in terms of human passions experienced during our life on earth. That is more or less exactly what we find in the *Phaedo* myth as well: a literal description of the underworld combined with the already half-buried traces of an allegorical interpretation along the same lines. In the *Phaedo*, the punishment of being immersed in the Pyriphlegethon is correlated with the passion of anger; in the *Gorgias*, the punishment of being forced to keep trying to carry water in a sieve is correlated with the passion of insatiable desire. In each case the basic idea is essentially the same.[21]

These parallels between the *Gorgias* and what we find in the *Phaedo* are plainly no coincidence, and the connection between them becomes even clearer when we note the other factors that both myths share in common. To begin with the underworld scene itself: in the *Gorgias*, Plato suggests with pointed vagueness that the literal level of the myth—its basic description of the underworld—is 'probably' the creation of 'a Sicilian or Italian'.[22] As far as the *Phaedo* is concerned, we have seen that the literal level of the myth—its basic description of the underworld—derives, precisely, from Sicily and Italy. And then there is the finished product: the complete combination of original myth and allegorical interpretation. Here, in almost identical terms in both the *Phaedo* and the *Gorgias*, Plato

---

[21] *Gorg.* 492e–3d (cf. Ch. 4 with the refs. in n. 6); *Phd.* 113e–114a.
[22] ἴσως Σικελός τις ἢ Ἰταλικός, 493a.

describes his source for this combination of myth plus interpretation as being a certain oral source, an unnamed 'someone': a person whom all the evidence tends to indicate was in both cases a Pythagorean.[23] At each level, the parallel between the *Phaedo* myth and the snippet of myth in the *Gorgias* is confirmed.

Analysis of the *Phaedo* myth in terms of this division into two levels can easily be extended to specifics, but here it will be enough just to mention the one crucial point around which all the other details revolve. At the basic, literal level of the myth we live on a spherical earth—an earth with its geography of volcanoes, hollows and caves, and subterranean rivers that sometimes rise to the surface to produce springs, streams, and lakes. As for the underworld, it is precisely that: another world hidden from us in the depths of the earth.[24] This basic situation—we on the surface of the earth and Hades below us in the realm of the dead—plainly applies at the literal level of the *Gorgias* myth as well.[25] But, in both dialogues, at the metaphorical level the whole situation has changed: everything is transposed. According to the *Gorgias* it is not a question of us being alive and the underworld being somewhere else. On the contrary, we are already dead (νῦν ἡμεῖς τέθναμεν), and the world that we think of as life on earth is really the underworld itself. The *Phaedo* myth makes exactly the same point in even fuller terms. Really, it explains, we are living in some small crevice or cave, one of the many little 'hollows' hidden in the depths of the earth as it truly is, 'but although we are living in the earth's hollows we are unaware of the fact and believe we are living up on the surface of the earth'. The Mediterranean is like a little pond in our dank cave, and we are like frogs around it.[26] If we bring everything back to its main point of origin in Sicily we can say that according to the literal level of the *Phaedo* myth we are living on the surface of the island, above its dark

---

[23] τοῦ ἔγωγε καὶ ἤκουσα τῶν σοφῶν (*Gorg.* 493a), ὡς ἔφη ὁ πρὸς ἐμὲ λέγων (493b), ὡς ἐγὼ ὑπό τινος πέπεισμαι (*Phd.* 108c); Ch. 8 with nn. 9–10; below, n. 36.

[24] 111d2 (ὑπὸ γῆν), 6, e5, 112c8–d1, 113a1, b4, c3; cf. *Phdr.* 256d6, *Rep.* 615a2–3, *Laws* 905a6–7 and, for the hiddenness of Hades, Ch. 4 with n. 43.

[25] Rohde 586–8; Burkert 248 n. 48.

[26] *Phd.* 109a–110a, esp. 109c3–4, 109c3–d1. For the association of 'hollow' (κοῖλον or ἔγκοιλον) and cave see Ch. 7 n. 14; *Od.* 12.84–93, Aesch. *Eum.* 22–3, Eur. *Andromache* 1264–5 and fr. 421, Diod. 3.13.1, etc.

hollows and caves opening into the depths of the underworld; but according to the metaphorical level we are already down in one of those caves, down in the underworld itself and only dreaming we are on the surface of the earth.

The relevance of this analysis of the *Phaedo* myth to Plato's other myths should need no emphasizing—in particular its relevance to our understanding of the allegory of the cave in the *Republic*. But as far as the *Phaedo* myth itself is concerned, there is one final point worth noting here. Analysis of the myth into its component levels puts an end once and for all to a debate which has clouded attempts at interpreting it for thousands of years: the debate as to whether the 'earth' referred to in the myth is to be understood literally as a reference to our own familiar earth, or metaphorically as an allusion to something else.[27] The simple answer is that it refers to both, depending on the level at which one chooses to read what Plato says. As is to be expected in the case of material subsequently reworked and supplied with an allegorical superstructure, the myth starts off at the literal level.[28] But then, as one would also expect, the metaphorical level rapidly begins to emerge. The result is that the *Phaedo* myth describes two earths, not one. The original

---

[27] For evidence of the disputes over the matter in antiquity see Dam. *Phd.* 1.503–22, 2.114–45. Damascius himself tried rather limply to persuade his students that the *Phaedo*'s 'true earth' is 'very simply the earth described by geographers' (1.504). Plainly he is taking sides here over an issue that had divided Platonists for centuries (cf. esp. 1.503, 2.114); in trying to provide a 'reasonable' solution to the problem he was following in the footsteps of Aristotle, who with typical down-to-earthness had already criticized the myth at its purely literal level and disregarded any allegorical meaning (*Meteor.* 355^{b}32–356^{a}33). In modern times the trend has been to follow Aristotle and Damascius, as e.g. J. Dillon, *California Studies in Classical Antiquity*, 4 (1971), 138–9; but for a rare insight into the allegorical level of the myth, and its relationship to the myth in Plutarch's *De genio Socratis*, see R. Heinze, *Xenokrates* (Leipzig, 1892), 135–6, and also de Santillana's comments, *Hamlet's Mill* (Boston, 1969), 186–8. The Platonic background to Plutarch's eschatological myths has been a matter of common knowledge for a long time: cf. e.g. Dieterich 145–9 and, more recently, Culianu 43–6. However, Culianu makes the normal mistake of claiming that in Plutarch 'Plato's data are constantly interpreted according to a modern, sophisticated, exegetical system'—the assumption being that there is neither sophistication nor exegetical system already implicit in Plato's myths.

[28] 108e–9a (sphericity of the earth); Ch. 8 with n. 5. This introductory passage, with its clear distinction between a central earth and the surrounding heavens, directly answers the questions raised earlier in the dialogue about the shape and position of the earth as we normally understand it (97d–98a; Sedley 360–3). For broadly comparable examples of subsequent interpretations attaching themselves to earlier material see Kingsley (1994*a*).

earth is our earth—or, to be more precise, the earth as viewed from a vantage-point in Sicily or southern Italy. And then, in addition to this literal earth we are also presented with another, 'true' earth as distinct from our poor imitation: an aitherial, celestial world inhabited by divine beings. They are the true living and, compared to their life, so-called life on the surface of our corrupting and corrupted earth is just a feeble shadow and a dream.[29] This juxtaposition of earths has, understandably, caused confusion; but when we stop trying to choose between the one level and the other and accept the existence of them both, we are in a position to start appreciating the real density and complexity of the myth at the end of the *Phaedo*.

It is time to sum up and bring a little order into the main points so far. Examination of the *Phaedo* myth has shown that it contains at least two different strata. At the lower level we have a narrative describing the earth's shape and geography, presented largely in terms of a detailed account of Sicily's visible terrain as well as of the vast rivers of water and fire streaming and burning under the surface of the island. This original narrative can be called Pythagorean not only in the sense that it was used and reinterpreted in Pythagorean circles, but also in the sense that it was almost certainly composed by a Pythagorean. Particularly significant here is the fact that the theory of the earth's sphericity—that is, the sphericity of the earth as we know it—already belongs to the literal level of the myth, prior to its allegorization.[30] There is nothing surprising in this. On the contrary, it provides yet another parallel with the fragment of myth in the *Gorgias*, because the evidence tends to suggest that the 'Sicilian or Italian' tentatively referred to there as the author of the original level of the myth was himself a Pythagorean.[31]

Then, one level up in the *Phaedo* myth, we have the reuse of

---

[29] *Phd.* 109b7–111c3, 114b6–c4; Ch. 8 with nn. 12–14, 16. For the reality of 'life' in the heavens see also *Tim.* 39e–40b, 41b–42d; *Epinomis* 981e–984b, esp. 984a6–b1; Arist. *De philosophia*, fr. 33 (22) Untersteiner; Jaeger 138–55.

[30] Ch. 8 with nn. 5, 11; above, n. 28.

[31] Burkert 248 n. 48; also Wilamowitz, ii. 89, Frank 299. For the expression 'Sicilian or Italian' cf. Cherniss (1935), 384–5, Sext. *Math.* 9.127 = DK i. 367.1–2, and Ch. 10 with n. 8.

this original narrative: its allegorical application to philosophical—or, to be more precise, mystical—ends. And with this allegorical interpretation we encounter something strikingly new: an interest in symbolically transposing into a cosmic dimension the details of Sicilian or Campanian geography. Later we will see another, even more vivid, example of how current this particular interest was among Pythagoreans before Plato's time.[32]

From this excavation of the *Phaedo* myth there are a number of conclusions to be drawn. One naturally concerns Plato himself, and the question of his originality. It has become an article of faith nowadays that his myths are the productions of his own creative genius. Expression of this belief is normally accompanied by acknowledgement that some of the basic material for his myths can be assumed to derive from Pythagorean sources; but in the same breath the qualification is inevitably added that it was Plato's own remarkable powers which were responsible for transforming these crude components, for combining and integrating them to produce the myths as we have them. Time and again one meets with the claim, stated in almost identical terms, that 'no one but Plato' has made the myths into 'the wonderful imaginative wholes' which they are.[33]

However, analysis of the *Phaedo* myth presents a very different side to the picture. Plato's supposedly original touches and creations turn out not to be his creations at all, and for the origins of the traditional material he used we do not have to look in various different directions because all the evidence points to one and the same source. This applies not just to the physical or geographical details, but also to the mythical and eschatological ones; and not only to these but to the allegorical interpretation of them as well. The structure and texture of the myth prove to be far denser and more complex than has so far been suspected: it consists of different levels of meaning, and Plato has reproduced—as opposed to created—not just the literal, mythological level but also the symbolic level of interpretation. In short, everything points to the conclusion that he

---

[32] Below, Ch. 13. Cf. also Virg. *Aen.* 6.887 with Ch. 10 n. 37; Plut. *De sera* 566a–e with Ch. 11 n. 26.

[33] Guthrie (1952), 168. So e.g. Frutiger 260, 266; Thomas 120–2, 157; Hackforth 172 and n. 2; Annas 122.

has done exactly what in so many words he himself says he has done: repeat the myth as he has heard it from someone else.

This is not to deny that Plato introduced changes of his own into the myth as he tells it. On the contrary, he very probably jumbled some of the details and blurred the edges of some originally fine distinctions—either through lack of interest or even through misunderstanding.[34] However, there is no evidence whatever that he contributed to the creation—or even the arrangement—of the mythical material in any significant way. We have already seen a good number of cases, starting with the intricate details of Sicilian and Italian geography, where precisely those features or touches which would seem at first sight most likely to be Plato's own original contribution prove to be nothing of the kind. To mention just one other example here, there is the seemingly personal touch where Plato quotes a line from Homer to illustrate the cosmological theory he is describing. But in fact, as we will see later, the particular passage in the *Iliad* which he refers to here happens to have had a very special significance for Pythagoreans before his time, and was already used by them to illustrate and formulate their cosmological theories.[35] Once again, the tall shadow of Pythagoreanism looms over every detail of the *Phaedo* myth that one chooses to look at carefully.

There is one other respect in which Plato appears to have made an original contribution: by putting down in writing what he heard from his oral source. This is a detail that, in our highly literary age, may not seem particularly important; but it was to prove a step of major significance. By presenting what he had heard in written form Plato was to make available to a far wider audience—even in the Greek world of his own time, let alone for ages to come—an oral tradition which had apparently been restricted to a very limited circle of Pythagoreans.[36] And

---

[34] See above, on Plato's lack of interest in the Pyriphlegethon/fire correlation; and below, Ch. 14.

[35] *Phd.* 112a, citing *Il.* 8.14; below, Ch. 13.

[36] Dodds (1959), 27 describes that other myth—complete with allegorical interpretation—which is preserved by Plato in the *Gorgias* as 'a text which he evidently does not expect his readers to know'. The point about unfamiliarity on the part of Plato's readers is clearly correct; but Dodds's description of the basic myth together with its symbolic interpretation as a 'text' (cf. also ibid. 298, 'Pythagorean writing') is directly contradicted by Plato's own repeated assertion that his source was oral, not written. Cf.

there is also a broader issue involved here. This particular substitution of literacy for orality is one aspect of a much larger phenomenon, and can only be appreciated fully when we remember that Plato was living at a time of crucial transition in the West from a primarily oral to a predominantly literary culture. He himself was only too aware of the limitations of the written text as a medium of genuine communication. But he was also a realist, and saw that his only hope was to beat the enemy at its own game: that the only way to counteract the new 'democratization' of the written word—which gave anyone and everyone the ability to find a mass audience regardless of the value of what they had to say—was to use literature himself to attempt to communicate and preserve some of the ideas he considered essential both for the individual and for the continuation of society as a whole.[37]

In the centuries after Plato, stories began and continued to circulate about his plagiarizing Pythagoreans by taking their writings and reproducing the contents of them in his own books; these stories were even taken over by later Platonists, who were happy to admit that Plato was essentially continuing in the line of Pythagorean tradition by 'making his own' (σφετερίσασθαι) the teachings of his predecessors. Nowadays these stories and ideas are automatically dismissed: the accusations of plagiarism are explained as malicious rumours concocted out of thin air to put a damper on Plato's reputation, while the Platonizing versions are condescendingly ascribed to the 'syncretistic' mood of later antiquity with its supposed fondness for creating non-existent filiations between different traditions.[38] But whatever the truth or untruth about the stories of buying or being given books, one thing has become quite

Gorg. 493a ('I myself have heard') and esp. 493b ('as the person who told me explained'); also *Meno* 81a ('I have heard from wise men and women'), *Phd.* 108c ('as a certain person has succeeded in persuading me'), *Rep.* 583b ('as I think I heard from a certain wise person'). For orality in early Pythagorean tradition see Burkert 140 n. 110; and for the *Meno* passage, below, Ch. 12.

[37] For various approaches to these issues see E. A. Havelock, *Preface to Plato* (Cambridge, Mass., 1963) and *The Muse Learns to Write* (New Haven, Conn., 1986), 7–8, 116; Brisson 42–9; Ferrari 204–22; M. Vegetti in M. Detienne (ed.), *Les Savoirs de l'écriture: En Grèce ancienne* (Lille, 1988), 387–419. Regarding the large audience at which the *Phaedo* in particular was aimed see Wilamowitz's comments, i. 324–6.

[38] For basic refs. and discussion cf. e.g. Festugière (1950), 17–25; Riginos 169–74; H.-R. Schwyzer, *Ammonios Sakkas* (Opladen, 1983), 90–3; Lloyd 169.

clear from examining the *Phaedo* myth: Plato was substantially indebted to Pythagorean oral tradition. Later, in an increasingly bookish age, it was only natural to translate the indebtedness into purely literary terms; and yet this was to transpose—not invent—the pattern of borrowing and reproduction. Circumstantial hints at Plato's links with Pythagorean tradition already existed, for example in what Cherniss tactfully described as Aristotle's 'tendency to treat Pythagoreanism and Platonism as closely related doctrines', or in the extent to which Pythagorean ideas appear to have infused and influenced the life of the early Platonic Academy; but there is no substitute for finding the evidence in Plato's writings themselves.[39] And any possible objection to the effect that the implications of this evidence should be confined strictly to the mythical interludes in Plato's works has little force. As we will see later, anything that has a bearing on the nature and origins of Plato's myths necessarily has a bearing on our understanding of his work as a whole.

---

[39] Cherniss (1935), 46. On this point Aristotle is quite specific: 'In most respects Plato's undertaking—ἡ Πλάτωνος πραγματεία—followed them (the Pythagoreans), but it did also have features of its own distinct from the philosophy of the Italians' (*Met.* 987ª29–31; cf. Burkert 30–1 and, for 'Plato's undertaking', *Phd.* 61c8, *Tht.* 168a8, *Letters* 7.341b7–d2). Even allowing for his ability to distort the history of ideas, Aristotle was—as Plato's closest pupil—in a much better position to judge the truth of the matter than we are; but this is no obstacle nowadays to describing 'the essential connection... which Aristotle claims exists between Pythagoreanism and Platonism' as 'apparently constructed by Aristotle himself' (Riginos 173 n. 36). For the Academy and Pythagoreanism cf. Burkert 83–95, and below, Ch. 20 n. 47; Wilamowitz's remarks on the subject (i. 246–8, 270–3) must be qualified in the light of J. P. Lynch's, *Aristotle's School* (Berkeley, Calif., 1972), 54–7, 61–2.

10
Plato and Orpheus
*

> Plato paraphrases Orpheus everywhere.
>
> Olympiodorus, *On the Phaedo*

FROM Plato we can now start to make our way back in time, via the myth in the *Phaedo*, towards Empedocles.

The reason for turning aside to look at the *Phaedo* myth in the first place was the possibility that it might throw light on Empedocles' strange equation of Hades and fire. As a result, a number of conclusions have emerged. The similarities between Empedocles' equation and what we find in the *Phaedo* myth, with its vast fires burning in the underworld, turn out to be no coincidence and are easily explained: just as in the case of Empedocles, the half-physical, half-mythical geography described by Plato has its origins primarily in Sicily. We have also seen that Plato inherited the structure and details of the *Phaedo* myth from western Pythagoreans. This provides an additional point of contact with Empedocles, whose own links with western Pythagoreanism cannot be denied.[1]

The next obvious question is how exactly Empedocles' poetry and the myth in the *Phaedo* are related. This is not the first time that connections between them have been pointed out: Burnet did precisely that, and assumed Plato took details of the myth directly from Empedocles. However, the reasons he cited in favour of his assumption were more than a little naïve. His major argument was that the *Phaedo* myth mentions hot springs and volcanoes while referring to Sicily, that Empedocles mentioned hot springs, so the details would appear 'to

---

[1] For these links cf. Bidez 110–24, Long 56–62, Burkert 133 with n. 72, 186 with n. 155, 220 with n. 12, 289 with n. 59, Zuntz 232–4, 264–6, Zhmud' 273–4, 289; below, Chs. 15–23.

come from the Sicilian Empedocles'.² But even apart from the obvious element of over-simplification here, we have already seen that the matter is far more complex than Burnet had allowed. The myth at the end of the *Phaedo* is not just one single, homogeneous entity; it consists of different strata, and if we are concerned about its origins we need to specify whether we have in mind the literal level of the myth or its subsequent, symbolic level. With regard to Empedocles the choice is straightforward: in the first instance we need to focus on the original level—on the level at which the earth's surface is still the earth's surface and Hades has not yet been projected allegorically into the world in which we live.³ The question, then, is whether the original level of the myth can in any sense be attributed to Empedocles.

Here we come up against much the same situation as the one faced by Olympiodorus in the sixth century, when he tried to identify the 'Sicilian or Italian' whom Plato was prepared to suggest had 'probably' composed the original stratum of the *Gorgias* myth. Olympiodorus' reasoning was just as simple as Burnet's: we are looking for a Sicilian or Italian who apparently had Pythagorean interests, Empedocles was both a Pythagorean and a Sicilian, so he is probably our man.⁴ In fact the correspondences noted earlier between the samples of myth in the *Gorgias* and in the *Phaedo* make Olympiodorus' question as to the identity of Plato's 'Sicilian or Italian' extremely relevant to the issue of who wrote the text that lies behind the *Phaedo* myth.

That the author in question was Empedocles can safely be excluded. To begin with the *Gorgias*, Plato describes the person in question not just as 'probably a Sicilian or Italian' but also as 'some ingenious myth-teller of a man'. Dodds saw in this description sufficient grounds to rule out Olympiodorus' choice of Empedocles: Empedocles is a philosopher, not a teller of myths.⁵ Dodds was a little too quick off the mark—Plato in his

---

² Burnet 134, justifiably queried by Hackforth (178 n. 1).
³ Above, Ch. 4.
⁴ Pl. *Gorg.* 493a6–7; Olymp. *In Gorgiam* 157.15–17 Westerink. It should be borne in mind that Olympiodorus was predisposed to finding specific allusions to Empedocles in the *Gorgias* for the equally unsubtle reason that the dialogue is named after a man who had Empedocles for his teacher: cf. ibid. 8.1–3.
⁵ (1959), 297.

more wistful and less polemical moments was quite prepared to lower the barriers between a philosopher and a teller of myths[6]—but, even so, he was clearly correct and Olympiodorus wrong. For Plato, Empedocles was not 'probably a Sicilian or Italian' but a Sicilian, very simply.[7] Plato himself will have known Empedocles' poetry much too well not to be able to recognize it behind the thin veil of its allegorical interpretation, and his audience in Athens will also have been familiar enough with it not to be taken in—let alone amused—by such a pointless equivocation. And we can in fact go one stage further. Zeller and Wilamowitz noted that the wording used by Plato—'some ingenious myth-teller of a man, probably a Sicilian or an Italian' ($\tau\iota\varsigma\ \mu\upsilon\theta\text{o}\lambda\text{o}\gamma\hat{\omega}\nu\ \kappa\text{o}\mu\psi\grave{\text{o}}\varsigma\ \grave{\alpha}\nu\acute{\eta}\rho,\ \ddot{\iota}\sigma\omega\varsigma\ \Sigma\iota\kappa\epsilon\lambda\acute{\text{o}}\varsigma\ \tau\iota\varsigma\ \mathring{\eta}\ \mathit{'I}\tau\alpha\lambda\iota\kappa\acute{\text{o}}\varsigma$)—is a humorous paraphrase of a line by the poet Timocreon about 'an ingenious Sicilian man' ($\Sigma\iota\kappa\epsilon\lambda\grave{\text{o}}\varsigma\ \kappa\text{o}\mu\psi\grave{\text{o}}\varsigma\ \grave{\alpha}\nu\acute{\eta}\rho$). As Wilamowitz also noted, the precise way in which Plato modifies and adds to his model clearly implies that, although the myth-teller in question must have had close links with Sicily so as to call to mind Timocreon's words in the first place, the man probably came not from Sicily but from mainland Italy.[8] That excludes Empedocles altogether.

When we turn back to the *Phaedo* myth we arrive at very much the same conclusion. Burnet's theory of borrowings from Empedocles was tolerable on the assumption that what we find in it is simply a pastiche with the odd detail taken from here and the rest borrowed from elsewhere. But as we have seen, the basic level of the *Phaedo* myth is a continuous, detailed narrative describing the nature of the earth, its surface and interior, and its accommodation of the dead. If written by Empedocles, Plato would have realized the fact immediately—and doubtless also have expected his readers to realize it.

So we must look elsewhere. As noted earlier, in terms of a twofold division of the *Phaedo* myth into its literal and allegorical strata, the literal level was almost certainly the work of a Pythagorean. This is not to forget, though, that in a work of

---

[6] So e.g. *Phd.* 61a3–e4 ($\phi\iota\lambda\text{o}\sigma\text{o}\phi\acute{\iota}\alpha\varsigma\ldots\ \mu\upsilon\theta\text{o}\lambda\text{o}\gamma\iota\kappa\acute{\text{o}}\varsigma\ldots\ \phi\iota\lambda\acute{\text{o}}\sigma\text{o}\phi\text{o}\varsigma\ldots\ \mu\upsilon\theta\text{o}\lambda\text{o}\gamma\epsilon\hat{\iota}\nu$). Cf. also *Sophist* 242d7–243a2 on the Empedoclean 'Muse', and—for the relation between song, 'music', and myth-telling—*Phd.* 61a–b, *Rep.* 392b.

[7] *Sophist*, loc. cit.

[8] Zeller, i/1. 558 n. 2; Wilamowitz, ii. 89 ('der Verfasser war also aus Italien, ein Pythagoreer'); Timocreon, fr. 6 Page (*PMG* § 732); below, Ch. 12 n. 73.

Plato and Orpheus

such strongly local character the author will almost inevitably have accommodated earlier mythical and religious traditions.[9] And here we come to another possibility which needs to be considered: that the original stratum of the myth was in the form of a poem ascribed to the prophet Orpheus.

As is well known, during the century or more before Plato's time there was an extremely close link between Pythagoreanism and the production of Orphic literature; it was plainly an established tradition for early Pythagoreans to attribute to Orpheus poems they wrote on gods and the cosmos, salvation and the soul.[10] Hardly surprisingly when we consider the geographical distribution of Pythagoreanism at the time, for the source of these Orphic productions we are drawn repeatedly to Sicily and southern Italy.[11] For example an Orphic *Descent to Hades*, now lost, is ascribed to 'Orpheus of Camarina'—clearly the name assumed by an author from this town on the south coast of Sicily, below the Palici and Etna, founded in 599 BC by Syracuse.[12] Again, a work called *Salvation* (Σωτήρια) is ascribed to a Timocles of Syracuse.[13] Syracuse, as we have seen, was an extremely important centre of the Persephone mysteries, and there is undoubtedly a connection here with the fact that Orphic literature itself was focused to a very large degree on the figure and fate of Persephone.[14] What is more, Orphic literary production exhibited a marked tendency to gravitate towards the mysteries of Demeter and Persephone, and assume the role in relation to them of sacred narrative texts—as clearly emerges in the case of the Eleusinian mysteries.[15] And here we come back to the *Phaedo* myth: with its painstaking description of Sicilian and Campanian landscape, plus a number of other more specific details already noted, it raises the strong suspicion that the original stratum of the myth prior to its allegorical interpretation was some kind of

---

[9] Cf. e.g. Burkert 112–13, with n. 21 *ad fin*.

[10] Burkert 125–33; Graf (1974), 92–4, 148–9; West 7–15. On the inappropriateness of the term 'forgery' in this context see Kingsley (1993b), 23.

[11] *OF* tests. 173–9; Nilsson, ii. 644; Burkert (1969), 17 and n. 37.

[12] *OF* test. 176; Rohde 336 and 349 n. 7; Nilsson, loc. cit. For the founding of Camarina see Thucydides 6.5.3.

[13] *OF* test. 178; Lobeck, i. 383–4.

[14] For some points of contact between Syracuse and Orphic literature cf. Loicq-Berger 61; Graf (1974), 143–4; F. di Bello in *OMG* 191, 193.

[15] Richardson 77–85; Graf (1974); West 24.

sacred narrative designed to serve much the same function in relation to mysteries in the West as the Homeric or Orphic hymns to Demeter served in relation to the mysteries at Eleusis. This automatically solves another problem: whereas Empedocles had to be ruled out as author of the original myth because his poetry was so well known, in this respect an Orphic text would fit the bill precisely. Many Orphic poems were plainly intended in the first instance for the use of circles of initiates: a fact which helps to explain why they appear to have had comparatively little public exposure and 'a very limited circulation'.[16]

But there is no need to be satisfied with these general considerations, suggestive though they are. There is also much more specific evidence which points in the same direction; and we can begin with the snippet of myth in the *Gorgias*. The idea that the original, pre-allegorical stratum of the myth derives from an Orphic poem about the underworld is nothing new, and is well founded. In broad terms, the evidence for literature ascribed to Orpheus emanating from Italy and Sicily has been noted for its relevance to Plato's vague mention of some 'Sicilian or Italian' as author of the fragment of myth in the *Gorgias*.[17] This possibility of a link with Orphic literature is also reinforced to an extent by a passage in Plato's own *Republic*.[18] What is more, a fresh light has been thrown on the matter by the dramatic discovery in 1962 of the Derveni papyrus—a papyrus containing a text that was probably written not many years before Plato wrote the *Gorgias*, and which consists of the allegorical interpretation of a poem ascribed to Orpheus.[19] As we have seen repeatedly, the structure of the fragment of myth

[16] West 79–83.
[17] Burkert 130. Cf. also ibid. 248 n. 48; Guthrie (1952), 161–3, 190, 241–2.
[18] The idea in the *Gorgias* of being forced to carry water in a sieve as a punishment in the underworld (φοροῖεν ὕδωρ τετρημένωι κοσκίνωι, 493a5–c3) recurs in the *Republic*, where it is attributed to 'others' in addition to Musaeus and Eumolpus (363c3–d8, κοσκίνωι ὕδωρ ἀναγκάζουσι φέρειν). In this direct connection with Musaeus the 'others' are most likely to refer primarily to Orphic writers who reinterpreted and adapted traditional Greek ideas (for the combination of Musaeus and Orpheus cf. *Apology* 41a, *Ion* 536b, *Protagoras* 316d, *OF* tests. 11, 15–18, 166–72, and *passim* and, for the reinterpreting of earlier traditions in Orphic poetry, below with nn. 54–5); this is further confirmed by Plato's mention almost immediately afterwards of the 'hubbub of books of Musaeus and Orpheus' (βίβλων ὅμαδον Μουσαίου καὶ Ὀρφέως, 364e3).
[19] For the discovery and dating of the text see West 75–82, Brisson, *RHR* 202 (1985), 397–8; for its relevance to the *Gorgias*, Burkert (1968), 100 and n. 13.

in the *Gorgias* mirrors down to the smallest details the structure of the myth at the end of the *Phaedo*: any light shed on it is automatically light shed on the *Phaedo* myth as well.

That brings us to the most direct evidence of all, the *Phaedo* itself. The first point to note is that already in the main part of the dialogue, and before embarking on the final myth, Plato punctuates the discussion with references to Orphic literature on the subject of the underworld.[20] That the references are to Orphic texts is undeniable, and this in itself is enough to prepare one for the possibility that when Plato comes to tell the final myth—on, of course, the subject of the underworld—he will again be basing what he says on Orphic material.

That, roughly speaking, is how the matter has tended to be left in previous discussions of the *Phaedo*: as distinct possibility, but no more. However this is not because of lack of evidence, as we will see, but because of a combination of lack of interest and changes in fashion. It is a century now since the last serious study was made of the relation between Orphic literature and Plato's myths, by Dieterich in his *Nekyia*. Gradually the subject fell out of favour, forced into the background by the new-found belief in Plato's originality as a myth-maker. During the 1930s Wilamowitz and Thomas dealt what seemed almost a death-blow to the whole subject of 'Orphism' by demonstrating that there never was such a thing as a single or unified body of Orphic doctrine, and no such thing as an Orphic 'church'. In doing so they performed a miracle: acknowledgement of the existence of a genuine, pre-Platonic Orphic literature virtually vanished from the scholarly scene.[21] One particular result has been that, when there is an apparent overlap between themes and images used in Orphic literature and ones found in Plato, it is nowadays considered more 'prudent' to speak in terms of the Orphic material as having a

---

[20] For the statement in 62b that 'according to an utterance in the mysteries' life on earth is a prison, see *Crat.* 400c where the same teaching is ascribed explicitly to 'Orphics' (οἱ ἀμφὶ Ὀρφέα); cf. *OF* frs. 7–8 with further refs., L. J. Alderink, *Creation and Salvation in Ancient Orphism* (Chico, Calif., 1981), 62, West 21–2. For the reference in 69c to non-initiates 'lying in the mud' see *Rep.* 363d6–8 with n. 18 above, *OF* fr. 5, Graf (1974), 100–7; also Olymp. *Phd.* 8.7.1–2 and, in general, P. Courcelle, *Connais-toi toi-même*, ii (Paris, 1975), 502–19. For the verse quotation immediately following (69c8), see Olympiodorus' comments and metrical restoration, *Phd.* 7.10.10–12, 8.7.1–4 and 10.3.13–15 = *OF* fr. 235; Dam. *Phd.* 1.170.8–9.

[21] Wilamowitz (1931–2), ii. 193–200, 251 n. 1; Thomas 51 n. 192, and *passim*.

Platonic 'model' or needing to be understood 'in a Platonic perspective'.[22] This is all very convenient; but the trouble is that in the long term it has meant running away from a historical problem, not solving it. The problem has made itself more acutely felt in recent times as a result of archaeological discoveries—especially the discovery of the Derveni papyrus, which has honourably reinstated the existence of an Orphic literature before Plato's time.[23] And yet even the Derveni find runs the risk of focusing attention away from the crucial evidence of Plato's own works.

On a number of occasions Plato quotes or paraphrases lines of Orphic verse, and more often than not makes a point of emphasizing their antiquity.[24] But the most telling evidence of all is to be found in the *Phaedo*. The few references to Orphic ideas in the main body of the dialogue are, quite literally, only the start of the story. For example, the most striking item of underworld symbolism in those earlier Orphic allusions recurs repeatedly in the underworld geography of the final myth: the mention of lying in the mud (ἐν βορβόρωι κεῖσθαι) is picked up three times in the context of the Pyriphlegethon, Tartarus, and Sicily's vast subterranean rivers of mud.[25] Some very strange suggestions have been put forward as to the origin of the expression 'lying in the mud', but it is quite clear from references to the idea elsewhere—by Plato as well as other writers— that the image intended is of punishment reserved for sufferers in the underworld. Aristophanes' description is especially illuminating: those who have committed crimes, in particular the crime of raising their hand in violence against their father or mother, are forced to 'lie in the mud' (εἶτα βόρβορον ... ἐν δὲ τούτωι κειμένους) in the underworld close to the Acherusian Lake. Apart from the identical idea here of being left to lie in the mud it will be noted that according to the *Phaedo* the rivers

---

[22] Brisson, *ANRW* ii.36.4 (1990), 2915, 2920.
[23] Cf. Boyancé's comments, *REG* 87 (1974), 94; Burkert (1977), 1–3; Mansfeld, XIV. 290.
[24] *Phd.* 69c3–d1 = *OF* fr. 5 ('ancient hint'): cf. 63c6, 67c5–6, 70c5–6. *Crat.* 402b–c = *OF* fr. 15: for the idea of antiquity here see Mansfeld (1990), 46–51. *Phlb.* 66c = *OF* fr. 14: cf. West 118 and n. 8. *Laws* 715e = *OF* fr. 21 ('ancient saying'): cf. fr. 21a, Pap. Derv. cols. 17–19, Boyancé, loc. cit., Mansfeld, XIV. 290 and n. 75, West 89 and n. 35.
[25] 69c5–6 (above, n. 20); 110a5–6, 111d5–e2, 113a6–b6, Frutiger 259 n. 2.

of mud form part of the Pyriphlegethon, which flows close to the Acherusian Lake and is reserved in particular for those who lay hands in violence on their father or mother.[26] What the Aristophanes passage and the other parallels do here is bring together the allusions to lying in the mud and to subterranean rivers of mud which are scattered through the *Phaedo*, and show that they are so many facets of one and the same fundamental idea. In other words, they make it unmistakably clear that the final myth in the dialogue explains the earlier, brief allusion to lying in the mud by placing it in its proper, mythical context—the context in which it makes sense. At this basic level of the myth Plato has not added anything in the way of embellishment or subtle exegesis, and neither has his source; on the contrary, he has simply presented the broader canvas in which the image has its rightful place. If the isolated idea of 'lying in the mud' is Orphic, then the mythical canvas in which it belongs is almost certainly Orphic as well—and that means not just the rivers of mud but the subterranean landscape of the *Phaedo* myth as a whole.

Thematically, the earlier Orphic references are related to the final myth: they give a brief foretaste of the kind of material to come, and the possibility of connections between the different parts of the *Phaedo* finds itself confirmed. The same pattern of anticipation and fulfilment also repeats itself in other, complementary ways. So, the image evoked earlier in the dialogue of the impure soul being left to lie in the mud had, as its logical counterpart, the vision of the pure soul freeing itself from Hades to go and live with the gods; Plato's phrasing of the dichotomy is typical of specifically Orphic doctrine as described by him elsewhere, and the very same contrast

---

[26] Aristophanes, *Frogs* 137, 145–50 with Graf (1974), 92–4; Pl. *Phd.* 113a5–b5, 113e6–114a8 (Ch. 9 with n. 17). See further Pl. *Rep.* 363d6–7 (cf. 533d1–2); Virg. *G.* 4.478–80; Luc. *True Story* 2.30–1 and *Alex.* 25; Plot. 1.6.6.3–5, 1.8.13.17–26. With the *Phaedo*'s description of the 'huge lake' formed by the Pyriphlegethon, 'blazing with much fire' and 'boiling with water and mud' (113a6–8), compare also the so-called *Apocalypse of Peter*, with its portrayal of sinners in the underworld trapped in 'a huge lake filled with blazing mud' (Dieterich 4.50–1, 6, 81, 221). The similarities between the imagery here and in the *Phaedo* are obvious, and Dieterich devoted a book to showing the extent to which the Christian *Apocalypse*—found only in fragmentary form at Akhmīm in Upper Egypt and in a version in Ethiopic—is dependent on Orphic and Pythagorean traditions: cf. Dieterich, *passim*; Cumont 245–8; C. Maurer in E. Hennecke, *New Testament Apocrypha*, ii (London, 1965), 667.

reappears—drawn in identical terms—in the final myth as well.[27] Here too, the Orphic allusions earlier in the *Phaedo* spread into the myth at the end, and become an invaluable key to understanding its structure, language, and meaning. The conclusion becomes increasingly unavoidable that the basic stratum of the myth—in particular the details and description of its underworld rivers—derives from an Orphic poem.

At this point corroboration arrives from a surprisingly obvious source. In his commentary on the *Phaedo* myth Damascius explains Plato's selection of four underworld rivers—unusual from the point of view of Homeric tradition—as a sign of his indebtedness to Orphism. He states that a similar arrangement of underworld rivers was described 'at some length' by Orpheus; and he adds, on Proclus' testimony, that according to 'Orphic tradition' (Ὀρφέως παράδοσις) it was interpreted symbolically by correlating each river with an element.[28]

What are we to make of this? During the wave of scepticism earlier in the century about all matters 'Orphic', Damascius and Olympiodorus suffered particularly badly: it was argued that, apart from living nearly a thousand years after Plato, they were gullible mystics and the value of their reports in helping to reconstruct early Orphic literature almost nil.[29] But there are a few elementary facts that need to be either stated or restated. Just like Aristotle and Theophrastus, Damascius and Olympiodorus can be very convenient sources of information when handled with care; and in contrast to Aristotle and Theophrastus, they have the advantage that they had no real taste for deviousness or wilful misrepresentation of their predecessors.

---

[27] 69c6–7 (ὁ δὲ κεκαθαρμένος ... μετὰ θεῶν οἰκήσει) ~ 111b6–c1, 114b7–c2, 114d3. For 69c6–7 κεκαθαρμένος τε καὶ τετελεσμένος and 114b8 ἐλευθερούμενοι see *Rep.* 364e5–365a3 ('Orpheus and Musaeus'); and for 114b8–c1 cf. *Crat.* 400c7 (οἱ ἀμφὶ Ὀρφέα).

[28] Dam. *Phd.* 1.497.3–5 and 541.1–6, 2.145.1–6 = *OF* frs. 123, 125. The *Phaedo* modifies Homeric tradition in making the Cocytus a river in its own right as opposed to a branch of the Styx, demoting the Styx from a river to a lake, and adding Ocean to the number of underworld rivers (but see *Od.* 10.508–11). Cf. Hackforth 182 n. 1; *OF* fr. 222; Ch. 9. For Damascius' reference to 'Orphic tradition' see esp. Procl. *Tim.* iii. 161.3–4 and 168.9–18 = *OF* test. 250; iii. 170.15–16 and 250.17–18 = *OF* frs. 104, 217; also G. Casadio in *La Tradizione: Forme e modi* (Studia Ephemeridis 'Augustinianum', 31; Rome, 1990), 190–7.

[29] So e.g. Wilamowitz (1931–2), ii. 193, 197, 198 n. 1; Dodds 148 and n. 90.

As for the question of sources, we know very well that Damascius in particular had access to Orphic literature dating back to Plato's time at least.[30] But, these general considerations apart, the accuracy of the information available to Damascius and Olympiodorus—and the authority with which they were able to speak about Plato and his literary antecedents—are matters that have hardly begun to be appreciated adequately. Olympiodorus could name a comic poet whose work Plato alludes to obliquely and in passing in the *Phaedo*, and even quote the lines from the poet which Plato has paraphrased; without Olympiodorus' explanation we would have remained in the dark about the allusion, and never have known not only that he was paraphrasing another writer's words but also who the poet in question was.[31] And to take another example, again from his commentary on the *Phaedo*, Olympiodorus is able to supply the original Orphic verse—metrically correct—which Plato has also paraphrased in passing.[32] In the light of evidence like this it is a rather high-handed approach not to take Damascius very seriously when he says that Plato derived his arrangement of underworld rivers in the *Phaedo* myth from a poem by 'Orpheus'.

Damascius' frequent eccentricities as an interpreter of earlier philosophers are one thing; but a liar he was not. If he says that the same basic arrangement of underworld rivers which we find in the *Phaedo* was described—and described at some length—in a poem by Orpheus, then it was described in a poem ascribed to Orpheus. The obvious question is whether the Orphic verses were really older than Plato, or whether they were later creations with the geography of the *Phaedo* myth as their direct or ultimate model. Here, all the probabilities point in the same direction. On the one hand, Orphic literature produced after Plato's time pays remarkably little attention to

[30] Cf. e.g. West 116–18; KRS 19 n. 2, 22.
[31] Olymp. *Phd.* 9.9.1–12 = Eupolis, fr. 386 Kassel–Austin; Pl. *Phd.* 70b10–c2. The lines are also quoted, in general less accurately and with no attribution, by Ascl. *Met.* 135.23–4 (cf. Westerink, i. 136 n.). Olympiodorus' scrupulous analysis of Plato's allusion is worth reading with care. On the other hand, when he was reduced to making guesses of his own about Plato's sources he was considerably less reliable (above, n. 4).
[32] *Phd.* 7.10.1–12, 8.7.1–4, 10.3.13–15 = *OF* fr. 235; Pl. *Phd.* 69c8–d1. Regarding the intellectual 'sport' of tracking down quotations in authors from even earlier authors cf. Whittaker 66–7; also Ch. 4 with the refs. in n. 27.

him. There is no evidence whatever that, beyond the occasional Platonic saying which slipped into the mainstream of Hellenistic ideas, the later Orphic poets showed any interest in his writings or his myths; what philosophical nourishment they needed they appear to have found elsewhere. On the other hand Plato himself, as we have seen, knew and used a body of Orphic literature which already in his days was considered ancient; and earlier in the *Phaedo* he alludes to this literature in a way which prepares us for the fact that he will be using Orphic sources in the culminating myth. Sometimes the presence of a poetic original behind the prose of the *Phaedo* myth is difficult to ignore.[33] Also, in discussing the rivers of the underworld Proclus quotes a few brief lines from an Orphic poem which bear marked similarities to the overall eschatological schema outlined by Plato but at the same time are decidedly simpler: clearly any borrowing is not by the Orphic poet from Plato, but the other way round.[34] Finally, as for Damascius' remarks on the 'Orphic tradition' about the Acheron, Cocytus, and Pyriphlegethon and on the allegorical interpretation of these rivers in terms of elements, the Derveni papyrus is vivid proof that even the allegorizing reinterpretation of Orphic poetry along these lines was already current before Plato's time—not to mention the original Orphic poetry itself.[35]

These general considerations would themselves weigh the balance heavily in favour of concluding that the Orphic poem Damascius refers to was older than Plato, even were it not for the fact that we have already arrived at the same conclusion on

---

[33] So e.g. at 113b3 (οὐ συμμειγνύμενος τῶι ὕδατι) Plato is almost certainly following a line of verse modelled on *Il.* 2.753 (οὐδ' ὅγε Πηνειῶι συμμίσγεται ἀργυροδίνηι . . .); see also below, with nn. 50, 53–5, 57.

[34] *Rep.* ii. 340.11–20 = *OF* fr. 222.

[35] See the refs. above, n. 19; and below, n. 41 *ad fin*. In objecting to the allegorical interpretation of each of the Orphic rivers as representing an element, Damascius specifically takes issue with his immediate source and predecessor: Proclus (*Phd.* 2.145.2, 7–14; cf. also 1.541.5–6, and note that in 1.497.2–5 'Orpheus' is presented as distinguishing between—rather than equating—underworld rivers and elements). And yet any possibility of ascribing this allegorizing interpretation simply to Proclus is out of the question. Proclus cites the same 'Orphic' tradition of interpreting the underworld rivers in *Tim.* ii. 49.17–21; and from the extraordinary mass of subtleties and clumsy attempts at refinement—inclusion of the fifth element, and so forth—in which the interpretation there has become immersed, it is clear that already in Proclus' time it had a long history behind it.

the grounds of the *Phaedo* itself. The convergence is precise. From the references to lying in mud and rivers of mud it emerged that the underworld landscape of the *Phaedo* myth as a whole—which means above all its arrangement of rivers—is almost certainly based on a work of Orpheus. Damascius simply provides the welcome corroboration. The truth of the matter was seen by Dieterich a century ago; but there is no accounting for fashion.[36]

There is one small detail which helps to confirm this conclusion and put it in a clearer perspective. Damascius, as mentioned earlier, distinguishes explicitly between the original mythology of underworld rivers ascribed to Orpheus and the allegorical interpretations—correlating each river with an element—to which it was subsequently subjected. He provides a specific example to illustrate this basic distinction: while the Pyriphlegethon was interpreted allegorically as the element of fire, Ocean as water, and Styx as earth, justification for equating the Acheron with the element of *aer* was found in the fact that 'Orpheus' himself describes the Acherusian Lake as *aerios*.[37] This presents a very interesting sequence. The original description of the lake as *aerios* was plainly intended to mean not that the lake was 'air-like' or 'up in the air' but that it was 'misty'. The simple yet visually impressive image of a misty lake is an obvious parallel to the use of *aer* and its derivatives by writers such as Homer, Hesiod, and Empedocles to describe the mist rising from and hovering over bodies of water.[38] The significance of this original meaning of the word emerges when we go back over the ground covered earlier in our analysis of the history of the term *aer*. Through the sixth and fifth centuries BC, the word along with its derivatives retained the basic sense of 'mist' or 'fog'. But during the same period the new, broader sense of 'air' started to appear, and by the early fourth century

---

[36] Dieterich 123-4; cf. also Eitrem, *RE* xx. 259.

[37] Dam. *Phd.* 2.145.3-6 = *OF* fr. 125: . . . τὸν Ἀχέροντα . . . ἀέριον, διὸ καὶ Ὀρφεὺς τὴν Ἀχερουσίαν λίμνην ἀερίαν καλεῖ. Westerink's interpretation of ἀερίαν as a proper name, 'Lake Misty' (ii. 363-4 n.), is unjustified and implausible. Norden overlooked Damascius' distinction between original poem and subsequent allegorization of it, with disastrous results: *P. Vergilius Maro Aeneis Buch VI*[4] (Darmstadt, 1957), 26.

[38] *Il.* 23.744, *Od.* 20.64, Hes. *Op.* 549-50, Emp. B38.3; Kahn 151; above, Ch. 3. For *aerios* in this sense in the Orphic passage see also Westerink, ii. 363-4 n. (above, n. 37).

this new meaning had taken over in everyday speech and literature. The old sense of 'mist' survived for a while as a literary relic in early Alexandrian poetry, but eventually disappeared so totally that even the greatest commentators on Homer misunderstood and reinterpreted the word when it occurred in its original sense.[39] The broad relevance of this evolution to the history of Orphic literature is straightforward. The later Orphic poetry composed during the Hellenistic or post-Hellenistic periods represents an amalgam of phrases and whole lines of verse taken over from earlier Orphic literature (as we now know for certain from the Derveni papyrus) combined with new material written in freshly contemporary language and style; it is important to note that, even in the case of passages very probably adapted from an older model, there are signs of deliberate attempts to substitute modern diction and terminology.[40] What is more, the predominant and pervasive affinities of these later Orphic texts—at the linguistic as well as philosophical level—are Stoic.[41] These general comments are strikingly borne out in practice when we look at how the word *aer* and its derivatives are used in the surviving Orphic literature. *Aer* itself is always, as in Stoicism, the element of air.[42] As for the adjective *aerios* (or *ēerios* in its epic form), it occurs repeatedly—but not in the sense of 'misty' or 'shrouded in a fog'. With one single exception it always means 'up in the air', 'airlike', or 'airy'.[43] The one exception, where the word is

---

[39] Above, Chs. 2–3. For examples of early Alexandrian antiquarianism in 'this respect cf. Aratus 349, Apollonius Rhodius 1.580.

[40] Cf. West 225, and, for the reuse and adaptation of older material, ibid. 34, 82–4, 89–90, 218, 229; *OF* pp. 93, 206; Mansfeld, xiv. 290 n. 75; F. Vian, *Les Argonautiques orphiques* (Paris, 1987), 8 and n. 1.

[41] Dieterich (1891), 83–6; West 36, 58–61, 193–4, 219, 224–6; Brisson, *RHR* 202 (1985), 409–10; Kingsley (1994a) with nn. 28, 52. The relations between Stoicism and Orphism are best considered in the first instance in terms of affinities rather than of one-way influences from Stoic onto Orphic literature, because there are definite signs of influence the other way round: cf. e.g. *SVF* ii. 316.11–22, Guthrie (1952), 76–7, K. Ziegler, *RE* xviii/1. 1201, and Boyancé's comments, *REG* 87 (1974), 93–5, 108. In this connection it is worth noting that already in the Derveni papyrus we find the tendency to interpret Orphic poetry allegorically by correlating particular things—including rivers—with the element of air (col. 23.1–3).

[42] *OF* frs. 168.25, 228b, 247.27; above, Ch. 2 n. 9, Ch. 3 with nn. 9, 31, 44.

[43] *Orphic Hymns*, proem 32, 11.16, 16.8 (cf. 6), 20.2, 21.1 (cf. 4), 59.17 (cf. 3), 69.9, 71.6, 81.6 (cf. 4). Compare also *Chaldaean Oracles*, frs. 61.5 and 10, 73.4, 91, 216.4, 219 Majercik; Lewy 139–44, 268 and n. 28, 271 and n. 41.

used in the sense of 'misty', occurs in a line of a fragment which is clearly pre-Platonic in date.[44]

Here, owing to the nature of the problem and the evidence at our disposal, it is impossible to be categorical; but from these various details some highly probable conclusions emerge. Damascius' report of Orpheus describing the Acherusian Lake as *aerios* or enveloped in mist very strongly suggests a poem of pre-Hellenistic date, hardly written much if at all later than the end of the fifth century BC. As for the subsequent deflection of a derivative of *aer* from its original sense of 'misty' to make it refer to the element of air, this is exactly the kind of misinterpretation to which other sixth- and fifth-century literature was routinely subjected.[45] As far as Damascius is concerned, whatever knowledge he may have had of the poem in which the word occurred is most likely to have come to him through the much later Orphic *Rhapsodies*: a vast poetic compendium synthesizing earlier Orphic verse with more recent material, and the basic source on which the Neoplatonists relied.[46] This has the practical implication that he is likely to have had no way of knowing—and probably no reason to want to know—the title of the poem in question. Otherwise, we are drawn once again to the conclusion that in general terms he knew what he was talking about: that the original Orphic poem which was subsequently interpreted allegorically was pre-Platonic, contained one of those eyewitness accounts of Hades which already in Plato's time were considered the special preserve of Orphic and Pythagorean literature,[47] and gave a description of underworld geography which formed the basis of the myth at the end of the *Phaedo*.

From our earlier conclusions there is nothing surprising in this at all. We have already seen that the *Phaedo* myth comes from the West, is Pythagorean in origin, and that it was common practice for western Pythagoreans to attribute poems they wrote—especially poetry dealing with eschatology and the underworld—to Orpheus. We have also seen (Ch. 7) that the

[44] *OF* fr. 72, cited by Proclus directly after fr. 66, which we will come to shortly.
[45] Above, Ch. 3 (Empedocles); ibid., n. 9 (Theagenes).
[46] K. Ziegler, *RE* xviii/1. 1362; West 69; Brisson, *ANRW* ii.36.4 (1990), 2886.
[47] Epigenes *ap.* Clem. *Str.* 1.21.131.3–5 = *OF* pp. 52 (test. 174), 63 (test. 222), 304–5; West 9, 12; Burkert 155–63. Cf. also Diod. 1.92.3, 1.96.4–6 = *OF* fr. 293; Servius on Virg. *Aen.* 6.392, 565 = *OF* frs. 295–6; Julian, *Orations* 7.11, 216d; Lobeck, ii. 810–13.

geography of underworld rivers provides the fundamental structure for the myth as a whole. What this means altogether is that while the physical foundation for the *Phaedo* myth is chiefly Sicilian, its mythical foundation now appears to be Orphic. In short, the myth arose out of the soil of Sicily and Italy and took the form of an Orphic poem written, used, interpreted, and eventually transmitted to Plato by western Pythagoreans. The significance of this sequence can hardly be overestimated. For one thing, the interrelation between the categories of 'Orphic' and 'Pythagorean' is graphically demonstrated. For another thing, habits die hard, and in spite of the evidence of the Derveni papyrus it is still normal to find the allegorizing of Orphic poetry and mythology presented as a primarily Neoplatonic phenomenon.[48] Here, however, we have the allegorizing interpretation of Orphic literature not only attested before Plato's time, but actually feeding into and creating the Platonic myths themselves.

After the mud and the mist, there is one final detail in the *Phaedo* myth that calls for a few words. Plato begins his outline of its underworld topography by describing a huge labyrinth of chasms, craters, and water channels, all grouped around one chasm in particular—the largest of them all. Part of this description was quoted earlier; here it will be worth giving it more fully.

One of the chasms (*chasmata*) in the earth happens to be vaster than any of the others, and is bored right through the entire earth. It is this chasm which Homer had in mind when he described it as

'far, far away, in the deepest abyss underneath the earth';

elsewhere both he and many other poets refer to it by the name of Tartarus. All rivers flow together into this chasm (*chasma*) and flow out of it again, and each of them takes on its own particular quality according to the nature of the earth it flows through. The reason why

---

[48] So e.g. Brisson, *ANRW* ii.36.4 (1990), 2926–7 ('Interprétations allégoriques'). Similar misconceptions surround the closely related issue of the allegorizing of Pythagorean *akousmata*: an issue where Burkert mistakenly assumed that 'allegorical tradition' is somehow synonymous with 'late tradition' (229 n. 55). In fact, as Burkert himself points out (166, 174), the allegorizing interpretation of *akousmata* can safely be dated back at least as far as the time of Philolaus. Cf. also Richardson (1975), 74–6.

all rivers flow out of this place and then flow back into it is the fact that the liquid has neither bottom nor fixed base (111e–112b).

This last expression, 'has neither bottom nor fixed base' (πυθμένα οὐκ ἔχει οὐδὲ βάσιν), does not seem particularly remarkable in the context of Plato's overall description; but there is more to it than meets the eye. One line of Orphic poetry which is quoted and referred to repeatedly by Neoplatonists describes the 'vast, gigantic chasm' (μέγα χάσμα πελώριον) formed at the start of creation, in which 'no boundary existed, and neither bottom nor set base' (οὐδέ τι πεῖραρ ὑπῆν, οὐ πυθμήν, οὐδέ τις ἕδρα).[49] The parallelism between this Orphic verse and the wording of the passage in the *Phaedo* has hardly received the attention it deserves; it was noted in passing by Guthrie but then, appropriately enough, seems to have sunk back into the oblivion out of which it emerged.[50] And yet even at a purely verbal level it is striking: the same word for 'bottom' (πυθμήν), the two synonyms for a fixed or set base (βάσις, ἕδρα), the same double negative construction. The two synonyms are in fact so exact that in his commentary on the *Phaedo* passage Aristotle substitutes the word found in the Orphic verse (ἕδρα) for the word used by Plato (βάσις).[51] The argument that these parallels could just be so many coincidences loses all its force when we note that, in both cases, the subject in question is the same: a 'vast, gigantic chasm' in the Orphic verse (μέγα χάσμα πελώριον) and 'the vastest of all chasms' in the *Phaedo* (ἕν τι τῶν χασμάτων μέγιστον).[52] What is more, it is clear that in both cases the

---

[49] *OF* fr. 66; Lobeck, i. 472–4.

[50] Guthrie (1952), 168–9, who correctly suggested that Plato should be considered the debtor. The *Phaedo* passage is not mentioned in *OF* ad loc., and the Orphic parallel is not noted in editions and translations of the *Phaedo*.

[51] *Meteor.* 356ª4 (οὐκ ἔχειν γὰρ ἕδραν).

[52] To argue that the two chasms are different because the chasm in the *Phaedo* exists now, whereas the chasm in the Orphic verse as given by the Neoplatonists was formed at the start of creation (*OF* fr. 66a), would be futile. The Orphic chasm as presented in the Neoplatonic context was created alongside *aither*, and there is no reason to suppose that *aither* at the start of creation was different from *aither* in the universe now. As it happens, the entire structure of the *Phaedo* myth is little more than an elaboration of this coupling of chasm and *aither*, with its fundamentally dualistic description of the fate of the soul on its passage down into the lower chasm of Tartarus (109a9–b7, 109c1–110a7, 111c4–114b6) and on its ascent to the higher reaches of *aither* (109b7–c1, 110a8–111c3, 114b6–c6).

chasm is being connected with Tartarus. This is made explicit in the *Phaedo* myth (112a4–5), while in the case of the Orphic verse the idea of a 'vast chasm' (μέγα χάσμα) is obviously meant to evoke the famous Hesiodic picture of the 'vast chasm' (χάσμα μέγ') mentioned in the context of a description of Tartarus.[53] Further, it will be noted that the emphasis both in Plato and in the Orphic verse on this vast chasm having no bottom—and, in the Orphic verse, no boundary as well—not only evokes but also deliberately corrects and modifies the chasm as portrayed by Hesiod, which for all its vastness does by implication have a definite bottom and definite boundaries. This kind of conscious mythological refinement of Hesiod's cosmology would be rather out of place in Plato, but it was a dominant keynote of Orphic cosmological poetry.[54] While these points serve to bring the passage in the *Phaedo* and the line of Orphic poetry even closer together, they also suggest the priority of the Orphic verse with regard to Plato; and the inevitable conclusion that the two passages are linked is further confirmed by the evidence noted earlier of substantial borrowing from Orphic literature in the *Phaedo* myth as a whole. We will find additional confirmation that this is what has happened here when later we come to look at a passage in Plutarch.

All that remains is to clarify the nature of the link between the Platonic and the Orphic passage. The balance of probability has already started to come down heavily on the side of Plato's indebtedness to an Orphic poem; but there are a number of other factors which need taking into account as well. To begin with, an Orphic versification of the *Phaedo* myth is extremely unlikely. Generally speaking, Greek prose was only converted into verse in the special case of famous apophthegms or canonical sayings, or when versification would prove practically useful as an aid to memory. Here we also need to

---

[53] *Th.* 736–45. Cf. also Parm. B1.18 with Burkert (1969), 8–13, Pellikaan-Engel 8, 10, 25, 28, 53–4, 73; Eur. *Phoen.* 1605; Pl. *Rep.* 614c2 with Adam ad loc., Bidez, *Eos, ou Platon et l'Orient* (Brussels, 1945), 45–6; Plut. *Moralia* 565e, 590f. For the *Phaedo* chasm and the Hesiodic chasm see also Solmsen 244–5.

[54] Cf. Hes. *Th.* 738 (boundaries), 741 (bottom), with Solmsen 242–3, 245–8, Pellikaan-Engel 25, 47. For the Orphic modification of Hesiod cf. Guthrie's comments, (1952), 83; and for the general poetic interest in 'rewriting' Hesiod during the 6th and 5th centuries, Bollack, i. 283–6. Regarding the obviously unsatisfactory nature of Hesiod's *chasma* description 'in the eyes of more sophisticated later generations' see Solmsen 246.

remember the point already mentioned, that Hellenistic and post-Hellenistic Orphic poetry showed a remarkable lack of interest in Plato's writings—certainly not the kind of interest needed to try versifying a complex section of narrative in the *Phaedo*. But, these broad considerations apart, there is one clear sign that the Orphic verse is primary. The end of the line—'nor any set base', οὐδέ τις ἕδρα—is taken over word for word from the end of a line in Hesiod. This is an obvious example of the general procedure so evident in Orphic poetry of adopting parts of lines from Hesiod and putting them 'to entirely new uses'.[55] In short, the model for the Orphic verse is not Plato but Hesiod, and the prose version in the *Phaedo* is based on it, not the other way round.

While the Orphic version must be primary, it is equally clear how Plato's very similar prose version came into being. The surviving works of Homer and Hesiod provide an ideal measuring stick for assessing not just the accuracy but also the style of Plato's paraphrases and quotations from earlier literature. For him, accuracy was not a matter of primary importance. On the contrary, he developed a deliberate art of casual allusion. In particular, he liked to alternate exact quotations with paraphrases—modifying lines of verse by adapting them to his prose, abbreviating them, preserving some of the original words and substituting synonyms for others.[56] Comparison of Plato's allusions to Homer or Hesiod with the poems themselves shows that even the smallest details in his manner of alluding to the Orphic verse have exact parallels elsewhere, and are typical of his normal style and procedure.[57] In short, not only are the parallels between the

---

[55] *OF* fr. 66b ~ Hes. *Th.* 386; Guthrie (1952), 83. See above, Ch. 4, for the same principle of imitation and variation in Empedocles.

[56] The most complete discussion and summary of the evidence remains Labarbe's, 257–360, 404–5, 420, and *passim*. Cf. also Whittaker 67.

[57] Beginning of verse omitted: Labarbe 260–2. Modification of grammatical case to suit the prose context: *Laches* 201b2–3 ~ *Od.* 17.347, *Protagoras* 309b1 ~ *Il.* 24.348, *Gorg.* 485d6 ~ *Il.* 9.441, Labarbe 259–62 and *passim*. More accurate quotation to begin with (πυθμένα ~ πυθμήν), followed by a drift into paraphrase (βάσιν ~ ἕδρα): ibid. 264–5. Substitution of synonyms: ibid. 332, 358, 420, Bluck 393. As for Plato's substitution of βάσις for ἕδρα, it is probably no coincidence that the word he uses occurs in Sophocles in a passage with a different sense but a very similar construction to his πυθμένα οὐκ ἔχει οὐδὲ βάσιν: οὐκ ἔχων βάσιν οὐδέ τιν᾽ ἐγχώρων κακογείτονα (*Philoctetes* 691–2). Plato in fact had a marked penchant for conflating from memory

Phaedo passage and the Orphic verse instructive, but so even are the divergences.

Once again, the way in which the Neoplatonists quote the Orphic line is no help in reconstructing its original context. If, as would seem to be the case, they found it in the *Rhapsodies*, there is no telling which Orphic poem it first occurred in: as noted earlier, lines of Orphic poetry were readily transferable property. But one thing we can be quite sure of is that it occurred in a poem which was pre-Platonic; and although this is not the first time we have found Plato paraphrasing Orphic literature in the *Phaedo*, what is especially interesting about the example this time is that it takes us to the very core of the final myth.

Even with the pitifully slender evidence at our disposal, we can begin to see that Olympiodorus probably knew rather well what he was talking about when he said 'Plato paraphrases Orpheus everywhere'.[58] Nowadays, the conviction among Neoplatonists that both they and Plato himself stood in a line of Orphic tradition has become a laughing matter. It is either ignored or dismissed with condescension as a typical example of the 'syncretistic' fantasizing so characteristic of later antiquity: as a delusion to be traced back to this or that authority in the Neoplatonic school.[59] Instead, modern scholarship has devised its own myth to live by: the myth of Plato as a kind of Prometheus or Palamedes to whom much of what is most important in antiquity can safely be ascribed. This myth has a certain practicality which makes it all the more attractive. Increasingly over the past fifty years attention has shifted to the study of philosophy and religion during the period of later antiquity, and in line with this shift Plato has become a very convenient hook on which to hang ideas. If a theme or idea can

entirely different passages—written either by different authors or the same one—on a purely associative basis: Labarbe 101-8, 189-92, 197-201, 342-3, 406-9, 420, West (1978), 181-2. For the usual Platonic 'method' of quotation and paraphrase see also Olympiodorus' comments, *In Alcibiadem*, 104.3-6 Westerink.

[58] παρωιδεῖ γὰρ πανταχοῦ τὰ Ὀρφέως, *Phd.* 7.10.10; πανταχοῦ γὰρ ὁ Πλάτων παρωιδεῖ τὰ Ὀρφέως, 10.3.13. For Olympiodorus the verb παρωιδεῖν meant simply 'to paraphrase, allude to'. Cf. *Phd.* 7.10.5-6, 8.7.1, 9.9.9-12; *In Alcibiadem*, 94.12, 104.3-6 Westerink, etc.

[59] See e.g. Brisson (1987), 43-53; below, n. 62.

be traced back to Plato then it comes from Plato; to try tracing it back even earlier would be too ambitious, an obvious case of crossing the dividing line between different disciplines and different periods of history. But things are not so simple. As we saw in the last chapter, there is a certain relationship of symmetry between Neoplatonic and pre-Platonic traditions; particularly where mythical and religious ideas are concerned, Plato himself is not so much a starting-point as an isthmus between two continuous bodies of tradition. Later periods in history, and later literary productions, are not chunks of material that can be carried away and studied on their own. Everything is interrelated, and so long as these complex interrelations remain unperceived everything goes wrong: we misjudge the historical accuracy of what Neoplatonists say, misunderstand their intentions, and criticize or simply ignore them when we should be listening to them most closely. In other words, the Neoplatonic evidence cannot be understood properly without understanding the real roots of Platonic tradition; and conversely, the Neoplatonic evidence has a very significant contribution to make in helping us understand the truth about those origins. The tendency nowadays is to focus attention on the ways in which the Neoplatonists supposedly departed from Plato, on the extent to which they differed from him and were unfaithful to him. But—quite apart from the fact that some of these supposed differences and departures are illusory[60]—it should be self-evident that before we can start to appreciate the discontinuity we need to understand the underlying continuity, even at the risk of letting a few basic preconceptions die. When Olympiodorus exclaims with an obvious touch of humorous exaggeration that 'Plato paraphrases Orpheus everywhere', we need to start by assessing how much truth there is in what he says, not how little. When Proclus says Plato received his knowledge of divine matters from 'Pythagorean and Orphic writings' (ἔκ τε τῶν Πυθαγορείων καὶ τῶν Ὀρφικῶν γραμμάτων), we may disagree with some of the picturesque details of initiatory succession which he provides, and with his emphasis on purely written transmission, but that is no justification for ignoring what is most

---

[60] Above, Ch. 9 with nn. 19–20 and 27.

important: the basic acknowledgement of Plato's indebtedness to Orphic and Pythagorean tradition.[61] And when Proclus claims that Plato 'took over from Orpheus the mythological details' in the *Phaedo* myth, or when Julian—writing a century earlier—says Plato followed the model of Orpheus in his descriptions of the underworld, we may object to their failing to distinguish between Orpheus and people writing under the name of Orpheus; but otherwise the evidence indicates that they were perfectly correct.[62]

[61] Procl. *Th. Pl.* 1.5; cf. e.g. *Tim.* iii. 160.17–161.6, 168.8–20, *Eucl.* 22.9–16. For Proclus' account of Plato receiving knowledge of the gods from Pythagorean and Orphic sources (*Th. Pl.* loc. cit.: Πλάτωνος ὑποδεξαμένου τὴν παντελῆ περὶ τούτων [sc. θεῶν] ἐπιστήμην . . .), see *Meno* 81a5–11 with Ch. 12 below.
[62] Procl. *Rep.* ii. 340.11–13, cf. also ibid. 21–2, 23–4 (*OF* fr. 222); Julian, *Orations* 7.11, 216d (out of context in *OF* p. 304). The Julian passage is not noted by Brisson (1987), 51 in his attempt to trace back to as early as Iamblichus Proclus' 'theory' that Plato was indebted to Orphic tradition: Julian was of course to a very large extent a mouthpiece for Iamblichus' ideas (Finamore 159–60 with refs.). For Iamblichus on Orpheus cf. also Julian, *Orations* 7.12, 217b–c (not in *OF*), and J. M. Dillon, *Iamblichi Chalcidensis in Platonis dialogos commentariorum fragmenta* (Leiden, 1973), 363.

11
The Mixing-Bowl
*

APART from the *Descent to Hades* by Orpheus of Camarina, and Timocles of Syracuse's *Salvation*, there is another Orphic poem worth mentioning in connection with the *Phaedo* myth—and for two separate reasons. This is a poem called the *Krater*: a word which originally meant a mixing-bowl used for mixing water and wine, but later acquired its by now more familiar sense through being applied metaphorically to hollows in the earth and volcanic 'craters'. It will be noted that even in Roman times the original meaning of the word was still very palpable behind the metaphorical usage.[1] We have already found this applied sense of the word in the myth at the end of the *Phaedo*,[2] and a few brief comments on it will not be out of place.

First, 'craters' or bowl-shaped hollows in the earth are mentioned by Greek and Roman writers as occurring in a number of places. But by far the most important and most often referred to were in two regions which were repeatedly compared with each other in antiquity because of the volcanic qualities they shared and which, as we have seen, were understood to be joined to each other along a single volcanic seam. These regions were, on the one hand, the area around the Bay of Naples in Campania and, on the other, the eastern part of Sicily extending from Hiera and Strongyle, just off the north coast of the island, down through Etna and—past Hybla Geleatis (modern Paternò) at Etna's southern face—to the lakes of the Palici and Syracuse.[3] Second, wherever in the

---

[1] Cf. Lucr. 6.701–2 (mixing of water and fire in Etna), with the commentators.
[2] 111d5; Ch. 7, with n. 11.
[3] Ch. 9, with the refs. in n. 15. The Bay of Naples was itself known as 'the *Krater*' in Roman times: Strabo 5.4.3 and 5.4.8, Cic. *Letters to Atticus* 2.8.2, J. H. D'Arms, *Romans on the Bay of Naples* (Cambridge, Mass., 1970), p. vii. The name is clearly meant to reflect the intense volcanic activity both immediately inland, and in the waters of the bay itself: cf. R. F. Paget, *In the Footsteps of Orpheus* (London, 1967), 27–8, 92; M. Guido,

ancient world these craters are mentioned as occurring they are almost always described as points of descent and entry into the underworld.[4] Third, through this special link with the other world they became 'places of power': sites of oracles, appropriate spots for the taking of oaths, places of ordeal where truth was revealed and untruth exposed, and locations associated with dreams and their interpretation.[5] And finally Sicily in particular, but Campania as well, were famous not just for their craters of fire but also for craters of water.[6]

Regarding the Orphic *Krater*, no verses specifically ascribed to the poem survive; its contents have remained a mystery in spite of—or rather because of—the fact that the *krater* was obviously an important Orphic symbol.[7] It recurs as a symbol elsewhere, but there is nothing specific enough to be of much help. Empedocles uses the imagery of mixing wine and water when he describes the universal swirl of creation as Love mixed the four elements to produce mortal things; and the *krater*

Southern Italy (London, 1972), 27–48, 64–7. For the 'craters' of Vesuvius cf. Strabo 5.4.8, and for Sicily see: Diod. 5.7.3–4, Strabo 6.2.10 (Hiera and neighbouring islands); Lucr. 6.701–2, Diod. 5.4.3, 5.7.3–4, Plin. *HN* 3.8.88, Luc. *DM* 6 (20). 4, *Dead Come to Life* 2, *Icaromenippus* 13, D.L. 8.69, 71, 75, etc. (Etna); above, Ch. 7 n. 11 (Palici, Hybla Geleatis, Kyane).

[4] Soph. *OC* 1593; cf. 1551–91, schol. on 1593 (above, Ch. 7 n. 14), R. C. Jebb on 1591 χαλκοῖς βάθροισι (correctly citing *Il.* 8.15 and Hes. *Th.* 811–12), 1593 κοίλου ... κρατῆρος (citing Pl. *Phd.* 111d) and Θησέως, and 1594 (citing Schol. Aristophanes, *Knights* 785). Pl. *Phd.* 111d2–113d4; Luc. and D.L., locc. citt., with Ch. 6 and n. 10; Macr. *Sat.* 5.19.18 and 24 = Aesch. fr. 6, Ciaceri 16–17, 23–6; Ov. *Met.* 5.424 (Kyane) with Ch. 9 above. For the region around the Bay of Naples cf. Burkert (1969), 17–18; McKay 1–2.

[5] For oracles see esp. Macr. *Sat.* 5.19.22 and 30 = Xenagoras, *FGrH* 240 F21, Freeman, i. 524, Ciaceri 31–2, J. H. Croon, *Mnemosyne*, 5 (1952), 119, 122–3 (Palici), Ael. *VH* 12.46, Hesychius, s.v. Γαλεοί, Stephanus of Byzantium, s.v. Γαλεῶται, Freeman, i. 514–17, Ciaceri 15–22 (Hybla Geleatis), Lyd. *Mens.* 4.86 (135.12–13), Freeman, i. 528–9, Ciaceri 151–2, and for Campania, Burkert, loc. cit., McKay 8–15, 130–2, Parke 92–3, 95 n. 5; the now fashionable but feeble argument that an oracle site at Lake Avernus is a purely literary invention (Hardie 279–86, Clark 68–71) is far less plausible than the alternative that an early—probably at first non-Greek—oracle by the crater eventually fell into disuse. For the exposure of truth and untruth and the taking of oaths cf. esp. ps.-Arist. *Mirabilia* 834[b]7–17, Diod. 11.89, Sil. 14.219–20, Macr. *Sat.* 5.19.19–21 and 28–9 = Polemon, fr. 83 Preller, Freeman, i. 167, 520–5, Ciaceri 24–6, 30–1, Croon, op. cit. 118, 120; no doubt the reference to oaths in Soph. *OC* 1593–4 is to be understood in this light. For dreams and their interpretation cf. Cic. *Div.* 1.20.39 and Paus. 5.23.6 = Philistus, *FGrH* 556 F57, Freeman, i. 160–2, 514–16, Ciaceri 15–16 (Hybla); Virg. *Aen.* 6.893–900.

[6] Ch. 7 n. 11; above, n. 3.

[7] Cf. Nilsson's comments, iii. 334–5.

occurs in Plato's *Timaeus* as a cosmic symbol of the vessel in which the creator-god mixes and produces the different kinds of soul.[8] There is also—not surprisingly, considering the links between the Hermetica and Pythagoreanism—a Hermetic work called the *Krater*; the importance of the symbol here can easily be gauged from the immense influence it subsequently exerted on the legend of the Grail.[9] But far from being able to clarify anything else, the Hermetic use of the symbol stands in need of clarification itself. It is a complex web of ideas, plainly due to the merging of different mystery traditions: the *krater* is not just the source of the liquid the initiate is expected to drink but also something he has to throw himself into so as to become pure and immortal, a 'perfect man' capable of reascending to heaven.[10] This second idea would seem to suggest a form of pagan initiatory ritual; commentators have plausibly pointed to Orphic and Pythagorean ideas and, in particular, to the legend of Empedocles throwing himself into the *krater* of Etna to achieve precisely the same results.[11]

However, there is a missing link to the puzzle: a text which directly associates the symbol of the *krater* with the name of Orpheus and, in so doing, promises to throw some real light on the contents and concerns of the lost Orphic *Krater*. The existence of the link, as well as its potential value, have been appreciated for a long time; but the obvious and also less obvious implications of the text appear not to have been drawn.

In one of his eschatological myths Plutarch gives a description of a visit to the other world during which the fictional character making the journey arrives at a 'vast *krater* with rivers

[8] Emp. B35.14–16: cf. *Il.* 9.202–3 with Gemelli Marciano 41–4, S. Eitrem, *SÖ* 5 (1927), 49. Pl. *Tim.* 35a–37a, 41d: cf. *Phlb.* 61b–c with Procl. *Tim.* iii. 250.17–22 = *OF* fr. 217, West 11.

[9] *CH* 4; B. P. Copenhaver, *Hermetica* (Cambridge, 1992), 131. For the Hermetica and Pythagoreanism see below, n. 11 and Chs. 20–1; for Pythagoreanism and the Hermetic *Krater* in particular, NF i. 56–7 n. 28. On the continuation of the *krater* symbolism in the Grail legend cf. H. and R. Kahane, *The Krater and the Grail* (Urbana, 1965).

[10] *CH* 4.4–6. Cf. Scott, ii. 140–2; Festugière (1967), 100–12; G. van Moorsel, *The Mysteries of Hermes Trismegistus* (Utrecht, 1955), 57–64.

[11] L. Ménard, *Hermès Trismégiste* (Paris, 1866), p. lxviii; Scott, ii. 141–2; van Moorsel, op. cit. 59. For continuity between Pythagorean, Empedoclean, Hermetic, and early Christian ideas of immortalization cf. W. Bousset, *Kyrios Christos*[2] (Göttingen, 1921), 342–3; for comments on the literary background, Norden 130–3, 279; and for Empedocles and Etna, below, Chs. 16–18.

pouring into it'. The character is then told by his spirit guide that this is as far as Orpheus managed to get on his descent to the underworld in search of his wife; that he suffered a lapse of memory and, as a result, when he returned to earth empty-handed he spread around the false story that the oracle at Delphi was shared by Apollo and Night. The guide goes on to explain that this *krater* is in fact the site of an oracle shared by Night and the Moon, and is the source of dreams.[12]

The passage is a classic example of how with the passing of time myth becomes contaminated and corrupted, mixed up with other traditions, and distorted by misunderstandings and polemic; but a genuine Orphic kernel to it there undoubtedly is. The oracle of Night, delivered by her from her mysterious sanctuary or cave, is an authentic and ancient feature of Orphic mythology.[13] To this we can add a passage in Proclus—the details show that he is not basing what he says on the fragment of myth in Plutarch—which associates the Orphic Night both with the Moon and with a vast cosmic *krater*.[14] But by Plutarch's time the original mythical material has been heavily modified through contact with outside factors and interests. A story of Orpheus' descent to the underworld has obviously been parodied, at the expense of Orpheus himself and to the decided advantage of Delphi. His mission is abortive, his memory fails him, he wrongly associates Night with Apollo at Delphi, and he is stupid enough to disseminate his lies and confusion among others. Possibly there is an allusion here to an actual statement by Orpheus to the effect that Night's role at Delphi had been ignored and suppressed; possibly a more universal statement by him expressing the primacy of Night's role as an oracular power was taken personally by Delphic propagandists and turned back against him.[15] But in the result-

---

[12] *De sera* 566a–c.

[13] As noted by Dieterich 147. Cf. Pap. Derv. col. 11.1–3; *OF* frs. 103–5, 144; the *Orphic Argonautica* 28; Burkert (1969), 17 and n. 37; West 71, 72–3, 84–8, 99–100, 213, 236, 256.

[14] *Tim.* iii. 169.15–170.16 = *OF* fr. 104. There is also a close connection between the Moon and Musaeus—Pl. *Rep.* 364e = *OF* fr. 3; Roscher 101 and n. 412; West 47–8—and the evidence suggests our *Krater* poem was addressed to Musaeus by Orpheus (Servius on Virg. *Aen.* 6.667 = *OF* test. 167 and p. 308).

[15] For Delphi and Night cf. Schol. Pi. ii. 2.5–6 Drachmann, West 101 n. 58; also Eur. *IT* 1262–7, 1277–9. Making Orpheus mistake the site of the *krater* for the site of the

ing comedy of errors two points at least are quite clear. First, the *krater* which Orpheus arrives at originally had nothing to do with Delphi. At the most obvious level, an oracle of dreams which gave individuals their own way of seeing into the past and future would be a threat and source of competition to the very principles on which the oracle at Delphi was based.[16] Also, Plutarch himself goes some of the way to set things right by making it plain that the oracle at the *krater* and the oracle at Delphi are two quite separate things.[17] Second, it is clear as well that the various mistakes and failures are features not of the original myth but of a parody of it produced in a spirit of blatant Delphic one-upmanship; the fact that Plutarch himself was a proud member of the Delphic priesthood is hardly a coincidence.[18] As for Orpheus' fundamental failure to bring back his wife, this is very probably a reflection of what appears to have been the widespread tendency to replace earlier stories in which Orpheus' descent to the underworld was successful with more morally 'acceptable' versions in which his superhuman journey proved too much for him.[19]

On the other hand behind the remodellings and reworkings there are those traces, still discernible, of the earlier stratum of myth; and here we come to the obvious point of significance in this passage of Plutarch. The vast *krater* which Orpheus arrives at on his descent into the underworld is not just some literal or

---

oracle at Delphi would be yet another effective way of ridiculing him by adding to the impression of his confusion.

[16] Eur. *IT* 1247–83; H. W. Parke, *Greek Oracles* (London, 1967), 35.

[17] It is an oracle of Night and the Moon, unlike the oracle at Delphi (566c); and in direct contrast to the solidly fixed site of the Delphic oracle (566d) it 'wanders everywhere among men' (566c). This whimsical idea of a wandering *krater* is obviously appropriate for a conveyor of dreams, but see also below, n. 22. The references to the Sibyl, Vesuvius, and Dicaearchia in 566d–e suggest an original location for the *krater* in Campania, and yet the details mentioned by Plutarch only date from his own lifetime: cf. P. H. de Lacy and B. Einarson in the Loeb edn. (Cambridge, Mass., 1959), 173–4. The earlier imagery of souls as birds drawing in their wings and making their way round a vast chasm instead of flying straight across it (565e–566a) is an obvious allusion to volcanic phenomena, and points especially either to Campania or Sicily: cf. Heraclid. fr. 128, Timaeus, *FGrH* 566 F57, Clark 67 and n. 88 (Avernus); Austin 109 (Solfatara and Agnano); Lycus, *FGrH* 570 F11a (Leontini).

[18] For a basic introduction to the ways and means of Delphic propaganda cf. J. Defradas, *Les Thèmes de la propagande delphique*² (Paris, 1972).

[19] Nilsson, ii. 637–8; cf. also Wilamowitz (1931–2), ii. 195–6; Guthrie (1952), 31; Boyancé in *OMG* 180–1; Kingsley (1994*d*).

metaphorical mixing-bowl, as it has loosely been assumed to be, but a specific physical phenomenon. Just like the geophysical 'craters' considered earlier, it is a point of descent and entry into the underworld. Just like them, it is associated with oracles and dreams; and the emphasis on Orpheus' lapse of memory and 'false declaration' (λόγον κίβδηλον) not only hints at the familiar motif of the underworld waters of Memory and Forgetfulness, but also tends to suggest that the *krater* he arrived at was—like other 'craters'—a place of ordeal where truth and falsehood are revealed.[20] Finally, any remaining doubts as to its geophysical aspect vanish when we set Plutarch's image of 'a vast "crater" with rivers pouring into it' (κρατῆρα μέγαν, εἰς δὲ τοῦτον ἐμβάλλοντα ῥεύματα) alongside the image in the *Phaedo* of the earth's rivers pouring into chasms 'which are like "craters"'—and especially into the one vast chasm or *krater* that is huger than them all.[21]

This is the obvious point that has been overlooked; but there is a less obvious detail which has also been missed. Plutarch explains the true nature of the *krater* which Orpheus arrived at in the following words: 'This is an oracle shared by Night and the Moon. It has no boundary anywhere on earth and neither does it have a single set base, but it wanders everywhere throughout mankind in dreams and visions.' That seems straightforward enough—except that Plutarch is referring here to the very same Orphic verse which Plato paraphrases in the *Phaedo*. Even if it did not occur directly after the mention of Orpheus, as it does, the reference would still be clear.[22] The striking similarities noted a moment ago between Plutarch's vast *krater* with rivers flowing into it and the vastest *krater* of all

---

[20] Above, nn. 4–5; *De sera* 566c. For the waters of Memory and Forgetfulness cf. e.g. Zuntz 358–83; Burkert (1977), 2 § 2.1.
[21] ... εἰς γὰρ τοῦτο τὸ χάσμα συρρέουσι πάντες οἱ ποταμοὶ ... εἰσρεῖν πάντα τὰ ῥεύματα (*Phd.* 111d2–112b2); *De sera* 566a–b.
[22] *OF* fr. 66b οὐδέ τι πεῖραρ ὑπῆν, οὐ πυθμήν, οὐδέ τις ἕδρα; *De sera* 566c οὐδαμοῦ τῆς γῆς περαῖνον οὐδ' ἔχον ἕδραν μίαν. Apart from the obvious echo of the Orphic πεῖραρ in Plutarch's περαῖνον, the expression γῆς περαῖνον also evokes the idea of 'boundaries of the earth', πείρατα γαίης: cf. *Il.* 8.478, 14.200 and 301, Solmsen 242 and n. 3, and for πείραρ–περαῖνον, DK i. 457.16, Onians 464–5, Burkert 250–9. The strange idea in Plutarch of a *krater* which 'wanders everywhere' (πάντηι πλανητόν) looks very much like a facetious extrapolation from the Orphic verse which describes the vast chasm as reaching 'here and there' (ἔνθα καὶ ἔνθα, *OF* fr. 66a2). For the links between the Orphic chasm and Night cf. *OF* frs. 1, 65–7; West 70, 111, 117; 208.

The Mixing-Bowl

in Plato's *Phaedo* had already suggested they were somehow connected; now we see that this is the case.

The relation between Plutarch's myths and Plato's is an interesting one. On the one hand, there are unmistakable similarities between what we find in Plato and what we find in Plutarch—a point long acknowledged. In fact the extent of these similarities has been considerably underestimated, due to the belief that Plutarch's projection of underworld geography into the cosmos represents a subsequent reworking and allegorizing of Plato's myths; the normal approach has been to attribute the decisive step in this process of reinterpretation either to Plato's disciple and eventual successor, Xenocrates, or to Posidonius, living centuries later.[23] As we have seen, though, this transposition of underworld geography is not only present in Plato's myths themselves but had already been carried out before his time; later writers were simply prolonging a tradition already well established. However, there is also a whole other side to the matter. In spite of the closeness of the parallels between Plutarch's myths and Plato's, Plutarch by no means derived his basic materials from Plato alone. Dieterich identified the general source for this additional material as a tradition which he described as 'Orphic–Pythagorean'[24]—a description that turns out to be fundamentally quite correct in terms of the evidence we have just examined. Both Plato and Plutarch paraphrase the same Orphic verse, but they paraphrase it in different terms; Plato omits or paraphrases words that Plutarch leaves intact, and vice versa.[25] This can only mean one thing: they are both independently dependent on the same ultimate source. That ultimate source is plainly an Orphic poem, and both Plato and Plutarch have inherited along with the poem a tradition of allegorical interpretation which is Pythagorean in origin.[26] As is only to be expected, the tradition has become

---

[23] So e.g. Bousset 60–2; cf. above, Ch. 9 n. 27.

[24] Dieterich 144–9.

[25] Plato omits πεῖραρ, preserves πυθμήν, and paraphrases ἕδρα (βάσις); Plutarch paraphrases πεῖραρ (περαίνον), omits πυθμήν, and preserves ἕδρα.

[26] For Plutarch's location of the *krater* in the heavens (*De sera* 566a–b, d–e) cf. Macr. *Somn.* 1.12.7 = *OF* fr. 241, Kahane and Kahane (above, n. 9), 15–17; for the colours of the rivers flowing into the *krater*—a white whiter than snow, purple, and other colours unspecified, 'each possessing its own unique brilliance when seen from a distance' (*De sera* 566b)—cf. Pl. *Phd.* 110b6–d3.

rather tattered and bedraggled by the time it reaches Plutarch, almost half a millennium after Plato. But the tradition still survives, which is also hardly surprising when one considers that late Neoplatonists writing almost as long again after Plutarch as Plutarch was writing after Plato had access to Orphic traditions now completely lost to us; or to be more precise, what little Orphic material has not been lost to us has survived almost entirely thanks to them.

The similarities between the *Phaedo* myth and the passage in Plutarch—the Orphic background in both cases, the same concern with underworld geography, the same central image of a huge crater with rivers flowing into it, the allusion to the same Orphic verse—all point to the conclusion that the same Orphic poem lies behind them both. In the case of the Plutarch passage, Dieterich gave grounds for supposing that the original poem was the lost Orphic *Krater*, and there are a number of factors which suggest he was right.[27] Here it will be enough just to add one little point by way of corroboration. For a couple of reasons, the Orphic *Krater* seems to have had close links with two other Orphic poems: the *Robe* and the *Net*. First, according to an ancient tradition all three works were written by one and the same author.[28] Second, the object mentioned in the title of each poem clearly represented a different cosmic or terrestrial symbol, in the sense that each of them appears to convey an image describing one major aspect of the structure of the cosmos or of the earth itself and the life upon it.[29] Plainly it is a question here not of reasoning or abstract argument but of the immediacy of visual imagery, capable of encapsulating a cosmic principle in one vivid symbol: the robe, for example, almost certainly refers to the patchwork garment 'symbolizing the surface of the earth'.[30]

As far as the *Krater* is concerned, the key to the symbolism lies in the conclusion we have already drawn both from Plutarch and from the *Phaedo*: that what the word means in these texts is not just a mixing-bowl but also a geological crater.

[27] Dieterich 145–8.
[28] *OF* test. 179 (Zopyrus); see also below, Ch. 12 n. 40.
[29] Lobeck, i. 379–82, West 10–11.
[30] West, loc. cit.; cf. *OF* frs. 33, 192–3, Eisler, i. 115–20. For Pherecydes' terrestrial robe compare Schibli 53–4, and for the relation between Pherecydes and Orphic literature, ibid. 37 n. 69, 126 n. 46, and *passim*.

The Mixing-Bowl

This of course points us towards southern Italy or Sicily as the place of origin of the Orphic poem in question, which is not at all surprising in light of the fact that precisely these areas are known to have been major sources of diffusion for Orphic literature; and more specifically, it points us to the two regions—the one around Cumae, the other in eastern Sicily—which we know provided the setting for the Orphic poem that lies behind the *Phaedo* myth. But the fundamental conclusion that the 'craters' in Plato and Plutarch are geophysical phenomena also takes us one step further. As we have seen, the vast crater in Plutarch into which the rivers flow corresponds in the *Phaedo* myth to the chasm of Tartarus, into which all rivers flow—keeping apart at first, then pouring together and being mixed with each other in one huge, terrestrial mixing-bowl.[31] In short, what the *Phaedo* myth together with the Plutarch passage provides us with is a remarkable symbolic image of the entire inner structure of the earth which is perfectly comparable, and exactly complementary, to the image of the earth's surface as one huge terrestrial garment. The evidence of Plato and Plutarch, and the titles of the Orphic poems, clarify and are illuminated by each other—with the clue lying, yet again, in the geography of Sicily and southern Italy.

So far we have looked at the Orphic *Krater* from the point of view of its title. But there is also another, entirely different respect in which the poem helps to throw light on the myth at the end of the *Phaedo*, and this has to do with the question of its authorship.

It has often been noticed that the intricate description in the *Phaedo* myth of the workings of the great terrestrial 'mixing-bowl', with its complex system of see-saw oscillations and balancing forces, displays all the signs of a writer possessing considerable mechanical flair and ingenuity.[32] On the basis of Plato's explicit reference to Sicily the tendency has been to attribute these aspects of the myth to Empedocles—and, in the process, compare Empedocles' famous use of the clepsydra as a

---

[31] Plut. *De sera* 566b; Pl. *Phd.* 109b6, c2, 111d4–5, 112a5, 113b3, c5–7, 114a3–7, b3–4.
[32] Cf. e.g. E. Oder, *Ein angebliches Bruchstück Democrits über die Entdeckung unterirdischer Quellen* (Leipzig, 1899), 49; Friedländer, i. 387 n. 7.

model for describing the principles of respiration.[33] However, as in the case of the other attempts to ascribe specific features of the *Phaedo* myth to Empedocles, the matter is far from being as simple or straightforward as has been assumed; and in this particular case there are several considerations that have been overlooked. To begin with, it is quite true that the oscillation system in the *Phaedo*, together with the imagery of the earth breathing through its subterranean channels and 'exhaling' during volcanic eruptions, is undoubtedly based on an analogy between microcosm and macrocosm: the earth is one vast breathing creature.[34] Similarly, it is also highly likely that Empedocles intended his clepsydra analogy to be understood as a model not only for the breathing process in humans but also for the breathing of the universe during alternating stages of the cosmic cycle.[35] In other words, both Empedocles and the *Phaedo* myth appear to agree in postulating the general idea of breathing processes at a macrocosmic level. But what is significant here is, first that the idea of cosmic breathing is a basic Pythagorean tenet which was invented by Empedocles no more than it was by Plato;[36] and second, that when we turn to the actual details we find Empedocles and Plato using fundamentally different models to illustrate this basic idea—on the one hand the model of the clepsydra and on the other a mechanical model based on an oscillation principle. In short, the similarities are to be explained in much the same way as the common concern with Sicilian geography: in terms of a Pythagorean background shared by Empedocles and by the author of the myth that lies behind the *Phaedo*. As for the differences, however, they argue against any direct indebtedness to Empedocles in the myth of the *Phaedo*; and we are left in the same situation as when looking, earlier, to see if Empedocles was the author of the original stratum of the *Phaedo* myth. He is not our man; it is to someone else, apparently writing under the name of Orpheus and to a considerable extent sharing

[33] So e.g. Burnet 127, 135; Dicks 97.
[34] *Phd.* 112b–c, 113b; cf. Burnet 135; Zaslavsky 206 n. 22; Dorter 165–6 (appropriately comparing *Tim.* 70a–d); Burger 198–9.
[35] F. A. Wilford, *Phronesis*, 13 (1968), 108–18; cf. the Introduction, n. 10.
[36] Arist. fr. 201 and *Ph.* 213^{b}22–7 with Simpl. *Ph.* 651.26–8, ps.-Plut. *Placita* 2.9.1 = Stob. i. 160.9–10 (*Dox.* 338), Sext. *Math.* 9.127; Olerud 43–57, Guthrie, i. 200, Philip 68–70, Burkert 35, 37, 271.

Empedocles' interests and ideological background, that we must look as the person responsible for portraying the vast system of subterranean rivers in specifically mechanical terms.

Here we come back, once again, to the Orphic *Krater*. Ancient sources give us a name for the author of the work: Zopyrus of Heraclea.[37] The name of the man and his place of origin raise two questions in particular. First, is he the same person as the Zopyrus from Tarentum mentioned in Aristoxenus' list of Pythagoreans; and second, which is the Heraclea that he comes from? The general tendency has been to answer the first question in the affirmative[38] and give at best an implicit answer to the second. In fact the two questions hang together very closely. To begin with the identification of the town: the fact that the historian Herodorus, from Heraclea on the Black Sea, is known to have written about Orpheus[39] might seem a possible indicator. However, the very close links between towns in the Greek West—Metapontum, Croton, Velia, Camarina, Syracuse—and the actual production of Orphic literature[40] point with much greater probability to two other cities of the same name: a Heraclea in Sicily and one in southern Italy. Of these two, on quite general grounds the Heraclea in southern Italy is the more likely candidate; it was significantly better known in antiquity and was the one more commonly referred to by classical writers simply as 'Heraclea' with no further specification.[41] But the single most decisive fact for our purposes is that this Italian Heraclea was actually situated on the gulf of Tarentum. It was founded by Tarentum, populated by Tarentines, and after its foundation became Tarentum's right arm and major source of muscle in its struggle for political

---

[37] Clem. *Str.* 1.21.131.3, *Suda*, s.v. Ὀρφεύς (iii. 565.4–5 Adler) = *OF* pp. 52 (test. 179), 63 (test. 222), 64 (test. 223d), 308. For Zopyrus cf. also *OF* test. 189 with West's comments, 248–51.

[38] So e.g. Rohde 349 n. 7, Kern at *OF* test. 179, West 10. For the mention of Zopyrus of Tarentum in what is evidently a list of Pythagoreans drawn up by Aristoxenus, cf. Iam. *VP* 267 with Burkert 105 n. 40, Zhmud' 273–4.

[39] *FGrH* 31 F12, 42–3.

[40] Ch. 10 with nn. 11–14; cf. also O. Kern, *Orpheus* (Berlin, 1920), 2–4, Loicq-Berger 61.

[41] Cf. e.g. the refs. in nn. 42, 45–7 below. For the ways of referring to Heraclea Minoa in Sicily see the *Suda*, s.v. Ἡράκλεια (ii. 580.34 Adler), K. Ziegler, *RE* viii. 437–8, and esp. E. de Miro, *Kokalos*, 4 (1958), 77–8.

and territorial power.[42] The resulting situation was rather paradoxical: Heraclea was able to achieve what Tarentum by itself was unable to do, but its success and fame were attributed to the founding city.[43] To refer to Heraclea was automatically to refer to Tarentum, and however important it became it remained 'Tarentine Heraclea' or 'Heraclea in the territory of Tarentum'.[44]

These considerations are, in themselves, enough to make it very probable that Zopyrus of Heraclea and Aristoxenus' Zopyrus of Tarentum were one and the same person; but there are other factors which serve to confirm the identification. This same Tarentine Heraclea is repeatedly mentioned by ancient writers for its connections with Pythagoreanism. We are told by Livy that Pythagoras himself set up schools (*iuvenum aemulantium studia coetus*) at Metapontum, Heraclea, and Croton. This is an obvious anachronism—Pythagoras died long before Heraclea was even founded—but is an equally obvious acknowledgement of Heraclea's importance, alongside Metapontum and Croton, as a centre of Pythagoreanism.[45] Again, in his *Pythagorean Life* Iamblichus gives Heraclea as the home of Philolaus and another Pythagorean, Clinias.[46] And the connection occurs a third time in relation to another Pythagorean: Plato's acquaintance Archytas, who during his rulership of Tarentum served as president of the Italian League at its famous congressional centre in Heraclea.[47] On two separate counts then—Heraclea's ties with Tarentum, plus the evidence for its very close links with Pythagoreanism—we can consider it almost certain that Zopyrus of Heraclea and the Pythagorean Zopyrus of Tarentum are one and the same person. So, in the figure of Zopyrus we are left with one further testimony to the intimate links between western Pythagoreans and the production of Orphic poems.

---

[42] Diod. 12.36.4; Strabo 6.3.4; E. L. Minar, *Early Pythagorean Politics* (Baltimore, 1942), 40, 87.
[43] See E. Ciaceri's comments, *Storia della Magna Grecia*, ii² (Genoa, 1940), 388.
[44] Ἡράκλεια τῆς Ταραντίνης: Strabo, loc. cit.
[45] Livy 1.18.2. For the date of Heraclea's foundation see Ch. 12 with n. 31.
[46] *VP* 266. Clinias is mentioned by Aristoxenus (fr. 131 = DK 54 A2 = D.L. 9.40) as a contemporary of Plato.
[47] Cf. Strabo 6.3.4 with the *Suda*, s.v. Ἀρχύτας (DK i. 422.6–7); Wuilleumier 62, 70, 85.

The Mixing-Bowl 145

Fortunately, there is one other cluster of details which settles the matter and confirms this conclusion beyond any reasonable room for doubt. As we have seen, in his *Pythagorean Life* Iamblichus refers to the Pythagorean Clinias as being based in Heraclea; but earlier on in the same work he describes Clinias as 'a Tarentine', and when he comes to reproduce what is evidently Aristoxenus' list of Pythagoreans he groups Clinias, again, among the Pythagoreans of Tarentum. Similarly, as was also mentioned, alongside Clinias Iamblichus cites Philolaus as being from Heraclea; but Aristoxenus' list, which he reproduces almost immediately afterwards, brackets Philolaus among the Pythagoreans from the region of Tarentum—alongside both Clinias and Zopyrus.[48] In other words there were clearly a number of writers in antiquity who, for reasons that are fully understandable from a political as well as geographical point of view, drew no distinction between Heraclea and Tarentum, and assimilated the smaller town to the general region and mother city of Tarentum. Among these writers was apparently Aristoxenus himself, a proud native of Tarentum who had every possible motive for glorifying his home town at the expense of one of its satellites.[49]

Apart from Aristoxenus and Iamblichus there is also another ancient writer who mentions Zopyrus of Tarentum—but in what is such an unphilosophical context that from the point of view of the history of Pythagoreanism the passage has been almost entirely ignored. Writing most probably in the mid-third century BC, little over half a century after Aristoxenus, Biton ends his work on war machinery with a section devoted to the mechanical designs and innovations of Zopyrus from Tarentum.[50] It was Diels who saw the significance of the passage and, in his *Antike Technik*, drew the inevitable conclusions. Biton's straightforward allusion to a 'Zopyrus of

---

[48] Clinias a 'Tarentine', *VP* 239 (also Diod. 10.4.1 = DK 54 A3); Clinias and Philolaus from Heraclea, *VP* 266, and from Tarentum, ibid. 267. Other references in classical authors to Philolaus as from Tarentum are listed by Burkert, 228 n. 48 (add Cic. *Or.* 3.34.139, Lyd. *Mens.* 2.12 = DK 44 B20a).

[49] For Aristoxenus and Tarentum cf. Wuilleumier 587–607; Kingsley (1990), 261.

[50] Biton, Κατασκευαί, ed. C. Wescher, *Poliorcétique des Grecs* (Paris, 1867), 61.2–67.3; ed. Marsden, ii. 74–6. For the date of Biton's work cf. A. Rehm and E. Schramm, *Bitons Bau von Belagerungsmaschinen und Geschützen* (Munich, 1929), 3; Marsden, i. 3, 13, ii. 5–6; B. S. Hall, *Isis*, 64 (1973), 529–30.

Tarentum' would, in itself, tend to indicate that the man in question is very likely to be the Zopyrus from Tarentum listed by Aristoxenus; but Diels raised two additional points which confirm this supposition. First, in quite general terms the early stages of innovation in the design of weaponry depended heavily on the practical application of mathematics, and Pythagoreans are known to have been pioneers in this very field; Diels drew attention as well to the evidence that during the late fifth and early fourth centuries southern Italy and Syracuse, which were centres of Pythagoreanism at the time, were also at the forefront in the innovative design and production of machines of war.[51] Second, and more specifically, Diels noted that the one person who during this period was associated most closely with applied mathematics and, in particular, with breakthroughs in the field of mechanics, was none other than the famous Pythagorean from Tarentum: Archytas.[52] To this we can add that according to a number of reports Archytas' mechanical interests were shared by other Pythagoreans in Tarentum.[53] So, Biton's references to the mechanical achievements of Zopyrus from Tarentum point us back to the school of Pythagoreanism; and even if there was no mention by Aristoxenus of Zopyrus of Tarentum as a Pythagorean, this is something that could with a fair degree of probability have been inferred from Biton's report alone. But there is also one other point, not given the emphasis it deserves by Diels, which clinches the matter decisively. Ancient sources link the Pythagorean school of Archytas—who was himself elected by Tarentum seven times as its military general—not only with general advances in mechanical science but also with innovations in warfare and the design of weaponry.[54] What is

---

[51] Diels (1920), 19–24; cf. also Marsden, ii. 44 n. 2, 98 n. 52.

[52] Diels (1920), 21–2. For Archytas and mechanics cf. Arist. *Politics* 1340^{b}26–8 (DK 47 A10), D.L. 8.83 (DK i. 421.36–7), Vitr. 1.1.17 (DK 44 A6) and 7 pref. 14 (DK i. 439.6–10), Gellius 10.12.9–10 (DK 47 A10a), Plut. *Qu. conv.* 718e–f (47 A15) and *Marc.* 14.5–6, Theophylactus, *Letters* 71 = *PG* cxxvi. 493a–b; Krafft, pp. x–xi, 3–5, 11, 144–54, and *passim*, Burkert (1989), 211; also D.L. 8.79 (DK i. 421.29–31) with Kingsley (1994c), § 1 and nn. 27–8.

[53] Vitr. 1.1.17 and Theophylactus, loc. cit. (Philolaus and Archytas), Plut. *Qu. conv.* 718e ('Archytas' school').

[54] In Plut. *Marc.* 14.3–6 the connection between Archytas' mechanics—'ignored for a long time by philosophy, it ended up as one of the arts of warfare'—and weaponry is made explicit. Plutarch's mention of ὀργανικὰς καὶ μηχανικὰς κατασκευάς in *Qu. conv.*

more, as Diels pointed out in another context, the remarkable passage in Vitruvius which draws an analogy between the role of an artillery commander—supervising the correct tension, tone, and harmony of his catapult springs—and the role of a musician is undoubtedly to be traced back to the Pythagorean school at Tarentum.[55] According to Heraclitus war is harmony; according to Pythagoreans, war is a harmony when correctly conducted.

When we draw all the various threads of evidence together, it emerges that the supposed author of the Orphic *Krater* was Zopyrus of Tarentum: a Pythagorean who was a mechanical designer and inventor. The implications of this with regard to the *Phaedo* myth should need no emphasizing. We have already seen that the myth is based on an Orphic work which was very probably the poem known in antiquity as the *Krater*. But one puzzling feature about the myth is that the author of the poem on which it is based evidently took a considerable interest not only in mythology but also in mechanics. This interest cannot simply be attributed to Plato himself. First, Plato was no more mechanically minded than he was a musician, astronomer, or mathematician: he took a great interest in all these subjects, but whenever one looks closely at his treatment of them one invariably gains 'the impression of the enthusiastic layman trying to catch science on the wing.'[56] And second, we have seen that the basic description of subterranean geography in the *Phaedo* myth derives not from him but from a Pythagorean text which was almost certainly transmitted to him via the school of Archytas—a school that happens to have had an unrivalled reputation in the field of mechanics. The fact that

718e is, again, an allusion to weaponry (cf. Onasander 42.3–4, Diod. 17.43.1), and the link recurs in Theophylactus, loc. cit. (στρατιωτικὴ καὶ γεωμετρική). For Archytas as military commander see esp. D.L. 8.82 = Aristox. fr. 48 (DK 47A1) with Iam. *VP* 197 = Aristox. fr. 30 (47 A7), D.L. 8.79 (47 A1), Ael. *VH* 7.14, *Suda* s.v. Ἀρχύτας (47 A2), Wuilleumier 67–74, 574, 581. For Archytas' interest in ballistics see also the next note.

[55] Vitr. 1.1.8; Diels (1920), 24 n. 3. This naturally throws a refreshing light on the passage in Archytas B1 (DK i. 434.5–8 = Bowen 80.26–9) which refers to missiles in the context of music, sound, and harmony. This part at least of the fragment is certainly genuine (Burkert 380 n. 46); on the authenticity of the fragment as a whole see now Bowen 83–99, Huffman, *CQ* 35 (1985), 344–8, and esp. Cassio 135–9.

[56] de Santillana 197.

the author of the Orphic *Krater* appears to have come from Tarentum and to have belonged to that rare breed of ancient specialist—the professional engineer and mechanic—is hardly a coincidence.[57] The evidence is remarkably consistent, and confirms the conclusion that the poem which lies behind the *Phaedo* myth was by Zopyrus of Tarentine Heraclea.

[57] More specifically, it is worth noting that the principle on which the see-saw model in the *Phaedo* myth is based—the maintenance of balance between equal and opposing forces—was also the chief principle behind the successful design of artillery and war machinery in general: for the fundamental requirement of 'counter-weight and delicate balancing' cf. e.g. Marsden's comments, ii. 92. Zopyrus' own innovations in mechanical pull-back systems depended on mastery of the laws of equilibrium and the 'requirement for equal force' (τὸ ἴσης δεῖσθαι βίας). Cf. Marsden, i. 14–15, ii. 74–7, 97–103 and, for the expression, Heron, Βελοποιικά 84.2 Wescher (Marsden, ii. 24).

12
'Wise Men and Women'
*

ABOUT Zopyrus' dates we can be fairly specific. Under Diels's influence the convention has been established of dating him to the period between 400 and 350 BC.[1] This is much too late. To begin with Aristoxenus: a date as late as 350 has been allowed on the basis of the general conclusion that the names recorded in his list of Pythagoreans, as preserved for us by Iamblichus, 'take us down to the first half of the fourth century'.[2] But we can be more precise than that. According to Aristoxenus the 'last' of the Pythagoreans were contemporaries of Plato and Archytas, and active in around 365.[3] Zopyrus is not mentioned among them, so we can safely assume he lived and worked even earlier. By 365 some at least of these 'last' Pythagoreans were already fairly old. Echecrates was a young man in 399, and Xenophilus possibly even older: according to Aristoxenus he was over 100 when he died at Athens, probably some time in the 330s or 320s.[4] If Zopyrus belonged to a generation earlier than these 'last' Pythagoreans, he will almost certainly have lived no later than Philolaus who—again according to Aristoxenus—was one of their teachers.[5] That takes us back before the fourth century altogether into the second half of the fifth.[6]

Significantly—because it provides yet further confirmation that the Pythagorean Zopyrus of Tarentum was the same person as Zopyrus the artillery designer—the information in

---
[1] Diels (1920), 23, 97; Kern at *OF* test. 179; Marsden, i. 13, 78, ii. 47 n. 15, 98 n. 52.
[2] Burkert 105 n. 40.
[3] D.L. 8.46 = Aristox. fr. 19; Iam. *VP* 251 = Aristox. fr. 18; Diod. 15.76.4; Burkert 198, 200.
[4] For Echecrates cf. Burnet 1, 82; Frank, *AJP* 64 (1943), 222 n. 4; Burkert 92 and n. 40. For Xenophilus see Aristox. fr. 20 (also frs. 1, 18, 19, 25, 43), and Kingsley (1990), 261 n. 98.
[5] Aristox. fr. 19; Burkert 198, 228.
[6] Wilamowitz, ii. 86; KRS 323.

Biton points to precisely the same period. Artillery had its grand début in the Greek world at Syracuse in 399 BC. Dionysius I stage-managed a remarkable scene by inviting expert craftsmen from Italy and elsewhere to come to his city and transform it into an immense workshop for the mass production of weapons of war.[7] Diodorus Siculus, probably basing what he says on an eyewitness account by the Sicilian historian Philistus,[8] describes how 'all kinds of catapults were constructed, plus a considerable number of other missile launchers'; these weapons are almost certain to have included the kind of non-torsion catapult invented by Zopyrus and described by Biton.[9] To put the matter in a different way, the kind of 'catapult' that Zopyrus designed—it was actually a type of crossbow supplied with a pull-back winch system and a base—transformed artillery into a stunningly effective weapon of war, and it is very difficult not to suppose that this type and level of development lie behind Dionysius' initial artillery success at his siege of Motya in western Sicily during the year 397.[10]

These points would in themselves incline one to date Zopyrus' activities to the period during and before 399–397 rather than afterwards—even if we were not provided, as we are, with more definite information. Biton mentions among Zopyrus' creations one particular catapult design which he produced at Miletus.[11] Here preconceptions, and generalizing theories about the early development of artillery, must yield in the face of some basic historical facts. In 412 BC Miletus freed itself from the yoke of Athens. Under the protection of the fleet that had arrived from Syracuse it immediately started arming itself, and during the ten years between 411 and 402 built up its own defences and fortifications; but then, as part of the bargaining conducted between members of the anti-Athenian

[7] Diod. 14.41.2–43.4, 47.1; Diels (1920), 20–1, Marsden, i. 48.
[8] *FGrH* 556 F28; Marsden, loc. cit., Caven 89, 93–5.
[9] Diod. 14.43.3; Kingsley (1995*b*).
[10] For the role of catapults at Motya cf. Diod. 14.49.3, 50.4, 51.1; Kingsley (1995*b*), with n. 17. Diels (1920), 22–3 saw the connection between Zopyrus' achievements and Dionysius' artillery successes, but failed to draw any specific conclusions about the dating of Zopyrus.
[11] Κατασκευαί 61.4–62.1 Wescher, ii. 75 Marsden: ὁ γαστραφέτης ὃν ἠρχιτεκτόνευσε Ζώπυρος ὁ Ταραντῖνος ἐν Μιλήτωι.

alliance, Sparta handed the city over to Persia. For almost seventy years, from 401 BC down to the time of Alexander the Great's unwelcomed arrival, Miletus remained willingly under Persian rule; it avoided contact with the Greek world, became the home of a Persian garrison, and concentrated on making carpets.[12] Prior to 412, a trip made by Zopyrus from Tarentum to Asia Minor to help the Milesians arm themselves is out of the question. Tarentum and Athens were sworn enemies,[13] and any attempt on Zopyrus' part to help Miletus eject the Athenians by working right under their noses would have been doomed to failure. As for trying to date Zopyrus' journey to some time after 402, this possibility must also be excluded. Once under Persian control Miletus required no intervention at all from Greeks in the West; any attempt by Zopyrus to equip it with weapons would have been superfluous and inexplicable, unasked for and unwelcome. This leaves us only with the ten-year period between 411 and 402 as an acceptable time for Zopyrus' trip to Miletus; and, what is more, during this same period it makes perfect sense. Miletus had just managed to liberate itself from Tarentum's long-standing enemy; straight away it appears to have taken the logical step of actively arming and fortifying itself, and Zopyrus was precisely the kind of professional whose services the city would want to call on in such a situation. Also, it happens that the last decade or so of the fifth century is exactly the kind of dating we had already arrived at on quite independent grounds as the most likely period of Zopyrus' activity.

Zopyrus' journey to Miletus is revealing for a number of reasons. First, it has been used to show that he 'travelled widely in search of employment'.[14] In this respect it certainly links him up with a well-attested and common—although until recently

[12] For the Athenian domination down to 412, see A. G. Dunham, *The History of Miletus* (London, 1915), 103–11, 137–8. Revolt and struggle against the Athenians: Thucydides 8.17, 24.1, 25–39, 57.1, 60.3–63.2, 75.3, 78–80, 83–5, 99, Xenophon, *Hellenica* 1.1.31, 1.2.2–3, 1.5.1, 1.6.2 and 7–12, Dunham 111–13. For the dating of the fortifications at Miletus see also S. Hornblower, *Mausolus* (Oxford, 1982), 331. Transferred to Persia: Xen. *Anabasis* 1.1.6–8, 1.2.2–3, Dunham 115–20. Persian garrison, and resistance to Alexander: Arrian, *Anabasis* 1.18–19, Dunham 119–20. Carpet-making: Dunham 107, 117.

[13] Wuilleumier 43, 59–63 with refs. For the Athenian garrison established in Miletus by the mid-5th century cf. Dunham, op. cit. 104, 137–8.

[14] Marsden, i. 78 n. 1.

underrated—phenomenon during much of the first millennium BC: the phenomenon of specialized experts and craftsmen, both Greek and oriental, wandering across the Mediterranean and Near East in the areas between Italy in the West and Ionia, Mesopotamia, and central Asia in the East.[15] Second, it gives the lie to the general assumption that, once Pythagoras had emigrated from Samos to southern Italy, Pythagoreanism effectively broke its ties with the East. This was, anyway, an extremely implausible assumption given the normal frequency and speed with which people as well as news and information travelled backwards and forwards between Italy and Sicily on the one hand, and Asia on the other.[16]

But, these points apart, in this particular case there would seem to be another, very simple factor involved. Miletus would not have been able to hold on to its new-found freedom from Athens during the year or two immediately following the summer of 412 BC without the presence and assistance of the fleet from Syracuse; the ships had arrived, under the command of Hermocrates, to protect the Milesians and 'to contribute to destroying once and for all what remained of the Athenians'.[17] This is already the second time that Syracuse crops up in connection with Zopyrus and his mechanical innovations, and it would anyway be very difficult not to suspect some connection between the presence in Miletus of a man from Tarentum and the apparently simultaneous presence of the Sicilian fleet off shore. In fact, on this specific occasion a link between Zopyrus of Tarentum and the forces of Syracuse is not only plausible but extremely likely. When, some fifty years later in the 360s, Plato tied the bonds of friendship—*philia*—between Dionysius II of Syracuse and the Pythagoreans of Tarentum[18] he was only re-establishing for a new generation of ruler a

---

[15] Burkert (1983), (1992); C. Nylander, *Ionians in Pasargadae* (Uppsala, 1970), n. 328 and *passim*; T. F. R. G. Braun, *Cambridge Ancient History*, iii/3² (1982), 22–4; S. Dalley, *Iraq*, 47 (1985), 31–48; Grottanelli 649–70, esp. 664.

[16] So e.g. Hdt. 3.131–7 (Croton to Samos to Susa, and back to Croton via Sidon and Tarentum), 6.22 (Sicily to Ionia), 6.23 (Ionia to Sicily), 6.24 (Sicily to Susa, back to Sicily, and back to Susa). Tarentum was a normal stopping-off place for travellers with Sicilian or other Italian destinations, regardless of their point of departure.

[17] ξυνεπιλαβέσθαι καὶ τῆς ὑπολοίπου Ἀθηναίων καταλύσεως, Thucydides 8.26; cf. 8.28.2, 29.2, 35.1, 45.3, 61.2, 78, 84–5, Xenophon, *Hellenica* 1.1.27–31.

[18] *Letters* 7.338c–d (ξενίαν καὶ φιλίαν ποιήσας). Cf. also Plut. *Dion* 18.2, Wuilleumier 71–2.

relationship of mutual sympathy and interest between Tarentum and Syracuse which went back a fair distance in time. Quite apart from the question of Zopyrus' involvement, there are good grounds for supposing that the craftsmen from Italy whom Dionysius I invited to Syracuse in 399 consisted to a large extent of experts from Tarentum;[19] it has often been noted that Tarentum was more or less the only major city which appears to have stayed on friendly terms with Dionysius during his subsequent incursions into Italy;[20] and the combined Syracusan and Italian fleet which sailed out under Dionysius' instructions against the Athenians in 387 almost certainly included ships from Tarentum.[21] But as soon as we move back further in time to the period of Hermocrates, the evidence for links between Syracuse and Tarentum becomes much clearer and more definite. When, not long before he led the Syracusan fleet out to Miletus, Hermocrates appealed for allied assistance in preparing against the Athenian threat, he looked to southern Italy as well as Sicily; he named Tarentum in particular as a friendly city—*philia chōra*—to be relied on for providing active assistance, and his prediction proved correct.[22] Again, immediately after the Milesian expedition, ships from Tarentum sailed out with ships from Syracuse during the summer of 411 in another excursion against the Athenians.[23] In short, nothing could have been more natural or understandable than a Tarentine accompanying the Syracusan fleet out to Miletus in 412 or early 411. Apart from the help that this gives us in dating Zopyrus' journey, it also serves to pull the threads linking him with Syracuse even closer together, and suggests that Sicily was very probably as well known to him as he was to Sicily.

Finally, we have one other means for dating Zopyrus. Besides the work he did in Miletus, Biton also mentions that he designed a special item of mountain artillery at Cumae in

---

[19] Diod. 14.41.3; Diels (1920), 21; Wuilleumier 192, 581.
[20] Harward 10, 16; Caven 136, 196.
[21] Xenophon, *Hellenica* 5.1.26; cf. Thucydides 8.91.2, with Wuilleumier 65.
[22] Thucydides 6.34.1 and 4–5 (ἐκ φιλίας χώρας ... ὑποδέχεται γὰρ ἡμᾶς Τάρας); 6.44.2 and 104; Diod. 13.3.4. For evidence of relations between Tarentum and Syracuse going back before the time of Hermocrates cf. Loicq-Berger 200, 215.
[23] Thucydides 8.91.2.

Chapter 12

Italy.[24] Marsden seriously suggested that Zopyrus designed this piece of weaponry for the city of Cumae in the mid-fourth century BC;[25] but this is out of the question. In 421 BC Cumae came under heavy siege from the neighbouring Italian tribes, or Sabellians. It was savagely defeated, depopulated, and converted into an Oscan colony before eventually passing under Roman control. The male population who survived fled to nearby Naples, where in due course they were offered help and protection by Tarentum. As for Cumae itself, by 420 it was no longer a Greek city. Nothing survived of its Greek heritage except some relics of its earlier customs and culture: 'the town no longer had any history of its own.'[26] Down until the time of its transfer to Rome in the mid-330s, for any western Greek to attempt to arm Cumae's new occupiers would have been sheer suicide—especially for someone from Tarentum, which during this period was involved in its own struggle for survival against the Sabellian tribes.[27]

So for Zopyrus' work in Cumae we are drawn back to the period before, not after, the late 420s. Here the picture is very different indeed. In 474 BC Cumae had come under imminent threat of occupation by the Etruscans and appealed to Syracuse for help, which was immediately given; the Etruscans were defeated, and a close and enduring relationship established between Syracuse and Cumae that was to last for over fifty years.[28] During the second half of the fifth century Cumae found itself in the situation of having to build up its defences against the neighbouring Sabellians, and the tension reached a climax in the mid-420s as the Sabellians started moving down in force from the mountains of the interior towards the

[24] Κατασκευαί 64.4–65.2 Wescher, ii. 76 Marsden: τὸν ὀρεινοβάτην γαστραφέτην ... οἷον ἠρχιτεκτόνευσε Ζώπυρος ὁ Ταραντῖνος ἐν Κύμηι τῆι κατ' Ἰταλίαν. For comments on the weapon cf. Marsden, ii. 101 n. 66.

[25] i. 78; Diels, too, seemed unaware of the chronological factors involved, (1920), 23.

[26] Diod. 12.76.4, Strabo 5.4.4, Livy 4.44.12, Dionysius of Halicarnassus, *Antiquit. Romanae* 15.6.4, Velleius 1.4.2; Beloch 150–1, E. T. Salmon, *Samnium and the Samnites* (Cambridge, 1967), 39. The quotation is from Beloch, 151.

[27] Active collaboration of any form between Tarentum and the Sabellians can be excluded until well after Cumae had already fallen into Roman hands (Wuilleumier 88–9, Salmon, op. cit. 121, 212); cf. also L. Pareti's comments, *Storia di Roma*, i (Turin, 1952), 584 and n. 7.

[28] L. Homo, *Primitive Italy* (London, 1927), 155 and n. 1 with refs.; M. Napoli, *Napoli greco-romana* (Naples, 1959), 16, McKay 144–6. For relations between Cumae and Syracuse prior to 474 cf. Loicq-Berger 199–202.

vulnerable Campanian plain.[29] During these last few years of its existence as a Greek city, Cumae had every possible reason to commission a special design of artillery from its allies in the West for easy assembly and use in mountainous terrain, so that it could carry its lines of defence and offensive out of the Campanian plain into the mountain regions that contained the Sabellian strongholds.[30] In this general scenario one notes, once again, the Syracusan connection running like a thread through the background of Zopyrus' professional activities. And to what we have already seen of his evidently close links with Sicily, we can now add his contact with Cumae: two connections which, as noted in earlier chapters, together provide him with the ideal credentials as an author of the Orphic *Krater*.

From the historical point of view, the information about Zopyrus which we are left with is remarkably consistent. All the evidence for his inventions, along with what can be deduced from Iamblichus and Aristoxenus, points to a period from the 420s down to around the turn of the century as the time when he was active. On the other hand, any much earlier dating can be excluded for the simple reason that Heraclea was only founded by Tarentum in 432 BC.[31] So, when the various pieces are put together we are left with a picture of the man living and working during the last third of the fifth century BC.

Restoring these details to their proper place has a number of important consequences. From the point of view of the history of warfare, it provides the vital material for a new, first chapter on the development of western artillery. One very basic point worth noting here is the fact that the struggles to which Zopyrus added his contribution were directed not only against Carthage and indigenous Italian tribes, but also against the

---

[29] McKay 151; Napoli, op. cit. 17. For a brief description of the geography—Cumae, Campanian plain, 'mountains of the Samnites and Osci' (ὄρη τά τε τῶν Σαυνιτῶν καὶ τὰ τῶν Ὄσκων)—cf. Strabo 5.4.3-4.

[30] Setting up buffer zones of this kind against the Sabellian tribes was no doubt a military tactic well known to Zopyrus. At least a part of Tarentum's motivation for founding Heraclea in the 430s was almost certainly the need it felt to establish an early warning and defence system against impending attacks by Sabellians from the interior (Wuilleumier 62; Salmon, op. cit. 37).

[31] Diod. 12.36.4.

power of Athens. This may not seem particularly significant, but the situation is in a sense unmistakably symbolic. There is not much doubt that the way in which the evidence about Zopyrus has been so badly treated—if treated at all—has more than a little to do with the fact that the study of the ancient Greek world is still fundamentally Athenocentric. The case of Zopyrus can be taken as a reminder that even at the height of Athenian power there were other centres of culture in the Greek world which were just as important in their own way, and which would have seen events in antiquity in a very different perspective from the one we have become accustomed to.

Secondly, we are given a fresh insight into the history of mechanics, and into the kind of Pythagorean interests and traditions inherited by Archytas. Predictably, Archytas himself has tended to steal all the limelight through his own achievements and innovations in the field of mechanics[32]—predictably, because of his direct relationship with Plato and because of the arbitrary way in which history is so often made. The Greeks could be as simplistic in their nomination of pioneers and inventors as they were in their reconstruction of history and chronology in general;[33] as often as not they either committed gross anachronisms by projecting discoveries back far too early in time, or they made the opposite—and equally understandable—mistake of describing as inventors people who simply presented to the public eye discoveries that had already either been made or at least well prepared for by others. The same pitfalls still await the historian today; especially when trying to reconstruct the development of early philosophy or science, modern scholarship sometimes outdoes the naïvety of the Greeks. It is very tempting to portray the history of ideas as an elegantly symmetrical zigzag of Hegelian dialectic, or as a piece of theatre written for a miniature cast; but the case of Zopyrus is an eloquent warning against the tendency to idolize the achievements of a few individuals while

---

[32] So e.g. Plut. *Marc.* 14.5 (τὴν ὀργανικὴν ἤρξαντο κινεῖν οἱ περὶ Εὔδοξον καὶ Ἀρχύταν); D.L. 8.83 = DK 47 A1 (οὗτος πρῶτος . . .). For Archytas' own acknowledgement of his debt to 'the wisdom of his predecessors' cf. DK 47 B1 with Bowen 83–4, Huffman, *CQ* 35 (1985), 346–7.

[33] Kingsley (1995*b*); (1990), 259–64.

forgetting about the people who remain behind the scenes and do much of the most invaluable work.[34]

Thirdly, from the point of view of politics we are provided with a new perspective on the history of relations between Tarentum and Syracuse. In particular, Zopyrus' activities present a strangely familiar *arrière-plan* to Plato's itinerary on his first crossing from Tarentum over to Sicily; and it also becomes clear that the 'friendship' or alliance ($\phi\iota\lambda\iota\alpha$) which Plato managed to secure between Archytas of Tarentum and Dionysius II of Syracuse was not quite as unprecedented as it is often assumed to be, but had a very definite history behind it.[35]

Fourthly, from the point of view of the history of philosophy we are presented with a very different picture of pre-Platonic Pythagoreans from the usual stereotype of them as impractical dreamers, their minds fogged and obsessed with number mysticism, who had no 'clear idea of the value of empirical research' because all that interested them was discovering metaphysical principles.[36] One is usually left with the image of Archytas as little more than an eccentric old man who liked to spend his time inventing toys for children, or of Eurytus—another Pythagorean from Tarentum—as an archaic gentleman with a penchant for 'arranging pebbles somewhat after the manner of esthetic stationmasters in our own time'.[37] But the example of Zopyrus suggests Pythagoreans could be far more

[34] Compare West's comments, still largely ignored, on 'the exaggeration of the achievement and influence of individuals' in the early history of philosophy and ideas, and on 'the implicit assumption that nobody contributed to the development of thought except the few whose writings survived them': (1971), 218–19.

[35] *Letters* 7.338c, Thucydides 6.34.4; above, nn. 18, 22. For Plato's first journey to Sicily see above, Ch. 8.

[36] Lloyd (1979), 146. Cf. also G. S. Kirk and J. E. Raven, *The Presocratic Philosophers* (Cambridge, 1957), 216 ('only secondarily ... interested in the material aspect of the world'), M. M. Sassi in *Un secolo di ricerche in Magna Grecia* (Atti del ventottesimo convegno di studi sulla Magna Grecia; Taranto, 1989), 232–41. Elements of this criticism go back to Aristotle (*Cael.* 293ª20–7), although in fact Aristotle's criticism here is not that Pythagoreans ignored the sense-perceived world but that they refused to base their conclusions on sense perception alone—concluding in this case, to Aristotle's horror, that the earth is not at the centre of the universe. Elsewhere he acknowledges that the Pythagoreans 'carefully observe' ($\delta\iota\alpha\tau\eta\rho\circ\hat{\upsilon}\sigma\iota$) the interrelation between parts of the universe, and echoes Plato's diametrically opposite accusation that they paid too much attention to material things: cf. *Rep.* 529a–531c with Plut. *Marc.* 14.6, *Qu. conv.* 718e–f; Arist. *Met.* 989ᵇ33–990ª5, Kingsley (1990), 250 n. 30. Either way, in the new arena of specialization the Pythagoreans were bound to be the losers.

[37] de Santillana 112.

practical than is usually supposed, sometimes deadly practical. It also suggests that for them ideas and principles were things to be applied at every level of existence, from the universal and cosmological down to the everyday—including the level of devising weapons of war. For Plato this emphasis on practicality remained a powerful ideal; and yet, ideals apart, practically speaking it went against the grain of his temperament, his abilities, and against the conditions of the times in which he lived. With him and Aristotle the philosophical life as an integrated combination of practice and perception fell apart at the seams, and another ideal came to predominate instead: 'a new type of man, the unworldly and withdrawn student and scholar'.[38] Certainly there had been partial precedents in the ancient Greek world for this new ideal. But it was only with Plato and Aristotle that it received its most decisive turn, so as to become the defining characteristic of what was to prove the most enduring Athenian contribution to intellectual history: instead of the love of wisdom, philosophy turned into the love of talking and arguing about the love of wisdom.

Apart from these general implications, there are also some more specific conclusions to be drawn with regard to both Empedocles and Plato. As far as Empedocles is concerned, ever since the fourth century BC—and especially during the last century or so—his multifaceted interests have tended to be viewed as self-contradictory and eccentric; his apparent ability to combine mystical leanings with scientific concerns, passion with observation, has often been explained on the assumption that he was some kind of a freak, 'the last belated example of a species which with his death became extinct in the Greek world'.[39] But the example of Zopyrus indicates that while Empedocles may—even in the fifth century in which he lived— have belonged to a breed apart, there were other horses in the same stable. Although not written by Empedocles, the original work which lies behind the *Phaedo* myth bears several of the marks of his cosmological, scientific, and mythological

---

[38] Jaeger 426; Jaeger's warnings (428–9) against projecting this ideal of the 'theoretical philosopher' back beyond Plato onto the Presocratics are strangely ignored, in the case of Empedocles, by G. Müller, *MH* 17 (1960), 121–4. See further below, Chs. 15, 21.

[39] Dodds 145. Cf. Guthrie's comments, ii. 123–7.

interests, just as it bears so many marks of the influence of the country in which he lived. The figure of Zopyrus helps us understand Empedocles as much as Empedocles helps us understand him.

Where Plato is concerned, Zopyrus' chief significance naturally lies in his ability to throw light on the origins of the Platonic myths. When so much has been written and assumed about Plato's supposed originality as a myth-maker, it is obviously invaluable not only to be able to talk in general terms of a 'Pythagorean background' but also to be able to mention specific names. And it is fairly clear that Zopyrus' relevance to Plato is not confined to the *Phaedo* alone. As was mentioned earlier, he is credited by ancient sources not only with the *Krater* but also with an Orphic poem called the *Net*.[40] It is to stretch the bounds even of the greatest scepticism too far to see no connection between this and the fact that two of the visually most impressive images which run through the *Timaeus* are the imagery of a *krater* and the imagery associated with the fabrication and structure of a net. This is not even to mention the fact that the two strands of imagery occur one immediately after the other, side by side, at the end of a passage about the gods which undoubtedly refers to Orphic theology.[41]

But however important he was, Zopyrus—like Plato—is not someone whose contribution can be appreciated out of context or in isolation. As we saw earlier, the Orphic poem on which the *Phaedo* myth is ultimately based went through considerable changes—amplifications, interpretations, allegorization—before finally reaching Plato. In the case of the *Timaeus* it is

---

[40] *Suda*, s.v. Ὀρφεύς (iii. 565.7–8 Adler) = *OF* tests. 179, 223d; Ch. 11 with nn. 28–30. For the alternative attribution to Brontinus see Eisler's comments, i. 115 n. 1. It is also worth noting that Brontinus soon assumed the role of a semi-legendary figure to whom it was natural to ascribe Pythagorean writings—Thesleff (1965), pp. iv, 54–6; Burkert 114—and that our probable authority in this case for the attribution to Brontinus (Epigenes) appears to have been rather poor at distinguishing between legend and fact: for Epigenes on Cercops see Burkert 114, 130 n. 60.

[41] *Krater* imagery: *Tim.* 41d, cf. also 35a–36b, 37a. Imagery of the weaving and structure of nets: 78b–79d, cf. 36e, 41c–d. For the juxtaposition of the images cf. 41c–d, and for the Orphic allusion in 40d–41a, F. Adorno in *OMG* 17–18, West 117. The fact that the net imagery at 78d–79d is cited to explain the process of respiration, and using terms virtually identical to what we find in the passage about breathing in the *Phaedo* myth, is also certainly no coincidence: see Taylor 557 on 78e7 διαιωρούμενον with Ch. 11 above. Regarding the wider connections between Orphic literature and the *Timaeus* see Olerud's comments, 101–14.

noticeable how top-heavy the images of *krater* and net have become, weighed down almost to sinking point by sophisticated elaboration and technical intricacies. The obvious temptation is to ascribe this removal of visual imagery from its mythical base to Plato himself, but the very similar situation of reworking and transfer of imagery in the *Phaedo* myth strongly suggests that—here too—the decisive part was played not by Plato but by his Pythagorean sources. Certainly the *Phaedo* makes it quite clear that when considering the influence of Zopyrus on Plato we have to make allowance for the generation in between, and for the contribution of Plato's immediate informants: men such as Archytas, or Archytas' Sicilian disciple Archedemus.[42] Whichever way we look at the matter, there was undoubtedly more than one person involved. This makes it possible to appreciate and start to map out what so far has only been suspected but never analysed in detail: the complexity, intensity, and shiftingness of Pythagorean tradition during the late fifth and early fourth centuries. It becomes easy here to understand the meaning of the word 'tradition', as indicating something neither rigid nor fixed but fluid and accommodating: a kind of receptacle allowing for the pooling and absorbing of individual resources, so that new contributions transform the old until they are transformed in turn.

This aspect of individual interaction within a group context, of individuals preserving and contributing their uniqueness in a framework larger than themselves, is reflected in a Platonic passage which has often been discussed but not always with the clarity it deserves. In the *Meno*, Plato has Socrates introduce the doctrine of reincarnation by describing what 'I have heard from both men and women who are wise about divine matters'.[43] For a long time it has been disputed whether these men and women were 'Orphics' or 'Pythagoreans', and the overriding tendency has been to decide in favour of the conclusion that they were Orphics rather than Pythagoreans.[44] But for

---

[42] Above, Ch. 8. For Archytas' role as immediate source for material embodied by Plato in the *Timaeus* cf. Lloyd 169 n. 18, and also Burkert 237 n. 95 *ad fin.* on D.L. 5.25 Τὰ ἐκ τοῦ Τιμαίου καὶ τῶν Ἀρχυτείων (Aristotle).

[43] 81a, ἀκήκοα ἀνδρῶν τε καὶ γυναικῶν σοφῶν περὶ τὰ θεῖα πράγματα.

[44] So e.g. Boyancé 85 n. 4; Bluck 274–6.

a number of reasons this is clearly wrong. First, Wilamowitz's scathing attacks on the modern idea of a unified Orphic church or movement have had the enduring effect of making it untenable to talk of 'Orphics' in the abstract. From a theoretical point of view it is valid to claim that all ideas, beliefs, and practices which people in antiquity put under the patronage of Orpheus were automatically invested with a common spiritual affiliation; but practically speaking we need to be more specific, and when it is a question of 'Orphics' we always need to ask *which* are the Orphics in question.[45] As far as Plato is concerned, it could hardly be plainer that the 'wise men and women' mentioned in the *Meno* are not the same people as the famous wandering Orphic priests on whom he pours his scorn in the *Republic*: far from being individuals from whom he would be likely to want to learn anything through personal contact—or have Socrates learn anything through 'hearing'— he gives the impression that they are people he would prefer to keep well clear of.[46] And then, even more importantly, there is Plato's mention of men *and women*. In twentieth-century terms the detail may seem unremarkable; and there is a very real danger for precisely this reason of failing to notice how remarkable it is in a culture where sexual discrimination and above all segregation were the rule, not the exception. As for the wandering Orphic priests (and Plato only mentions men), the broader evidence would tend to suggest that although they were not necessarily celibates they were quite probably misogynistic.[47] It has been thought that this general conclusion needs modifying in the light of a passage where Plutarch mentions so-called 'Orphic' rites to which women were admitted.[48] But this is to misunderstand the evidence. Another passage in Plutarch indicates that the rites in question were open *only* to women.[49] In other words, we have here yet another sign of the sexual

---

[45] Cf. Kingsley (1993b), 23; West 2–3.

[46] *Rep.* 363e–365a. For the background to Plato's description of them as itinerant beggars cf. E. Fraenkel, *Aeschylus: Agamemnon* (Oxford, 1950), iii. 590–1; Burkert (1982), 4–12.

[47] Burkert (1985), 302 and n. 13. Cf. also Pl. *Rep.* 620a = *OF* test. 139; *OF* tests. 76, 115, fr. 234; Guthrie (1952), 49–50; M. Detienne in *OMG* 55, 70–9.

[48] Plut. *Alexander* 2 (*OF* test. 206); Guthrie (1952), 50. Cf. Demosthenes, *De falsa legatione* 281.

[49] Plut. *Caesar* 9; cf. Linforth 244.

polarization and instability which seem to have been so marked a characteristic of many 'Orphic' phenomena.

Any attempt to identify the men and women in the *Meno* by following this line of inquiry leads to a dead end. On the other hand, everything falls into place as soon as we remember that to attempt to distinguish on principle between 'Orphics' and 'Pythagoreans' is to create a false dichotomy: as we have seen, for the origins of ideas and literature that can be described as Orphic we are drawn repeatedly to Pythagorean circles in the West.[50] And in fact there can be no real doubt that the 'wise men and women' mentioned in the *Meno* were indeed Pythagoreans. One of the—by classical standards—most extraordinary features of ancient Pythagoreanism was, precisely, 'the equal status of women side by side with men': 'the fact that there were female as well as male Pythagoreans is often stressed in the sources'.[51] Here, plainly, lies the explanation of Plato's pointed reference to both sexes.

We could also have arrived safely at the same conclusion by following a quite separate route. The 'wise men and women' from whom Plato has Socrates receive teachings about the other world through 'hearing' (ἀκούειν) bear an unmistakable resemblance to the Pythagorean informant, 'one of the wise', from whom Plato elsewhere has him receive teachings on the

---

[50] Chs. 10–11. In the light of this evidence, Burkert's strange insistence (229 n. 55) that 'Philolaus' and 'Orphism' are mutually exclusive terms is historically unsustainable; and the relation between what can be described as either 'Orphic' or 'Bacchic' religious phenomena and Pythagoreanism was no doubt much closer than is generally supposed. For Archytas see below, n. 57; and for the interrelationship at Tarentum between Orpheus, Dionysus, and Bacchic mysteries, West 24–6. As for Philolaus, the fact that he is credited with a work called *Bacchae* which almost certainly contained authentic material—DK 44 B17–19; Burkert 268–9; Huffman 16 and 418 with Kingsley (1994*f*)—has caused considerable embarrassment. Burkert suggested the name was 'a late, "romanticizing" substitute' for an original title *On Nature* (269 n. 148), but even if this were correct the question still remains what there was in the work itself or in the Philolaic tradition to justify such a title: nobody ever seems to have thought of renaming Aristotle's *Physics* the *Bacchae*. There can in fact be little doubting that this connection between the Philolaic and the Bacchic needs taking seriously: see G. Pugliese Carratelli, *PP* 29 (1974), 143, Feyerabend 21, and, for Bacchic mysteries and Pythagoreanism, Ch. 17 below.

[51] Burkert (1982), 17–18, with refs. Cf. also C. J. de Vogel, *Pythagoras and Early Pythagoreanism* (Assen, 1966), 238 and n. 2; Dodds 165 n. 59. For the relevance of this social background to the *Meno* passage cf. e.g. K. Ziegler, *RE* xviii/1. 1381 n. 1; Long 69.

same subject and in exactly the same way.[52] But whereas in those other places Plato focuses attention on the individual and away from the others who share his wisdom, here he focuses it away from the individual and onto the group to which he belongs. This simultaneously helps to explain why, in the *Phaedo* as elsewhere, Socrates simply refers to his informant as a 'someone' and gives no name. Partly that was no doubt a literary device, a convenient way of avoiding the anachronisms involved in having the historical Socrates learn from people whom Plato himself learned from years after Socrates' death; but we must remember that there were other dramatic options open to Plato apart from leaving the person anonymous, and that he managed to get around the problem in other ways in other similar situations. Clearly the lack of a name served, as well, to lend the conversation an air of mystery. And yet, much more specifically, the anonymity also serves to remind us that Pythagoreans saw themselves as parts of, and mouthpieces for, a tradition greater than their own personalities. In practice this meant that they often appear to have seen it as a spiritual discipline, an act of modesty and self-effacement, to preserve their anonymity by attributing their individual achievements not to themselves but to the founding spirit of the tradition to which they belonged.[53] There was wisdom in leaving divinity nameless, Pythagoras nameless, and Pythagoreans nameless as well—all for different but related reasons.

After mentioning the 'wise men and women' in the *Meno*, Plato immediately goes on to describe them in greater detail: they are 'priests and priestesses, who are concerned to be able to explain what it is that they do'.[54] This description is significant for a couple of reasons. First, the mention of priests and ritual activity has been assumed to be incompatible with 'members of Pythagorean societies' and a proof that Orphics,

---

[52] 'As I have heard from one of the wise' (*Gorg.* 493a), 'as I seem to remember hearing from one of the wise' (*Rep.* 583b); Ch. 9 nn. 23, 36. For the dating of the *Meno* in relation to Plato's first journey to Italy and Sicily see R. W. Sharples, *Plato: Meno* (Warminster, 1985), 3. The Pythagorean identity of the men and women could also be inferred with a fair degree of cogency from the immediate context of the *Meno* itself (Cameron 69–76).
[53] Cf. esp. Iam. *VP* 88 and 198, with Burkert 90–1, 135.
[54] 81a, εἰσὶ τῶν ἱερέων τε καὶ τῶν ἱερειῶν ὅσοις μεμέληκε περὶ ὧν μεταχειρίζονται λόγον οἵοις τ' εἶναι διδόναι.

not Pythagoreans, are meant.[55] This shows little understanding of the ritual background to Pythagoreanism, and even less awareness of the evidence indicating the role assumed by Pythagoreans as priests in the mysteries—especially mysteries associated with Demeter and Persephone, where the priesthood was largely in the hands of women. It is hardly a coincidence that, immediately after introducing the subject of reincarnation as taught by these same priests and priestesses, Plato quotes some lines from Pindar specifically on the mysteries of Persephone.[56] And approaching the matter from an entirely different point of view, it is difficult to suppose that during his period of prominence and eventual rulership at Tarentum Archytas will not have been expected to preside over the mysteries of Persephone in much the same way that rulers at Syracuse were expected to officiate in their own city at the mystery ceremonies associated with the goddess.[57]

Second, there is Plato's emphasis on the priests and priestesses going to special trouble to 'explain' their activities. If we take this in the strictly Socratic sense of wisdom as being the ability to offer a philosophical definition of the essence of one's work, Plato's portrayal is obviously unhistorical and would have to be dismissed as simple literary fiction. But, as the Derveni papyrus dramatically shows, there is more to the matter than that. In highly vivid and critical terms the author in the papyrus attacks wandering Orphic priests—the details of the description show that these priests are indistinguishable

[55] So e.g. W. J. Verdenius, *Mnemosyne*⁴, 10 (1957), 294.

[56] *Meno* 81b–c = Pi. fr. 133. For Pythagoreans, priesthood, and ritual cf. e.g. Hdt. 2.81; Isocrates, *Busiris* 28; Eudoxus, fr. 325 Lasserre; D.L. 8.33 *ad fin.*; Iam. *VP* 85; Hipp. *Ref.* 1.2.18; Boyancé 136–7; Burkert 127–31, 153–61, 174 and (1969), 23–4. For priestesses of Persephone cf. e.g. Zuntz 96 and n. 14. In line with the recent revival of interest in Bacchic mysteries it has become common to describe Plato's priests and priestesses as 'Bacchic'—so Feyerabend 17, comparing Henrichs, *ZPE* 4 (1969), 237–8; Graf 97–8. First, however, it is important to define exactly how this term is to be understood: for the 'eschatological' wing of Bacchic religion, which focused on the relation between Dionysus and Persephone, see now *TGL* 3–16 (with the refs., ibid. 12, to Plato's *Phaedo*), and for the involvement of Pythagoreans in this aspect of Bacchic religiosity, above, n. 50, below, n. 57, and Ch. 17.

[57] Pi. *Ol.* 6.93–5, Hdt. 7.153, Griffith 172. It will be noted, as well, that Archytas refers casually but knowledgeably to the Dionysiac mysteries in fr. 1 (DK i. 35.2–5 = Bowen 80.38–40; for the genuineness of the passage see Ch. 11 n. 55); cf. also A. D. Trendall in *OMG* 175, Bowen 91, and above, n. 50. For Archytas' rulership at Tarentum and his active involvement in all aspects of the city's life see Wuilleumier 70; for the mysteries of Persephone and Dionysus in Tarentum, ibid. 496–501, 502–11.

from the ones mocked at by Plato in the *Republic*—for going about their business and earning money performing their rituals without being able to explain either to themselves, or to anyone else, what they are really doing.[58] What is particularly interesting here is that what the writer of the papyrus means by being able to give an explanation is the ability to interpret allegorically the Orphic literature which accompanies the mysteries.[59] Applied to the *Meno* passage, the implication of this is simple, obvious, and extremely meaningful. The 'wise men and women' are presented as doing precisely what Socrates' oral informant has done in the *Phaedo* myth: take a sacred Orphic text about reincarnation and the other world, and interpret it allegorically.

With these men and women, 'wise' about what they do, we are also brought back once again to the 'wise man' in the fragment of myth in the *Gorgias*, who is wise because he is so skilful at explaining allegorically the real meaning of accounts about the other world.[60] Yet again, it is striking how harmoniously the details fit into place. However, the very fact that there are so many of them, scattered through different dialogues, demands a few final comments on Plato's attitude both to myth and to the allegorical interpretation of it.

Platonic myth has had a poor reputation during the nineteenth and much of the twentieth century. But although the assumption that his famous attacks on Homeric myths in the *Republic* were attacks on the genre of myth as a whole is superficially tempting, it is fundamentally mistaken; and recent research has done a great deal to rehabilitate the Platonic myths from the inferior position to which they had most often been

[58] Pap. Derv. cols. 18.3–20.12 (cf. Pl. *Rep.* 364b–c); Burkert (1982), 5 ('... invective against people who "make a craft of the holy rites", who just take the money of their clients and do not give any explanation while they make them "see the holy things"').

[59] Cf. esp. Pap. Derv. cols. 18.3–19.7, 23.1–3, and for the role of the literature accompanying Orphic ritual, Burkert (1982), 8–9, (1983), 116, 119.

[60] *Gorg.* 493a1–c3. For 'wisdom' as the ability to explain passages (allegorizing) and words (etymologizing) by bringing out their true meaning cf. e.g. *Crat.* 396c–e (σοφία), *Rep.* 509d (σοφίζεσθαι), Iam. *VP* 86 (ἐπισοφίζεσθαι: Burkert 174). For the broad sweep of meanings and connotations that Plato was prepared to give to the terms *sophos*, *sophia*, and their derivatives cf. Ferrari's comments, 234–5 n. 12. By Plato's time allegorizing and etymologizing were in practice inseparable: see Woodhead 44–51; Buffière 60–5; Brisson 157–8; T. M. S. Baxter, *The Cratylus* (Leiden, 1992), 115–16.

confined as some kind of philosophically irrelevant appendage.[61] It is impossible to overemphasize the significance of the fact that, after attacking and re-attacking the moral degeneracy of traditional Greek myths in the *Republic*, Plato could bring the work as a whole to a final dramatic crescendo with a spiritually edifying myth about the other world.[62] And to mention one other approach to the matter, there is a great deal to be learned from the parallels suggested by Plato between the role of the myth-maker and the role of creator of the universe.[63] What is more, as we have already seen, Plato repeatedly tends to set up the two apparently opposing categories of myth and logic only to end up merging and demolishing them. Recent studies have shown increasingly that the radical distinction once assumed to exist between Platonic myth and argument is not nearly as clear-cut as it had seemed; instead, there has been a growing emphasis on uncovering the subtle but very significant ways in which the mythical and non-mythical in his dialogues interconnect and interact.[64] The essence of this 'new' tendency had already been expressed quite plainly by Julian in the fourth century AD: 'The mythical passages have been incorporated not just as incidental accessories but with a definite harmoniousness'.[65] In other words, there can be no clear distinction between Plato's attitude to myth and his attitude to philosophical writing; any conclusions about his own myths inevitably affect our understanding of his work as a whole.

As with myth, so with the allegorizing interpretation of myth. There is still a general belief that Plato was resolutely opposed to allegory as a method of explaining literary texts.[66]

[61] Cf. e.g. K. Gaiser, *Platone come scrittore filosofico* (Naples, 1984), 125–52.
[62] S. Halliwell, *Plato: Republic 10* (Warminster, 1988), 2–3. For the *Republic* itself as a myth cf. 376d9–10 and 501e4 with Zaslavsky 183 n. 50.
[63] Cf. L. Brisson in *Le Texte et ses représentations* (Études de littérature ancienne 3; Paris, 1987), 121–8. This naturally helps to explain why in the *Timaeus*—which is concerned precisely with the creation of the universe—the theoretical distinctions between myth, science, and argument prove so impossible to maintain. See also *Tim.* 92c and *Crit.* 106a with W. Welliver's comments, *Character, Plot and Thought in Plato's Timaeus–Critias* (Leiden, 1977), 58–9.
[64] So e.g. Annas 119–22; H. R. Scodel, *Diaeresis and Myth in Plato's Statesman* (Göttingen, 1987), esp. 165–6; Ferrari 113–39 and *passim*; Sedley. See also above, Chs. 7, 10; and below, n. 70.
[65] *Orations* 7.11, 217a, οὐ παρέργως ἀλλὰ μετά τινος ἐμμελείας ἡ τῶν μύθων ἐγκαταμέμικται γραφή.
[66] So e.g. Brisson 152–9 ('Refus de toute interprétation allégorique').

However, the fact is that his attitude to it was as ambiguous and complex as his attitude to myth itself. The *Republic* shows how ruthlessly he condemned allegorizing interpretations when used as a method of attempting to divert attention away from the obvious surface meaning by finding 'higher' truths in myths that are morally objectionable, harmful, and blasphemous; and as the *Phaedrus* shows, he could object to any form of allegorizing—and to the closely related practice of etymologizing—if it ended up dissolving the meaning of a myth into fragments of interesting, but useless, facts which distract the individual from the search for the truth of himself.[67] And yet Plato's attack on allegorizing in the *Republic* is not an attack on the method but only on its use in certain circumstances; while the *Phaedrus* also shows that mythical allegory, when practised with real wisdom, can direct a man towards the truth of himself just as much as it can direct him away.[68]

No discussion of Plato's attitude to myth and allegory, and to his predecessors in these fields, would be complete without a few words about his sense of humour. In a recent paper devoted to the same passage from the *Gorgias* which we have kept coming back to, D. Blank has understood Plato's distinction between an original myth-teller and the subsequent allegorizing interpreter of the myth as implying his approval of what the allegorizer has to say but his disapproval for the myth-teller himself.[69] Blank's analysis is plainly wrong, and yet he touches on an issue which has some important implications.

Blank begins by trying to conjure up the ghost of the view that for Plato myth was something by definition inferior to dialectic; about that enough has already been said.[70] He then goes on to claim that Plato's description of the mythical author as 'some ingenious myth-teller of a man' (τις μυθολογῶν κομψὸς ἀνήρ) must be understood as derogatory; and here we

[67] *Rep.* 378d3–e3; *Phdr.* 229b4–230a6. For the relation between allegorizing and etymologizing see above, n. 60.
[68] *Rep.*, loc. cit.; Ferrari 11–12, 235 nn. 14–15.
[69] *Hermes*, 119 (1991), 22–36.
[70] Ibid. 24, 35. Blank's simple assertion that Socrates views the concluding myth in the dialogue 'as a *logos* not a *mythos*' (cf. *Gorg.* 523a1–3; similarly Zaslavsky 195) misses the fundamental point that at the most obvious level the final story *is* a myth: Plato's point is simply that it is not *just* a myth. The effect here is not to reinforce the distinction between myth and logic, but to undercut and undermine it—as Plato does repeatedly elsewhere. Cf. Dodds's comments, (1959), 376–7.

come to the heart of the matter. First, although the word 'ingenious' (κομψός) is neutral in connotation and can be used by Plato in a clearly positive or complimentary sense, it is also true that he uses it sarcastically; here it is difficult not to detect a touch of irony.[71] But secondly, Blank has completely overlooked the detail pointed out by Zeller and Wilamowitz long ago: in having Socrates describe the author as 'some ingenious myth-teller of a man, probably a Sicilian or an Italian', Plato is humorously half-quoting, half-paraphrasing a line by the poet Timocreon about 'an ingenious Sicilian man'.[72] Of course, as we have seen, by Plato's time this kind of literary borrowing and variation had come to be prized and applauded as a type of skilful ingenuity all of its own; Plato himself was a master of this form of allusion, using it several times elsewhere in the *Gorgias* with considerable effect.[73] In other words, the very act of alluding to the myth-teller as ingenious is itself an example of literary ingenuity. Plato has, light-heartedly but quite deliberately, locked the figure of Socrates in a perfect circle of irony by linking him with the very person whom it might seem that he is ironically depreciating. We can only allow that Plato was criticizing the myth-teller if we are prepared to allow that he was condemning both Socrates and himself.

This may seem too fleeting a paradox to be of any real significance; but in fact Plato goes on to do exactly the same thing in the very next words. Immediately after describing the myth-teller as 'ingenious', he presents him as 'leading away' (παράγων) through his myths by using etymological procedure to explain the meaning of words.[74] Blank quite correctly points

---

[71] Blank, op. cit. 24–5, with refs.
[72] Zeller, i/1. 558 n. 2; Wilamowitz, ii. 89; Timocreon, fr. 6 Page (*PMG* § 732); above, Ch. 10 with n. 8.
[73] Cf. *Gorg.* 449a7–8, 513c2, and 516c3, with Labarbe's comments, 310–11 ('... nuance d'ironie', 'il lui a fallu ruser avec le poète ...'). For Plato's style of quoting and paraphrasing poetry see Ch. 10 with nn. 56–7; for literary borrowing and variation in general, Ch. 4 with nn. 25, 27. Dodds (1959), 301 objected that the context of the original phrase in Timocreon 'is very different and I see no reason to assume that Plato had Timocreon in mind'; but first, the paraphrase is so close as to be unmistakable (cf. Page *ap. PMG* § 732), and second, the effectiveness of Plato's allusions to Homer elsewhere in the *Gorgias* depends precisely on this incongruity between the original context and his own (Labarbe, loc. cit.). The incongruity is a normal feature in displays of ingenuity of this kind: cf. Bollack, i. 284–6.
[74] 493a6–7, παράγων τῶι ὀνόματι διὰ τὸ πιθανόν τε καὶ πειστικὸν ὠνόμασε πίθον . . .

out that on Plato's lips the verb here must have something of the sense of 'lead astray' or 'mislead'.[75] What he does not point out is that the verb in question also has the standard, semi-technical sense of 'deriving' a word from another through the use of etymology;[76] all other connotations apart, that is unmistakably the primary meaning of the word here. What is more, when we look at the passage as a whole it also becomes clear that the etymological procedure being referred to forms the very basis for the allegorical interpretation of the myth-teller's tale.[77] In other words, although the allegorizing and etymologizing explanation of the myth is *ascribed* to the 'ingenious' myth-teller himself, in actual fact it belongs not to him but to the separate 'wise man' who is interpreting the myth. That may seem strange, and yet it is perfectly normal procedure in allegorizing contexts. The interpreter generously attributes his own interpretation not to himself but to the writer whose work he is interpreting: 'he already intended these subtle meanings, and I am simply bringing out into the open what he left implicit'.[78] Even more important, though, than the fact that this interpretative process is the work not of the myth-teller but of the 'wise man' explaining the myth, is the fact that elsewhere Plato himself unashamedly uses the same etymologizing procedure—and even the very same etymologizing example—which on a surface reading he seems here to be dismissing as 'misleading'.[79] Once again, Plato implicates himself well and

---

[75] Op. cit. 25–6, with refs. Cf. also Burkert (1968), 100, and add the account in *Phdr.* 262d of 'how the man who knows the truth can, by playing with words, lead astray (παράγοι) his hearers', together with Ferrari's comments on the 'dizzy irony' of the passage (45–59).
[76] Cf. Pap. Derv. col. 23.1; *Crat.* 398c–d, 400c, 407c, 416b; LSJ s.vv. παράγω II.5, παραγωγή III.1, παραγωγός II.2a; Burkert (1968), 95 n. 4. Blank's statement (op. cit. 25) that as a grammatical term the verb simply means 'change slightly' is only partly true, and misses the essential.
[77] *Gorg.* 493a6–c7.
[78] For exactly the same attribution of hidden meanings by a commentator to the poet whose work he is interpreting allegorically, and expressed using exactly the same verb, cf. Pap. Derv. col. 23.1–3 (τοῦτο τὸ ἔπος παραγωγὸμ πεπόηται, καὶ τοῖς μὲν πολλοῖς ἄδηλόν ἐστιν τοῖς δὲ ὀρθῶς γινώσκουσι εὔδηλον ὅτι . . .). For the relevance of the Derveni papyrus to Plato's myths—and their interpretation—see Ch. 10 with the refs. in nn. 19, 23, 41.
[79] After using etymology to explain the allegorical meaning of the mythical 'jar' that refuses to hold any water (above, n. 74), he does exactly the same with the word for 'Hades' (τῶν ἐν Ἅιδου, τὸ ἀιδὲς δὴ λέγων, 493b4–5)—just as he does at *Phd.* 81c10–11;

truly in the web of his own irony. He appears to distance himself both from the myth-teller and from the allegorizer, only to identify himself with them even more closely at another level. This is not the Plato we are generally used to: the image of the serious philosopher rebuking the poets and aiming to cleanse the state and the soul of all ambiguity. The irony of the situation is inescapable; the more we try to extricate ourselves, the more we end up trying to extricate Plato from himself. The resulting complexities are ones that would tend to have been better understood in the Renaissance, when there was a greater general appreciation of the arts of humour and philosophers 'had learned from Plato that the deepest things are best spoken of in a tone of irony'.[80] And what they had learned from him was not just something they imagined they had learned: Plato makes it clear that he viewed not only myth-telling but also dialectic as a game—which is not surprising considering the overlap between them in his writings.[81] Or as he said at the end of his life, the only thing worth being serious about is God;

but as for man, he is just made as God's plaything and that really is what is best about him. So every man and woman should live in accordance with this aspect of themselves and spend their time playing the finest of games, thinking the exact opposite of what they do now.[82]

This is enough about Plato, and about the *Phaedo* myth. Wilamowitz once dismissed it as 'unimpressive' compared to Plato's other myths. He considered its lengthy description of underworld geography 'unusually wordy', and was able to find 'no satisfactory explanation' for its mass of apparently insig-

---

cf. also *Crat.* 403a5–7. For Plato's use of and attitude towards etymology see Woodhead's fundamental discussion, 49–73.

[80] Wind 236, quoting e.g. Ficino, *Op.* ii. 1137.38–41, 1425.50–1. For Socrates and 'serious jesting' see already Cic. *Or.* 2.67.269–70; Vlastos 28–9.

[81] Myth-telling as a game: *Phdr.* 276e, Zaslavsky 183 n. 50. Dialectic as a game: *Parmenides* 137b, Fic. *Op.* ii. 1137.41, and cf. also Gorgias, *Helen* 21 (DK ii. 294.19–20), Huizinga 148–51.

[82] *Laws* 803c; discussion and further refs. in Huizinga 18–19, 26–7, 48, 145, 211–12.

nificant details.[83] But, like the proverbial rejected corner-stone, we have found that it allows us to reconstruct a whole prehistory of Platonic myth. For our specific purposes it has been especially helpful in bridging the gap between Empedocles and pre-Platonic Pythagoreanism on the one hand, and Plato and Platonic tradition on the other. There is much more that could still be said both about the literal level of the myth and about its level of symbolic transposition—but nothing that will be of any help to us in our immediate task, which is to understand the factors lying behind Empedocles' equation of Hades and fire.

[83] i. 329 and n. 1.

13
Central Fire
*

FROM examining the *Phaedo* myth one point has become clear: the idea of massive fires inside the earth was held not just by Empedocles but also by western Pythagoreans living later on in the fifth century BC. This conclusion, in turn, brings us face to face with one of the most vexed questions in the history of Greek philosophy—the problem of the Pythagorean 'central fire'.

The earliest specific reference to the idea of a central fire occurs in Aristotle. There it is just vaguely attributed to 'Italians' or 'Pythagoreans'.[1] However, later reports—doubtless stemming from Theophrastus—associate the idea specifically with the name of Philolaus, and there is no reason whatever to suspect that the attribution is incorrect.[2]

The doctrine of the central fire, as preserved by Aristotle and as ascribed in later reports to Philolaus, is very precise in its outlines. Earlier cosmologies had, as a matter of course, placed the earth at the centre of the universe. But this system gives the central place to a fiery 'hearth', surrounded by a number of bodies revolving around it in circular orbits at various different distances from it; and one of those revolving bodies is our earth. The closest body of all to the central fire is the so-called 'counter-earth' (*antichthōn*). This, just like the central fire itself, has the interesting property of always remaining invisible to the inhabitants of the earth on which we live. The next body out from the central fire is the earth itself; as for the name 'counter-earth', it can at least in part be explained as due to the fact that

---

[1] *Cael.* 293ᵃ20–ᵇ21; cf. *Met.* 985ᵇ23, 986ᵃ8–12. For the allusiveness of the attribution see Burkert 236–7; also above, Ch. 12 with the refs. in n. 53.

[2] Stob. i. 186.24–7 = Philolaus A17b; i. 196.18–25 = A16; ps.-Plut. *Placita* 3.11.3 = A17a; 3.13.2 = A21; Stob. i. 189.16–18 = Philolaus B7. Cf. Guthrie, i. 287 and n. 1, 289–90; Burkert 337–42; Philip 112–16, 118–22; Huffman 242–3.

this mysterious celestial body lies 'opposite' (*anti-*) or 'over against' our earth while revolving in its own smaller orbit, always keeping level with the earth and so remaining directly between us and the central fire.[3] Further out from the earth in the other direction comes the moon, then the sun, the five planets, and finally the sphere of fixed stars.

The most striking feature of this scheme from our point of view is of course the fact that it displaces the earth from its central position, and in this respect seems to anticipate the modern, heliocentric theory. Copernicus himself named Philolaus as one of his predecessors, and even recently the Philolaic scheme has been described as requiring 'a bold leap of the scientific imagination'.[4] However, on closer analysis it becomes clear that the scheme has little pretension to be 'scientific' in the modern sense of the term. Knowledge of the five planets is something for which Philolaus was indebted to Babylonia, along with the astrological doctrine also associated with his name in the ancient sources.[5] But there is no Babylonian equivalent to his counter-earth or central fire; there appears to be no empirical justification for positing their existence in the first place; and the very fact that they are both so emphatically invisible to us tends to relegate them—along with

[3] Arist. *Cael.* 293ᵃ23–4, ᵇ19–21; Alex. *Met.* 40.31–41.1; Simpl. *Cael.* 511.28–30 and 511.33–512.1. In discussing the scheme a few scholars—e.g. Buffière 569–70 and fig. 11; G. B. Burch, *Osiris*, 11 (1954), 267–94—assume that the name 'counter-earth' is meant to imply a body revolving at the same distance from the central fire as the earth, and in the earth's own orbit, while remaining diametrically opposite to the earth itself. However, we must be guided here by the repeated emphasis on the part of our sources that each of the revolving bodies occupies a different orbit, with the orbit of the counter-earth closer in to the central fire than the orbit of our own earth. Cf. Simpl. *Cael.* 511.26–30 = Arist. fr. 204; Stob. i. 196.18–25 = Philolaus A16; ps.-Plut. *Placita* 3.11.3 = A17a; Alex. *Met.* 39.1–3; A. Boeckh, *Kleine Schriften*, iii (Leipzig, 1866), 276–83, 320–36. For the sense of the term 'counter-earth' see also Dicks 67, plus the diagram on p. 69; Zeller, i/1. 528 n.; Burkert 348 n. 58; and below, with n. 48.

[4] Copernicus, *De revolutionibus orbium caelestium*, ed. A. Koyré (Turin, 1975), 18.8–24, 54.28–56.3; Guthrie, i. 282.

[5] Knowledge of the planets: Burkert 300–1 with the refs. in n. 9, 310, 313; Lloyd (1979), 177; M. L. West, *JHS* 100 (1980), 208; and cf. also Kuhrt 150. The *Epinomis* (986d–987d) is quite explicit about the oriental origin of the planetary doctrine which is attributed in our sources to Philolaus; cf. also Arist. *Cael.* 292ᵃ7–9, Cic. *Div.* 1.1.2, 1.42.92–3. For the references in the *Epinomis* to 'Syria' (i.e. Assyria; 987a2, b4) cf. Tarán (1975), 304, 384, and A. Kuhrt in H. W. A. M. Sancisi-Weerdenburg and Kuhrt (eds.), *Achaemenid History*, iv (Leiden, 1990), 177; for its mention of Egypt (987a2), Burkert 299 n. 3 and Kingsley (1995*c*). Regarding Philolaus and Babylonian astrology see Kingsley (1994*a*), with n. 30; (1994*f*); Ch. 18 and n. 33.

this whole side of his cosmological theory—to the realms of 'mystery' and myth.[6] To describe the theory as scientific is not only misleading but self-deceptive, because it begs the fundamental question as to how and why this aspect of the theory came to be formulated.

Since antiquity, more or less serious attempts to answer this fundamental question have repeatedly been made; but what is most noticeable about them is their half-heartedness and their failure to assess the evidence with the care it deserves. Already Aristotle tried, in his *Metaphysics*, to explain the lack of any empirical justification for positing the existence of a counter-earth or a central fire by suggesting that these entities were simply introduced to bring the total of moving celestial bodies up to the mystically significant number of ten. At first sight there would seem perhaps to be something to his suggestion. But on closer analysis it emerges that Aristotle is adopting his usual facetious stance when dealing with the Presocratics, and that there is little in the way of a serious understanding of the system to be extracted from his words.[7] What is more, even if this were not the case it would still be only too clear that his proposed explanation fails to explain anything at all. The Philolaic system presents us with, in all, eleven celestial bodies (counting the sphere of fixed stars as one), not ten; and considering the significance in Pythagoreanism of the number nine as well as the number ten,[8] from the point of view of mystical symbolism there would be much more to be said for having nine moving bodies revolving around a stationary tenth than ten moving bodies revolving around an eleventh. In other words, the number-symbolism approach could possibly be used to help account for a planetary system which incorporates either the counter-earth or a central fire, but fails completely to explain a system that includes them both.

---

[6] Burkert 337–48; von Fritz (1973), 471; Philip 114–16. Huffman (247) misrepresents these writers in claiming that for them Philolaus' system was nothing but '*just* myth'; see also above, Ch. 7 with n. 1, Ch. 8 with n. 11.
[7] Arist. *Met.* 986^{a}8–12; Burnet (1930), 305, Cherniss (1935), 45.
[8] Cf. e.g. Speus. fr. 28, Arist. *Met.* 1092^{b}26–30, 1093^{a}28–^{b}1 = DK 58 B27, *Cael.* 268^{a}10–13 = DK 58 B17, ps.-Arist. *Problems* 910^{b}31–8 (DK 58 B16), Eudemus, fr. 142 Wehrli = DK 58 B18, Sext. *Math.* 7.94–5 and ps.-Plut. *Placita* 1.3.7 = DK 58 B15; Boyancé 254–5, Burkert 309, 467–8, 474–5, KRS 233–4. For the moving 'heaven' of the fixed stars in Philolaus cf. Burkert 340–2 with n. 11.

Another clue to the peculiar structure of the scheme has been found in a statement made by Aristotle elsewhere. In his *On the Heavens* he raises the issue of the earth's location in the universe, and cites the view of the 'Italians' or 'Pythagoreans' to the effect that the midpoint of the cosmos is occupied not by the earth but by a central fire. He then goes on to offer the following line of reasoning in support of this view: the centre is an honourable place, what occupies an honourable place should be worthy of honour itself, fire is more honourable than earth, and so fire—not earth—should occupy the centre of the universe. This logic has seriously been accepted by a number of modern writers as expressing Pythagorean ideas, and as reflecting the kind of motivation which led to the Philolaic scheme being formulated in the first place. However, from Aristotle's own words one point is perfectly clear: he is ascribing this belief in the honourableness of the centre not to Pythagoreans at all, but to 'others' whom he specifically distinguishes from them. So, after first outlining the basic Pythagorean scheme of counter-earth and central fire, what he goes on to say is:

There are many others (πολλοῖς ἑτέροις) who might agree that it is not right to allocate the central position to the earth—people who look for proof to theoretical argument rather than to phenomena. *For they believe* that what is most worthy of honour deserves the most honourable place; and as fire is more honourable than earth, a bounding position more honourable than an intermediary one, and the outermost limit and the centre are both boundaries, they arrive by way of conclusion at the opinion that fire, and not the earth, lies at the centre of the spherical universe. *But as for the Pythagoreans themselves*, they also go on to say . . . (ἔτι δ᾽ οἵ γε Πυθαγόρειοι . . .).[9]

There can be no doubt that what we have here is a straightforward parenthesis in which Aristotle turns briefly aside from the Pythagoreans to bring into his discussion ideas held by other philosophers. He neatly indicates, at the start of the digression and at the end of it, both when he is turning aside from the Pythagorean view and when he is returning to it.[10]

---

[9] *Cael.* 293ᵃ27–ᵇ1.
[10] For the resumptive force of δέ . . . γε after a digression or diversion see J. D. Denniston, *The Greek Particles*² (Oxford, 1954), 154; and for an example of exactly the same construction, Procl. *Eucl.* 90.11–14 (ἄλλοι δὲ λόγοι λέγουσιν . . . οἱ δέ γε Πυθαγόρειοι . . .). On an unprejudiced reading of the passage it is quite clear that the

There is nothing at all here that tells us anything about the Pythagorean theory itself—except in the negative sense that we can safely assume the specific details mentioned in the parenthesis do *not* represent ideas which Aristotle considered Pythagorean.[11]

Who, then, are those 'many others'? Already in antiquity this question was something of an embarrassment. The Aristotelian commentators knew of nobody apart from Pythagoreans who, prior to the time of Aristotle, so much as considered displacing the earth from the centre of the cosmos. As a result, some of them proposed the solution that the 'others' mentioned by Aristotle did not exist at all, and that the question as a whole was fruitless: he had simply formulated certain propositions which he attributed to others—apart from Pythagoreans—on a purely conjectural and hypothetical basis. The same solution has been revived in recent times, and most forcefully by Cherniss.[12] However, while it is certainly true that Aristotle's initial statement, 'there are many others who *might* agree that . . . ', is no more than hypothetical, there is nothing hypothetical at all in his following remarks about them believing that what is most worthy of honour deserves the most honourable place. That he is referring to definite people here cannot be denied.[13] On the other hand, we must always be wary of trusting Aristotle when he outlines the views supposedly held by earlier philosophers: only too often when discussing his predecessors, he indulges his habit of committing them to dogmatic positions which they did not necessarily

---

subject of οἴονται in 293ᵃ30 and of ἀναλογιζόμενοι and οἴονται in ᵃ33 is the same as the subject of ἀθροῦσιν in ᵃ29—which, of course, refers back to the πολλοῖς ἑτέροις of ᵃ27–8. Philip's introduction of 'the Pythagoreans' as subject of οἴονται at ᵃ30 (112) is not only gratuitous but plainly wrong.

[11] For the tendency to draw wrong conclusions from the passage by mistranslating or interpolating it, cf. e.g. Philip 112, 114, with the note above. The attempt to make Aristotle's 'many others' mean 'many other Pythagoreans'—so e.g. F. M. Cornford, *Plato's Cosmology* (London, 1937), 126—is equally gratuitous; if that is what he had meant, that is what he would have said.

[12] Simpl. *Cael.* 513.8–13; Cherniss (1935), 394–5 and (1944), 560–1, followed e.g. by Tarán (1975), 106–7 and (1981), 445. Themistius' commentary on the passage (*De caelo*, 83.1–3 in Landauer's Hebrew text, 124.3–7 in his Latin translation) is simply a rough paraphrase of the text of Aristotle and—contrary to Cherniss's claim, (1944), 560—makes no attempt at all to solve the problem on the lines mentioned by Simplicius.

[13] Correctly Heath 187; Burkert 327 n. 16 *ad fin.*

hold by extrapolating seemingly logical conclusions from views to which they did more or less firmly subscribe.[14] In this case his peculiar mixture of hypothesis and assertion is instructive, and helps us delimit what can safely be inferred about the people in question. Plainly they held, as a matter of doctrine, to the belief that what is most worthy of honour deserves the most honourable place. And yet all we can say in addition is that they may well have flirted with the possibility that this requires displacing the earth from its traditional position at the centre of the universe, although without necessarily being prepared to commit themselves to a non-geocentric viewpoint as a definite item of doctrine.

In fact the identity of these people is not at all hard to guess. Time and time again in his works Aristotle couples Pythagoreans with Platonists, and turns from describing Pythagorean ideas to outline Platonic ones. As it happens, this is precisely what he does in the passage of *On the Heavens* immediately following the one we have been considering.[15] In his mind this was no arbitrary procedure, but reflected his perception of the extent to which Platonism was historically indebted to Pythagoreanism.[16] Also, and more specifically, Aristotle repeatedly combines allusions to Pythagorean views with a passing reference to ideas held or expressed by Speusippus—who, as Plato's nephew and successor, was an extremely important figure in the early Academy.[17] Here too, there is nothing arbitrary in his procedure. Speusippus had direct contact with Pythagoreans, wrote a work on Pythagorean numerology which places him in a line of tradition linking him closely with Philolaus, and played a crucial role for the Academy in transforming and modernizing Pythagorean doctrine by bringing it into harmony with the Platonism of his time.[18]

[14] See e.g. Kingsley (1990), 250 n. 30; (1995a), § IV.

[15] *Cael.* 293^{b}18 ὅσοι μέν (= Pythagoreans) . . . ^{b}30 ἔνιοι δέ (interpreters of the *Timaeus*). Cf. esp. *Ph.* 203^{a}4–16, *Met.* 996^{a}6, 1001^{a}9–10, 1053^{b}12–13; also *Met.* 987^{a}29–988^{a}1, 989^{b}29–993^{a}10, 1083^{a}20–^{b}19, 1090^{a}2–31, Burkert 30–1.

[16] Above, Ch. 9 with n. 39.

[17] Arist. *Ethica Nicomachea* 1096^{b}5–8, *Met.* 1072^{b}31, 1080^{b}14–21, 1083^{a}20–^{b}19, 1090^{a}2–1091^{b}22; Zeller, ii/1. 999 nn. 1 and 3, Burkert 52 and n. 118, Tarán (1981), 309, 316–17, 335–6, 348–50.

[18] Burkert 47, 63–4, 69, 93 n. 43, 95 n. 52. Cf. von Fritz (1973), 464–5; M. Isnardi Parente, *Speusippo: Frammenti* (Naples, 1980), 369–76; Tarán (1981), 259–65; Huffman 23–4. Huffman's doubts (362) about the genuineness of the title of Speusippus' book,

Other indications point in exactly the same direction. It should hardly need emphasizing that Aristotle's mention of 'people who look for proof to theoretical argument rather than to phenomena' fits Platonists better than it fits anyone else.[19] And then there is the significant matter of two reports about Plato's Academy which are preserved by Plutarch. According to one of them, for which Plutarch cites the authority of Theophrastus, towards the end of his life Plato regretted ever having placed the earth at the centre of the universe because he no longer felt it deserved to occupy such an important position (ὡς οὐ προσήκουσαν τῆι γῆι τὴν μέσην χώραν τοῦ παντός). The second report, which is obviously related to the first, repeats the same information in virtually identical words while adding the specification that Plato became reluctant to place the earth at the centre because it was not among 'the most honourable' parts of the universe (οὐ τῶν τιμιωτάτων τοῦ κόσμου μορίων ὑπάρχειν).[20] These two reports have for a long while proved a major bone of contention, and scholars have done their best to dismiss them by arguing that either Plutarch or Theophrastus, or both, had got their facts wrong.[21] However, the evidence is not disposed of so easily. Plutarch was no fool, and knew his Theophrastus well;[22] as for Theophrastus himself, he was in a position to know more about changes of heart and other goings-on in the early Academy than we can ever hope to. To the sceptics one point only can be conceded: due allowance must be made for the gap of twelve years between Plato's death and Theophrastus' arrival in Athens.[23] During that time the reinterpretation of Plato by Platonists had grown apace, and

On Pythagorean Numbers, are unfounded; for Philolaus and number theory cf. Kingsley (1994f).

[19] Cf. G. E. R. Lloyd's comments, *Aristotle: The Growth and Structure of his Thought* (Cambridge, 1968), 46–7, 54–7, 65–7, 205–6, 286.

[20] Plut. *Qu. Plat.* 1006c = FHSG i. 432–3 § 243; *Numa* 11. Tarán's exaggerated hesitancy, (1975), 106 and n. 477, in ascribing the second report as well as the first to Theophrastus is hypercritical. What is more, the attribution can be considered confirmed by Theophrastus' own use of extremely similar language in *Met.* 11ª10–12, 22–4.

[21] So e.g. Cherniss (1944), 561–4; Tarán (1975), 105–7; and further refs. in Burkert 327 n. 16.

[22] No less than 69 entries in the latest edition of Theophrastus fragments and testimonia derive from Plutarch (FHSG ii. 683–6).

[23] For the chronological details cf. Jaeger 115 n. 1, 311–13.

this inevitably created difficulties. For obvious reasons the Academy—including Speusippus in particular—appears to have been keen to use every opportunity to cite the authority of Plato in defence of its new ideas, which in themselves show a marked tendency to bring the Platonic writings into an ever closer harmony with Pythagoreanism.[24] All this has implications in helping us determine the precise extent to which Plutarch's reports can be taken as reflections of historical truth. While they could possibly indicate that towards the end of his life Plato did express doubts about the traditional geocentric view during oral discussions in his school, the most that can safely be deduced from them is that the Philolaic scheme had its more or less serious defenders in the Academy of the late 340s and early 330s.[25] That, of course, has a direct bearing on the identification of Aristotle's 'others'. Even the Aristotelian commentators' bewilderment about the identity of the people in question is now easy to understand: the severe lack of information available to them about ideas floated in the early Academy—and about Speusippus' philosophy in particular—is well known.[26]

This is not quite the last word in the matter, however; more definite confirmation of the identity of the people whom Aristotle has in mind in this section of *On the Heavens* is also close to hand. First, at the very end of the *Critias* Plato himself describes the centre of the universe as 'the most honourable' spot: precisely the view that lies at the heart of Aristotle's parenthesis.[27] Second, and even more significantly, we also have another passage which does not just assume—as in the *Critias*—that the centre of the universe is a place of special honour, but makes this theory a specific point of doctrine. The writer in question this time is not Plato, but Speusippus; our authority for what Speusippus says, Theophrastus. The parallel with the passage in Aristotle's *On the Heavens*, even

---

[24] Cf. e.g. Burkert 63–6, 71; Philip 10–12.
[25] So for example Heath 186–7; Burkert 327 n. 16.
[26] Tarán (1981), pp. xxii, 3, 225–6, 406–8.
[27] *Crit.* 121c2–3. The passage is far from clear as to what actually occupies this central spot; but Cherniss (1944), 564, is no doubt right in concluding that Plato's perspective, if he had any conscious perspective at all, was geocentric. See further below, Ch. 14 with n. 17.

down to the exact wording, is unmistakable.[28] 'How', it has been asked, 'does it happen that a doctrine which Aristotle attributes to the Pythagoreans ... is attributed to Speusippus by Theophrastus?'[29] The answer, very simply, is what we saw to begin with: Aristotle does not attribute the doctrine to the Pythagoreans at all. He ascribes it to 'others' who appear to have been Platonists—and primarily Speusippus himself, Platonic authority and reinterpreter of Pythagorean doctrine.[30]

So far we have come to learn something about ideas and interpretations circulating among the early Platonists; but about the motivations that led to the Philolaic scheme being formulated in the first place we are none the wiser. However, there is one other line of approach to the matter which would seem to be more promising—although, as we will see, it soon peters out for lack of enough solid evidence.

This approach begins from the observation that the peculiar cosmological scheme of a central fire at the heart of the universe starts to become easier to understand as soon as we posit the existence of an earlier, geocentric system preceding it: a system where the central fire still occupies the middle of the universe but is at the same time contained inside the earth itself. Support for the existence of such an earlier cosmology has been found in Empedocles. He, as we have seen, posited the existence of fire in the centre of the earth and also had close affiliations with Pythagoreans in the West prior to the time of Philolaus. If this appeal to Empedocles were right, that would mean being able to trace the Pythagorean—as well as the Empedoclean—idea of a central fire back to its roots in the

[28] Theophr. *Met.* 11ª22–4 (cf. 11ª12) = Speus. fr. 83. The parallel with the *De caelo* is noted e.g. by Philip, 118 n. 4. Compare esp. Theophrastus' τὴν τοῦ μέσου χώραν, *Met.* 11ª24, with *De caelo* 293ª28, and cf. also Plut. *Qu. Plat.* 1006c τὴν μέσην χώραν τοῦ παντός. The word for 'honourable' (τίμιος) is identical in both Aristotle and Speusippus as well as in Plato's *Critias*, and recurs again in Plut. *Numa* 11 (above, n. 20). Cherniss's view—(1944), 559; Tarán (1981), 447–9—that the Speusippus fragment refers not to cosmology but to the ethical doctrine of the 'mean' is refuted not only by the parallels in the *De caelo* and Plutarch but also by the context in Theophrastus: οἱ περὶ τῆς ὅλης οὐσίας λέγοντες (*Met.* 11ª22–3) can only mean 'those who speak about the nature of the universe'. M. van Raalte, *Theophrastus: Metaphysics* (Leiden, 1993), 560–1, notes this last point but is otherwise unhelpful.

[29] Philip, loc. cit.

[30] The Aristotle passage is correctly understood e.g. by H. Bonitz in the Berlin *Index Aristotelicus* 599ᵇ48–9 ('Platonici'); Heath 186–7; Burkert 70 n. 115 ('Platonists').

volcanic phenomena of Sicily and southern Italy.[31] Following this line of approach one arrives at the possibility that an earlier cosmological theory of a central fire inside the earth was reformulated by someone—presumably Philolaus himself—who projected this central fire out of the earth and into the surrounding heavens in much the same way that, in Pythagoreanism before as well as after Plato, we find traditional features of the underworld being projected from their original location into their new domain in the skies.[32] In fact, as we will see later, the comparison between central fire and underworld turns out to be more than just an analogy.

To suppose that the central fire which we encounter in the Philolaic system first belonged inside the earth definitely has probability on its side. It has been disputed whether Pythagoreans ever agreed with Empedocles in positing the existence of fire inside the earth;[33] however, through what we have learned from the *Phaedo* myth it has become clear that the idea of vast fires at the centre of the earth was held not just by Empedocles but also, indeed, by western Pythagoreans in the late fifth century BC.[34] Further than that it seems impossible to go. This is not to forget the evidence of a much-discussed passage in one of our commentators on Aristotle: Simplicius. There, immediately after citing Aristotle's account of the Philolaic scheme, Simplicius goes on to state that 'the more genuine' Pythagoreans held to the same basic idea and terminology of a central fire while maintaining a geocentric view with the central fire occupying the core of the earth. Proclus and Damascius, too, speak of the Pythagorean central fire as burning in the middle of the earth.[35] However, Burkert has pointed out that what Simplicius tells us here about the 'more genuine' Pythagoreans does not rest on the authority of Aristotle; on the contrary, it is self-consciously worded as a deliberate modification of the

---

[31] So e.g. C. Plésent, *Le Culex* (Paris, 1910), 158 and n. 2; H. Richardson, *CQ* 20 (1926), 120; Guthrie, i. 292.

[32] Cumont 189–218 and (1942), 182–8; Buffière 117–22, 446–7, 490–9; Kingsley (1994a) with nn. 21, 46, 51; above, Ch. 4.

[33] In a memorable passage W. Wiersma even doubted the existence of fire inside the earth according to Empedocles: *Mnemosyne*³, 10 (1942), 32.

[34] Chs. 7–9. The relevance of the *Phaedo* eschatology to Pythagorean ideas about a central fire was already hinted at by Cumont, 225 with nn. 2–3.

[35] Simpl. *Cael.* 512.9–20; Procl. *Tim.* iii. 143.24–144.8, Dam. *Phd.* 1.534–6.

Philolaic scheme which Aristotle himself describes. Burkert has also drawn attention to the fact that writers such as Proclus or Damascius had a vested interest in modifying earlier Pythagorean ideas so as to harmonize them with their own view of a geocentric universe; and he concluded that the geocentric version of the central fire theory cannot possibly be older than the Philolaic scheme but must be post-Aristotelian.[36] And yet this, too, is to go beyond the evidence at our disposal. In discussing the chronology of Pythagorean cosmological ideas it is essential to bear in mind that there does not ever appear to have been such a thing as one orthodox position to which all Pythagoreans adhered. Particularly during the fifth century BC, the rule seems to have been a plurality of different—and often mutually exclusive—interpretations of certain fundamental cosmological topics or themes: interpretations which for all their differences coexisted side by side.[37] That makes it very dangerous to speak of a straightforward genealogy of 'older' and 'younger' Pythagorean ideas. In the case of Simplicius' report, or Proclus' or Damascius', there is no reason why what they say should not be post-Aristotelian reformulations of theories which in themselves antedate Aristotle, Plato, and also Philolaus; to deny this would be to subscribe to a hopelessly over-simplistic view of the history of Pythagoreanism. And then there is the point already noted a number of times in the last few chapters: that Pythagorean ideas cited by Neoplatonists and normally dismissed as 'late' can turn out, on closer examination, to be much earlier in origin than has been assumed. In short, the overall probability must remain that in origin the Pythagorean central fire did occupy the middle of the earth. However, the evidence produced so far is not strong enough to allow us to be more specific; and certainly it is not strong enough to allow us to pinpoint the reason why this central fire should have been projected out of the earth and into the heavens. The nuts and bolts of the Philolaic scheme, its fundamental whys and wherefores, remain as great a mystery as ever.

This is as far as discussions and examinations of the Philolaic system have reached: basically, nowhere. The poverty of the

---

[36] Burkert 232–3; cf. also Zeller, i/1. 529 and n. 3.
[37] See above, Ch. 8, with the refs. in n. 15; and Burkert himself, 321–2.

results obtained in spite of the immense literature on the subject, plus the apparent insuperability of the problems involved, is so striking that it was even claimed not so long ago 'that perhaps new insights are more to be feared than hoped'.[38] In fact, however, the answer to the problems has been staring us in the face all along. It was simply a matter of not looking in the right direction.

Writing at Alexandria in the third century AD on the symbolic properties of the numbers up to ten, Anatolius starts off with a series of comments on the property of the monad, or number one. He mentions that, in spatial terms, the Pythagoreans equated the monad with the centre while in terms of time they equated it with the present moment, and so on. Then he adds:

They also said that at the centre of the four elements there lies a certain unitary fiery cube, and that Homer was also aware of its central location when he said

'As far below Hades as heaven is from the earth'.

It would seem that the Pythagoreans were followed on this point by Empedocles, Parmenides, and indeed the vast majority of the ancient sages, in the sense that they all describe the monadic nature as occupying the central position like a hearth and keeping the same place thanks to its equilibrium (διὰ τὸ ἰσόρροπον). And Euripides too, speaking as a disciple of Anaxagoras, refers to the earth in the following way:

'Those who are wise among mortals consider you the hearth.'[39]

The sequence of thought in this passage may seem woolly to us. Anatolius passes nonchalantly from a Pythagorean theory of fire as occupying the central position in the cosmos, via Empedocles and Parmenides, to Euripides' view of the centre of the universe as occupied by earth. But it is essential to appreciate the nature and purpose of Anatolius' work rather than judge it for what it is not. He was not writing a continuous piece of logically consistent argument, but presenting an anthology of philosophical theories about each of the numbers he discusses; it is left to us to extract from what he says the

[38] Philip 114.
[39] Anatolius, *De decade*, ed. J.-L. Heiberg, *Anatolius sur les dix premiers nombres* (Mâcon, 1901), 6.3–13 = *Theologumena arithmeticae*, 6.11–20 de Falco.

information relevant to our needs. To criticize his account as 'superficial and confused'[40] is to misunderstand the genre in which he was writing; and this misunderstanding becomes particularly dangerous when it results in attempting to impose a consistency on Anatolius' evidence which it simply does not possess. Repeatedly the mistake has been made of assuming that the Pythagoreans, Empedocles, Parmenides, and Euripides are all being cited together as sharing the same belief in an earth with a fiery core occupying the central position in the universe.[41] In fact it is clear from Anatolius' own words that for the Pythagoreans the central substance is fire ('at the centre of the four elements'), not earth, while for 'the majority of the ancient sages' the central substance is earth and not fire: what according to these sages 'keeps the same place thanks to its equilibrium' is plainly the central earth[42] and not some fire trapped inside it—which would be kept in the same place thanks to the containing earth, not thanks to some force of equilibrium. These discrepancies in doctrine were quite irrelevant to Anatolius, who was concerned to bring together as many views as he could about the existence of a single substance at the centre of the universe—regardless of whether that substance is earth or fire. In short there is no obstacle at all to accepting the simple and necessary conclusion that, alongside the traditional view of a geocentric universe, Anatolius has also introduced the peculiar Pythagorean theory of a central fire replacing the earth at the middle of the cosmos.

This naturally means that the ideas held by Empedocles, Parmenides, and Euripides cannot help us a great deal to understand the non-geocentric scheme which Anatolius so briefly refers to. And yet there is one other piece of information which obviously can: the quotation from Homer. Within the constraints of his literary medium Anatolius could hardly have given the quotation more prominence or significance, stating as he does that 'the Pythagoreans' actually cited the line of Homer in defence and explanation of their theory. However, no

[40] Guthrie, i. 292.
[41] So e.g. H. Richardson, *CQ* 20 (1926), 121; Guthrie, i. 293; Burkert 268 n. 139 ('The point of view is geocentric').
[42] Cf. ps.-Plut. *Placita* 3.15.7 = Parm. A44b; Arist. *Cael.* 295^{a}13–21 = Emp. A67, with Philo, *Prov.* 2.60, 86.26–30 = Kingsley (1993a), 53–5; Pl. *Phd.* 108e4–109a6; Furley 23–6.

modern scholar seems ever to have asked what the quotation is doing here and what its presence implies.[43] We only have to ask the question seriously for the answer to present itself. 'As far below Hades as heaven is from the earth' is Homer's way of locating and describing Tartarus.[44] So, if these words are applied—as they are in Anatolius—to the Pythagorean central fire, the inevitable implication is that this central fire *is being identified with Tartarus*. The two are interchangeable: Tartarus is the central fire, and the central fire is Tartarus.

One observation needs to be made here straight away: the identification of Tartarus and central fire which is to be inferred from Anatolius is far more than some unintelligent or arbitrary guess. As we have seen repeatedly, Greeks in the West associated Tartarus very closely with fire—the fire deep inside the earth. We find this association in Pindar, in the context of the geography of Campania and Sicily, and we find it in Plato's *Phaedo* in connection with Etna; while from the *Phaedo* we can safely deduce that it was well known to Pythagoreans in the West before Plato's time. The one difference in the Anatolius passage, which is also the difference in the Philolaic cosmological scheme, is that the central fire appears to burn no longer inside the earth but alone at the centre of the universe. It is this crucial difference which remains to be explained. And as we will now see, this difference is precisely what the Anatolius passage also allows us to understand.

Tartarus, according to Anatolius' quotation from Homer, is 'as far below Hades as heaven is from the earth'. That implies the following cosmic arrangement, starting from above: heaven, earth, Hades, and finally Tartarus. When we apply this arrangement to the Philolaic scheme, an interesting series of correspondences emerges. Letting the Homeric heaven and earth correspond to the Philolaic heaven and earth needs no justification. As for the Homeric Tartarus, we have already seen that it evidently corresponds to the central fire. And as for the Homeric Hades, above Tartarus but below the earth, only

---

[43] Delatte in his chapter on Pythagorean exegesis of Homer simply noted Anatolius' allusion in passing, but failed to draw any conclusions: (1915), 123. Guthrie, i. 292–3, omits the Homer quotation altogether when translating the Anatolius passage, clearly discounting it as mere poetic decoration.

[44] *Il.* 8.13–16.

one possible correspondence remains: the Philolaic counter-earth, above the central fire but below the earth. The correspondence between central fire and Tartarus makes, as we have seen, excellent sense; but it may seem less easy to make sense of the remaining correspondence between counter-earth and Hades. In fact, however, there is nothing difficult here at all. First, it will be noted that the most prominent single characteristic of the Philolaic counter-earth is its *invisibility*— also an essential characteristic of Hades.[45] Second, like Hades the counter-earth also has the peculiar feature of possessing inhabitants whom we are unable to see from our normal perspective on the surface of the earth.[46] And third, the very name 'counter-earth', *antichthōn*, plunges us even further into the substratum of mythical ideas underlying the seemingly 'scientific' façade of the Philolaic system. It cannot—as has sometimes been assumed—have been coined to describe a planetary body that acts as a counter-balance to our earth, for the simple reason that the body in question manifestly fails to perform any such role in the Philolaic scheme.[47] The name can, to some extent, be explained by the detail mentioned earlier: that the 'counter-earth' would seem to maintain a position 'over against' or 'opposite' our earth during its circuit around the central fire. Yet this explanation ultimately falls flat, and fails to satisfy or account sufficiently for how both the word and the idea itself came into being. On the other hand, though, in its literal sense of 'anti-earth' the word also evokes the idea of an earth in reverse, a kind of shadow-earth, a reflected or looking-glass earth which represents the Other World: the world of the dead. Burkert has well brought out this connotation of the term, tracing the motif of a 'reversed' or 'opposite' world of the dead in Greek mythology and emphasizing the relevance of this comparative material to our understanding of the Philolaic counter-earth; he is not the first to have been struck by the obvious correspondences between features of the counter-earth

---

[45] Invisibility of counter-earth: Arist. *Cael.* 293^{a}23–5, ^{b}22–3; Alex. *Met.* 40.28–41.1 = Arist. fr. 203; ps.-Plut. *Placita* 3.11.3 = Philolaus A17a; Simpl. *Cael.* 511.33–512.1. Invisibility of Hades: Ch. 4 with n. 43, Ch. 12 n. 79.
[46] Ps.-Plut., loc. cit.; Burkert 348.
[47] Burkert 348 n. 58; above, n. 3.

and the traditional Greek concept of Hades.[48] To these observations we can add one further detail. A Pythagorean variant of the Philolaic scheme, known to us from Alexandrian times, maintained a geocentric view of the universe by identifying the counter-earth not with some invisible body but with the moon; and yet the same sources that inform us about this variant also explain that the lunar counter-earth was identified with Hades.[49] The counter-earth may have become transposed, but where it moved the mythical Hades moved too.

Once again, as with the correspondence between Tartarus and central fire, Anatolius' Homer quotation appears mysteriously to hit the mark and, in the process, reveal an entire hidden dimension to the Philolaic system. However, this explanation of central fire as corresponding to Tartarus and of counter-earth as corresponding to Hades may seem all just a little too neat, too perfect. So, before going any further, some definite confirmation of it will be in order; and that means returning to Aristotle.

From our sources—which in this case means evidence either provided by or filtered via Aristotle—it emerges that the Philolaic system we have been considering gave two alternative names to the fire at the centre of the universe. One of them means either 'prison of Zeus' or 'Zeus' sentry-post' ($\Delta\iota\grave{o}s$ $\phi\upsilon\lambda\alpha\kappa\acute{\eta}$); the other, 'Zeus' defence-tower' ($Z\alpha\nu\grave{o}s$ $\pi\acute{\upsilon}\rho\gamma o s$).[50] To start with the title 'prison of Zeus' or 'Zeus' sentry-post': the question seems never to have been asked what it could have been intended to convey. The answer could hardly be simpler. The idea of Zeus having a sentry-post conveys nothing, but in

---

[48] Burkert 347–8; cf. e.g. Buffière 570. For the theme of the world of the dead as an inversion of this world add to Burkert's references Harva 347–9 (inverted world of the dead as 'another world' or 'another earth': cf. Ch. 8 with n. 12); see also Ch. 6 with n. 27, and J. Freccero's comments, *Dante: The Poetics of Conversion* (Cambridge, Mass., 1986), 180–5. For Huffman's mistaken view of the issues see above, n. 6 with refs.; Kingsley (1994*f*).

[49] Cf. esp. Plut. *De fac.* 944c (lunar 'counter-earth of Persephone'), with the further refs. collected by Cumont (1942), 184–8.

[50] Arist. *Cael.* 293ᵇ3, and fr. 204 = Simpl. *Cael.* 512.12–14; Calcid. 166.9 (*Iovis custos*); Procl. *Tim.* i. 199.3, ii. 106.21–3, iii. 141.11–12, 143.26, *Eucl.* 90.17–18; Dam. *Phd.* 1.535; Simpl., op. cit. 513.21, 26, 29, *Ph.* 1355.9. For the form $Z\alpha\nu\acute{o}s$ instead of $\Delta\iota\acute{o}s$ (Arist. fr. 204; Proclus and Damascius, locc. citt.; Simpl. *Ph.* 1355.9), see Burkert 222 n. 24 with further refs. The form recurs in Pythagorean tradition at Porph. *VP* 17.

Greek mythology Zeus' prison or guard-house (Διὸς φυλακή) is Tartarus. The crucial role of Tartarus as the well-guarded prison into which he throws his traditional enemies, the Titans, is emphasized at length in the text of the *Theogony*.[51] There his prison-guards or wardens are mentioned by name, and the Hesiodic expression 'Zeus' prison-guards' (φύλακες Διός) corresponds exactly to the Pythagorean expression 'Zeus' prison' (φυλακὴ Διός).[52] Pherecydes offers an alternative version, in which Zeus' prison is guarded by different mythical beings. But the prison, again, is Tartarus, and the verb he apparently used in this context for 'guarding' or 'keeping in prison' (φυλάσσουσι) is from the very same root as the nouns we find in both the Pythagorean and Hesiodic contexts.[53] Whichever mythical account we follow, there can be no doubting what the concept of 'Zeus' prison' will have meant for a fifth-century Pythagorean immersed in the imagery of Greek mythological tradition.

Then there is the alternative expression, 'Zeus' defence-tower'. Such a defence-tower (πύργος) is naturally to be understood as part of a defensive wall or enclosure; the Greek word in question is often used as a virtual synonym for the regular terms denoting a fortified enclosure or defensive wall (τεῖχος, ἕρκος).[54] Once again, there can be no doubting what is meant by the Pythagorean allusion to Zeus' fortified wall. The very same passage from the *Theogony* just mentioned describes in considerable detail the defensive wall as well as the famous

---

[51] Hes. *Th.* 715–35, 811–20.

[52] Gyges, Kottos, and Obriareus as φύλακες πιστοὶ Διός: *Th.* 735. That φύλακες here is to be taken in the sense of 'guardians' or 'prison-wardens' rather than in the weaker sense of 'defenders' or personal bodyguards must be considered clear from the context (cf. esp. 718 δεσμοῖσιν ἔδησαν, 726 ἕρκος, and 732–3), and is confirmed by the formal parallel between φύλακες at 735 and φυλάσσει (of Cerberus) at 769. The idea becomes a literary commonplace: cf. e.g. Virg. *Aen.* 6.395 *Tartareum custodem* (again of Cerberus), with R. G. Austin's note ad loc. on *in vincla*; Apollodorus, *Bibliotheca* 1.2.1 τὴν φρουροῦσαν αὐτῶν ... ἐν τῶι Ταρτάρωι φύλακας; *The Book of Thomas*, 142.32– 143.22 *tartarouchos* ... *shteko* = M. Krause and P. Labib, *Gnostische und hermetische Schriften* (Glückstadt, 1971), 99–101; Cumont 216.

[53] φυλάσσουσι δ' αὐτὴν (sc. τὴν ταρταρίην μοῖραν) θυγατέρες Βορέου Ἅρπυιαί τε καὶ Θύελλα· ἔνθα Ζεὺς ἐκβάλλει ...: Pherecydes, fr. 83 Schibli. Cf. Schibli 40 and n. 77, 100–1, 118–19, 137. For Pherecydes' links with Pythagoreanism, see West (1971), 2–4, 7, 77; Schibli 6–13, 20 n. 15, 104–9, 122–7, 131, plus (with specific regard to conceptions of Tartarus) 119 and n. 35.

[54] For πύργος and τεῖχος cf. e.g. *Il.* 7.436–7, 15.736–7, and for πύργος used in the loose sense of 'wall' rather than 'tower', *Od.* 6.262, Hesychius, s.v.

Central Fire

gates which safely enclose Zeus' fortified prison-area: the area known as Tartarus.[55] Each of the two Pythagorean expressions is simply an alternative way of alluding to the same mythical region. And finally, to bring all the threads together, we can note that the Homeric line which Anatolius quotes for its relevance to the Pythagorean central fire forms the climax to Zeus' famous boast about his ability to banish anyone he wants down to the impenetrable prison of Tartarus.[56] In short, the Anatolius quotation no longer stands alone as evidence for the mythical ideas underlying the Philolaic system, but links up directly with the names for the central fire preserved by Aristotle. The two sources—Aristotle and Anatolius—corroborate, confirm, and clarify each other. And from wherever we start we arrive at the same conclusion: the Philolaic central fire is none other than the Tartarus of Greek mythology.

It is time to look at the implications of what we have seen. The explanatory power of the Homeric line so informatively cited by Anatolius is no accident or coincidence. On the contrary, the verse provides the key to understanding the mythical background to the Philolaic system and how it came into being. That means in turn that the appeal to Homer cannot simply have been thought up by some poetically minded commentator on the original system, but must have played an essential role in the evolution of the scheme itself. This may seem a surprising conclusion, and yet it becomes anything but surprising when we reconstruct the genesis of the system from the evidence at our disposal.

For a consciousness influenced enough—as Philolaus certainly was—by the recent influx of information about Babylonian planetary learning to feel the need for radically revising traditional Greek ideas about cosmology,[57] and for a consciousness prepared to view the Homeric poems as inspired texts which contain passages capable of revealing a profound significance when interpreted with due care and respect, the

---

[55] *Th.* 726 ἕρκος, 733 τεῖχος.
[56] *Il.* 8.10–17. Cf. Pherecydes, fr. 83 Schibli φυλάσσουσι... ἔνθα Ζεὺς ἐκβάλλει θεῶν ὅταν τις ἐξυβρίσηι (above, n. 53).
[57] For Philolaus and Babylonian doctrine see above, n. 5; and on the part played by oriental ideas in contributing to the projection of the Greek underworld into the heavens, see also Frank 197–8, Burkert 357–60.

Iliad's description of Tartarus was capable of speaking volumes. Zeus' mention of it as being as far below Hades as our earth is below heaven was straightforward and explicit enough to inspire—and authorize—a dramatic new vision of the universe: a 'planetarization' of the cosmos which did not even stop short at the earth itself. It is important to appreciate that there was no need to read any fanciful explanation into the passage. On the contrary, just as in the most striking examples of allegorical and metaphysical interpretation of scriptural texts—whether Greek, Jewish, Christian, or Muslim—it was simply a matter of taking the words in the passage perfectly literally, at their exact face value. Tartarus is described as lying a vast distance below Hades, while Hades itself is routinely described in Greek poetry as situated not in but *under* the earth.[58] The sheer immensity of the distance between the earth and Tartarus as portrayed by Homer or Hesiod meant it was almost inevitable that at some stage, and sooner rather than later, the attempt would be made to explain such huge stretches of space in terms of a planetary model. To conclude that the fires of Tartarus so familiar to Pythagoreans in the West must in fact be burning not just under the earth's surface but way beneath the earth itself could hardly be more logical, or more faithful to the words of the poets. And of course, the further the earth is removed from Tartarus the more impossible it becomes to situate both of them at the centre of the universe. Keeping the earth at the centre was no longer a real possibility. Logically that would require an orbiting Tartarus; but the very fact of its being in orbit around the earth would mean it could not always remain 'beneath' us—as according to the poets it certainly does.[59] It would also tend to require an orbiting Tartarus which, for the very same reason, ought sometimes to be visible to us; but it never is. On the other hand, to transfer the fires of Tartarus at the core of the earth to their new position, alone at the middle of the universe, was to preserve for them their fundamental role as the invisible, central 'hearth'—

---

[58] Ch. 4 with nn. 41–2. In its less complex cosmic geography, the *Theogony* simply omits Hades and has Tartarus lie as far below the earth as the earth is below the heavens (720–5, cf. 727–45).

[59] ἔνερθ': *Il.* 8.16, *Theog.* 720–5, cf. 740–5.

the source not only of destruction but also of life and creation.[60] And inevitably, the more distant the earth becomes from that central, terrible but life-giving fire, the more it is reduced in importance to the role of a mere orbiting planet. The reasoning is ruthlessly simple, the vision that inspired it even simpler. Viewed from this perspective the Philolaic system as a whole can be defined as a product of the acute orientalizing of Homer, in the sense that a passage from the *Iliad* provides the opportunity for pushing to its logical extreme a planetary awareness inherited from Babylonia. The first gesture at displacing the earth from its position at the centre of the universe was, ultimately and indirectly, the result of eastern influence; and yet it was the result of eastern influence modified by western ways of thought. More specifically, it was the result of oriental influence enlivened by a strange gift of visionary imagination which took its point of departure not in abstract theorizing but from a verse in Homer.

This imaginative aspect of the Philolaic system needs no further comment here; the oriental contribution starts to become understandable as a fuller picture continues to emerge of the extent to which Babylonian influence shaped Greek astronomy and cosmology during the late fifth and fourth centuries BC;[61] but the significance of the role played by the poetry of Homer deserves a few extra words. It is no exaggeration to say that the line from the *Iliad* appears to have been the single most important factor in the genesis of the scheme as a whole. That may seem remarkable: to find a philosopher not just turning to the authority of Homer for confirmation of an idea, but actually deriving a major aspect of his cosmological system from a passage in the *Iliad*. And yet this simply gives us a taste of how serious writers were when, both before and after Philolaus' time, they kept insisting that in Greek society Homer

[60] With the role of fire inside the earth according to Empedocles (Ch. 6) cf. Stobaeus' description of the central fire in the Philolaic system (i. 196.18–20 = DK i. 403.14–15) and the descriptions of the central fire interpreted geocentrically in Damascius (*Ph.* 1.534–5) and Simplicius (*Cael.* 512.10–12). Central fire as 'hearth': Philolaus A16, A17, B7; Alex. *Met.* 38.23 and 40.30–1 = Arist. fr. 203; Dam. *Phd.* 1.535–6; Simpl. *Ph.* 1355.8; Boeckh 95.
[61] For an assessment of certain aspects of this influence see A. C. Bowen and B. R. Goldstein in E. Leichty *et al.* (eds.), *A Scientific Humanist: Studies in Memory of Abraham Sachs* (Philadelphia, 1988), 39–81; also Kingsley (1990), (1992), 345, (1995c).

was the authority 'from whom all men have learned from the beginning'.[62] We have already seen, from the example of Empedocles, how extensive and plastic a role the Homeric poems played in shaping not just the medium but also the message of other fifth-century philosophers. As for Pythagoreans, the view sometimes expressed in the past that before Plato's time they were totally hostile to Homer and refused to have anything to do with interpreting his poems has become indefensible; evidence that pre-Platonic Pythagoreans concerned themselves with the detailed interpretation of Homer is clearly discernible, even though by the very nature of our sources it is thin on the ground.[63]

Here we can be more specific. The myth at the end of the *Phaedo* appeals in a cosmological context to the very same passage from the *Iliad* which is cited by Anatolius. As we saw earlier, the *Phaedo* myth in its entirety, even down to the smallest of details, derives from a Pythagorean source.[64] The fact that the Philolaic system and the *Phaedo* give two divergent interpretations of the same Homeric passage simply testifies yet again to the lack of any rigid orthodoxy in pre-Platonic Pythagoreanism, and emphasizes the extent to which Pythagoreans evidently felt free to follow their intuition and develop their own systems of understanding—especially when it came to ways of interpreting poetry. And, finally, we come to Philolaus himself. Strong suspicions that he was in fact involved in the allegorical interpretation of Homer have already been raised;[65] while more recently, what little evidence survives has been used to support the claim that although Philolaus' 'use of mythology as a basis for physical cosmology' would not seem specifically to 'involve the reinterpretation of Homer', it did create 'a suitable background for such interpretation'.[66] We can now go a step further. Philolaus' use of mythology as a basis for physical

[62] Xenophanes, DK 21 B10; cf. e.g. Pl. *Rep.* 606e, Detienne (1962), 18, R. Pfeiffer, *History of Classical Scholarship*, i (Oxford, 1968), 8–12, 42–5, 69–73.

[63] Detienne (1962); Burkert 140–1; Richardson (1975), 74–6; R. Lamberton, *Homer the Theologian* (Berkeley, Calif., 1986), 31–43.

[64] *Phd.* 112a3 = *Il.* 8.14; Chs. 7–12. It is also worth noting that the lines about the 'golden chain' immediately following our passage in the *Iliad* (8.18–27) had become a routine subject for philosophical interpretation by the end of the 5th c. BC. Cf. Eur. *Or.* 982–4 and Pl. *Tht.* 153c-d with Richardson's comments, (1975), 70.

[65] Detienne (1962), 80 and n. 3.

[66] Richardson (1975), 75–6.

cosmology plainly *did* involve the interpretation of Homer—and to such an extent that without the text of Homer the cosmology would never have assumed its present shape at all. The Philolaic evidence rounds out and, for the first time, places on a firm footing our understanding of the textual use of Homer in Pythagorean circles before the time of Plato. While Philolaus' interpretation of Homer is remarkable enough in itself, it also sheds a clear light on the earlier stratum of ideas from which his process of reinterpretation took its point of departure. Behind the Philolaic scheme lies the basic idea of Tartarus as a vast mass of fire. But this association of Tartarus with fire is not to be found in Homer; and, far from being a commonplace in ancient literature, it points us to a very specific part of the Greek world. Before Philolaus' time Tartarus is associated with fire by Pindar, and afterwards by Plato—both times in the context of the mythical geography of Sicily and southern Italy. Bearing in mind that according to Greeks of the fifth century any distinction between Tartarus and Hades was more or less optional, the Sicilian background to Empedocles' association of fire and Hades also points in precisely the same direction.[67] Centuries after Philolaus, the idea of a fiery hell or Gehenna came to have a much wider circulation in Jewish, Christian, and Islamic literature—to a considerable degree as a result of the influence exerted by Italian and Sicilian Pythagoreanism.[68] In Philolaus' time, however, such a broad literary base did not yet exist. In short, the original background or initial starting-point for his own interpretation of Tartarus as the central fire around which our earth revolves was, undoubtedly, the volcanic fires of his home in southern Italy.[69] When Proclus and Damascius insisted on interpreting the Pythagorean central fire as a name for the fires

[67] Ch. 6 with the refs. in nn. 9–14.
[68] Cf. Dieterich 196–202 and (1891), 35–6; Kedar 997; Ch. 10 n. 26; and for the influence of Greek ideas of the other world on Judaeo-Christian literature see also A. M. Kropp, *Ausgewählte koptische Zaubertexte*, iii (Brussels, 1930), 89–90; E. Peterson, *Vigiliae Christianae*, 9 (1955), 1–20. On the significant but limited role of fire in the Egyptian underworld see Zandee 24–5, 133–46, 320–3. Zoroastrian ideas are often cited in this context; but the role played by fire in the Zoroastrian underworld was almost nonexistent—cf. *Ardā Vīrāz Nāmag* 55.1 and 3, ed. P. Gignoux (Paris, 1984), 101, 192—and the purifying fire that will flow into hell at the end of time is by no means at home there: *Greater Bundahišn* 34.31, M. Boyce, *A History of Zoroastrianism*, i² (Leiden, 1989), 243–4.
[69] For Italy as Philolaus' base see Burkert 228 and n. 48; Ch. 11 with n. 48.

of Tartarus burning deep inside the core of the earth, they were simply restoring the imagery of the Philolaic scheme to its original context. And when Simplicius insisted on calling the adherents of this same view the 'more genuine' Pythagoreans he was quite right, in the sense that the view in question was the original one from which Philolaus so dramatically and ingeniously departed.[70] Those who have argued for a geocentric cosmology as preceding the Philolaic one were correct. With the mythical key in our hands, understanding the chronological sequences involved could not be easier.

[70] Refs. above, n. 35.

14
A History of Errors
*

In the last chapter we saw how the Philolaic system needs to be approached and understood. However, something also deserves to be said about the misunderstanding of it. As one might expect, the problems begin in antiquity.

In addition to 'prison of Zeus' and 'Zeus' defence-tower' we find two other names for the central fire mentioned—both of them just once—in the ancient sources: 'home of Zeus' (Διὸς οἶκος) and 'Zeus' throne' (Διὸς θρόνος).[1] In modern times, the tendency when discussing the Philolaic system has been to group all four terms together indiscriminately: Zeus' home, prison (always mistranslated as 'watch-post'), throne, and defence-tower are so many mythical ways of saying the same thing.[2] But from what we have seen, this is far from being the case. Zeus' prison is precisely that: the secure gaol into which he throws his enemies and those who threaten him. It is not his home, let alone where he has his throne. In the same way, Zeus never lived in or ruled from Tartarus; on the contrary, his home and throne were up high—as far as possible from Tartarus—on Olympus. To call the Pythagorean central fire the throne or home of Zeus is incomprehensible from the standpoint of Greek mythology; and, what is most important, it is contrary to the logic and inspiration of the Philolaic system. The two terms betray a reinterpretation of the system which has completely failed to grasp its mythical resonance and subtlety. In short, there is only one way of accounting for the

---
[1] 'Home of Zeus': Stob. i. 196.19 (Philolaus A16). 'Zeus' throne': Simpl. *Cael.* 512.14.
[2] So e.g. Burnet (1930), 298, Burkert 37 and n. 43. The one exception appears to be Boeckh (95–6; followed, with some confusion, by Huffman 396–7). Already in 1819 he saw that the name 'Zeus' home' did not have Aristotle's authority, and suggested that the more general or commonplace idea of a home had been substituted by some later writer for the originally more specific idea of a watch-post. Boeckh was not familiar with the reference to 'Zeus' throne' in Simplicius.

alarming discrepancy in the names given by our sources to the central fire: somewhere along the line the original Philolaic system has been sifted through the mental filter of some person, or persons, who entirely misunderstood the scheme on which they thought themselves qualified to comment.

Who could this person or persons have been? We already have enough information at our fingertips to be able to answer the question. As we saw earlier, the first generation of the Academy—Plato and Speusippus in particular—considered no position more worthy of honour than the centre, and argued that the central point in the universe must be occupied by the most honourable entity of all.[3] That certainly harmonizes with the idea that the central fire is the throne or home of the king of the gods. However, it does not harmonize at all with the idea that the centre of the universe is occupied by a blazing, hellish dungeon. In other words, 'home of Zeus' or 'Zeus' throne' would have been suspiciously appropriate terms for Speusippus to use in elaborating on the Philolaic system, and indeed would have been entirely understandable modifications of the original names if the aim was to bring the system itself into line with Platonizing views about the centre. It is also important to realize that there was nothing at all special about the terms. Both of them—throne of Zeus and home of Zeus—were stereotype expressions which were readily available to anyone who had such a purpose in mind.[4]

Fortunately, this does not need to remain a conjecture. For the information that the Pythagorean central fire was given the name 'Zeus' throne' we are indebted to a very instructive passage in Simplicius' commentary to Aristotle's *On the Heavens*. Simplicius starts by citing Aristotle as authority for the fact that the Pythagoreans called their central fire 'Zeus' defence-tower' and 'prison of Zeus'. But then he continues: 'and it is also called "Zeus' throne", *as others say*' (οἱ δὲ Διὸς θρόνον, ὡς ἄλλοι φασίν).[5] This is not the first time we have come across the view of 'others' apart from the Pythagoreans

---

[3] Pl. *Crit.* 121c2–3, Speus. fr. 83, Arist. *Cael.* 293ᵃ27–ᵇ1; Ch. 13 with nn. 27–8.
[4] For 'Zeus' throne' cf. e.g. Theocr. 7.93 (Ζανὸς θρόνος), Soph. *Antigone* 1041, *OC* 1267, Eur. *Cyclops* 579, *Helen* 241–2, *IT* 1271; for 'home of Zeus', Eur. *Hippolytus* 68 (Ζανὸς οἶκος), Emp. B142.1, above, Ch. 4 n. 12.
[5] Simpl. *Cael.* 512.12–14 = Arist. fr. 204.

cited in the context of the Philolaic system and the fire at the centre. Aristotle himself pointedly distinguishes between the teaching of the Pythagoreans about the position at the centre and what 'others' ($\H{\epsilon}\tau\epsilon\rho o\iota$) have to say, and in that case it is clear who the people in question were. They were the Academy in general, and Speusippus in particular.[6] That Aristotle's 'others' and Simplicius' are in fact the same is strikingly confirmed by the point just noted: the idea of the centre of the universe as 'honourable' and the idea of its being the throne or home of Zeus are both extrinsic to the Philolaic system, but are also very obviously compatible with each other. And we have one last piece of corroborating evidence regarding the identity of Simplicius' 'others'. The passage at the end of Plato's *Critias* which, in agreement with Speusippus and Aristotle's 'others', refers to the centre of the universe as 'most honourable' also describes it in the very same sentence as 'home of the gods'— and of Zeus in particular: 'And he gathered all the gods together in their most honourable home ($\epsilon\grave{\iota}\varsigma$ $\tau\grave{\eta}\nu$ $\tau\iota\mu\iota\omega\tau\acute{\alpha}\tau\eta\nu$ $\alpha\grave{\upsilon}\tau\hat{\omega}\nu$ $o\H{\iota}\kappa\eta\sigma\iota\nu$) which stands steady at the centre of the universe.'[7] Here we have the notion of the centre of the universe as 'most honourable' presented side by side, and inextricably intertwined, with the notion of it as home of Zeus and the gods. Little doubt can remain as to the source of that peculiar reinterpretation of the Philolaic system which was to prove so successful in colouring and confusing our ideas about what the system originally meant.

To recapitulate: a cosmological scheme rich in mythical nuances and resonances has—no doubt with the best intentions—been covered over and replaced by a pious, ethically simplistic theory of the 'nobility' of the centre in which the original point of the scheme has been missed completely. There is little to be gained from avoiding some of the more obvious implications of this development. It is usual to assume, in line with an equally simplistic theory of historical evolution, that what Plato inherited from the Pythagoreans in the way of mythical material he invariably improved on. But this particular example suggests how much was probably lost during the transfer of myth and symbol from Pythagoreanism in Italy

[6] *Cael.* 293ª27–ᵇ1, discussed in Ch. 13.
[7] *Crit.* 121c2–3.

to the Academy in Athens; we already arrived at the same conclusion earlier on, when examining the myth at the end of the *Phaedo* (Ch. 9). Such a development was only to be expected. Certainly mythology continued to play an important role for Plato and the early Academy, yet it was a role that had become subordinated to other aims and interests. Mythical complexity and paradox had little chance of surviving in a climate where ambiguity was a quality that needed to be purged from ethics, and where interest in the intricacies of the cosmos had been made subservient to the attainment of a transcendental Good. As far as the possibilities for understanding the dynamics of the Philolaic system were concerned, with Aristotle the situation only changed from bad to worse. By his time the Academic reinterpretation of the system had already been sanctioned, sanctified. He does take the trouble, as we have seen, to distinguish between it and the ideas of the Pythagoreans themselves; but as soon as he turns to his inevitable 'refutation' of the scheme, historical distinctions go by the board as he attacks and dismisses it in its Academic dress.[8] This, too, was only to be expected. Apart from wanting to demonstrate his superiority to his predecessors, Aristotle had no real interest in or respect for the historical niceties of the situation. He was even less concerned than the Academics with tracing the Philolaic system back to its roots. All he cared about was to refute the scheme, and refutation was made all the easier by removing it from its historical background and context.

After Plato and Aristotle, the outlook for a real understanding of the Philolaic scheme was bleak. Aristotle's own report contained some vital clues; Anatolius' preserved the crucial line of Homer. However, to expect anyone in Hellenistic or post-Hellenistic times to make use of this information would be to demand of them the impossible. One of the greatest lessons of western intellectual culture had already been learned: argument is more important than appreciation, reinterpretation an easy substitute for understanding. For any real understanding of the Philolaic system, a grasp of its mythical dimension was absolutely vital; but, in the philosophical mainstream, mythology had either ceased to be an issue at all

---

[8] *Cael.* 293^{b}2 (τὸ κυριώτατον τοῦ παντός), 11–13 (τίμιον ... τιμιώτερον); see further below, with n. 11.

A History of Errors

or had degenerated into the superficial pastime of endlessly elaborating artificial new schemes and systems of correspondences. Proclus is a perfect example. He, like Damascius after him, introduces the Pythagorean idea of Zeus' defence-tower or prison into his discussion of the fires of Tartarus; and he even mentions in the same context the punishment meted out to Zeus' enemies, the Titans. But both he and Damascius were so absorbed in the intellectual game of juggling with as many mythological and metaphysical systems as possible that they were quite incapable of realizing anything so simple as the fact that here—in this connection between central fire, Tartarus, and the Titans—they possessed the key to understanding what the Pythagorean terms had originally meant.[9] What is more, Proclus had also accepted without any hesitation the Platonic interpretation of the centre as 'most honourable', which made any genuine understanding of the expressions even more impossible.[10] But, to be fair, it must be noted that Philolaus himself had already contributed to this general trend. By reinterpreting Homeric mythology, by transposing it into a new system of his own, he had only added to the ingenious confusion and set yet another precedent for what was to come. One side of having the freedom to be able to formulate theories and systems of one's own is undeniably positive in its implications; but the other side to it is the creation of a chaos of different schemes and mental structures in which no one is able to see things the same way as anyone else. Ironically, through his very creativity Philolaus had in a sense doomed himself to be misunderstood by western philosophical tradition in the centuries and millennia to come.

By far the easiest point of access to understanding the Philolaic system and its mythical dimension lay in the single expression 'prison of Zeus' or 'Zeus' guard-post' (Διὸς φυλακή). With this expression understood, the essential had been grasped; all the other details were secondary in importance. And here, too, we can see the extent of the damage inflicted, so rapidly and completely, by Greek philosophical tradition.

[9] Procl. *Tim.* iii. 143.26–144.9; Dam. *Phd.* 1.534–8. The perspective of both writers is geocentric (Ch. 13 with n. 35).
[10] πολυτίμητον, *Tim.* iii. 143.25–6.

Already with Aristotle the key had been lost—and with an almost malicious abandon. With the greatest of assurance he not only quotes the expression but also explains why it was used:

> because the supreme place in the universe (τὸ κυριώτατον τοῦ παντός) deserves to be guarded the most—and the centre is such a place—the Pythagoreans call the fire at the centre 'Zeus' guard-post'.

In other words, he has imported into his discussion and interpretation of the Pythagorean system the belief in the 'nobility' of the centre which a moment ago he had specifically attributed to others apart from the Pythagoreans.[11] Then, as if this were not enough in the way of careless exegesis, Aristotle goes on to dispose of the Pythagoreans using the weapon of mockery:

> there is really no need for them to get into a confused panic about the universe and call in a guard to protect its centre.[12]

But the confusion was Aristotle's—as, inevitably, it was to become the confusion of the Aristotelian commentators after him. So in a memorable passage Simplicius chases himself round in circles debating whether the Pythagorean expression should be taken to mean, as Aristotle implies, that the central fire is somehow watched over and protected or, on the contrary, that it somehow performs a protective role itself.[13] With the loss of the mythological framework, no amount of theorizing was capable of working the problem out.

While the idea that the central fire itself needs guarding because of its 'supremacy' goes back to Aristotle and, ultimately, beyond him to the Academy, an alternative interpretation of the name 'Zeus' guard-post' also springs eventually into view. And this one, too, points us back to the beginnings of the Academy. In the fourth century AD, Calcidius translated the Pythagorean name into Latin as *Iovis custos*, 'Zeus' custodian'. In the following century Proclus explained the original Greek expression with the utmost vagueness as referring to Zeus setting up his 'demiurgic watch-post' at the

---

[11] *Cael.* 293ᵇ1–4. He does the same thing even more overtly towards the end of his discussion (293ᵇ11–14).
[12] 293ᵇ8–10.
[13] *Cael.* 513.15–32, cf. 514.18–23.

centre of the universe.[14] Both Calcidius and Proclus are quite open about the fact that they are interpreting the Pythagorean name in the light of the passage in Plato's *Timaeus* where the earth, 'the first and most venerable of the gods created inside the heavens', is placed at the centre of the universe as 'custodian and creator of night and day' (φύλακα καὶ δημιουργὸν νυκτός τε καὶ ἡμέρας, 40c).

To point out that these Platonists tried to understand the Pythagorean expression in the light of the *Timaeus* is to state the obvious. But there would seem to be more to the matter than that. Plato's own choice of the word *phylax*—'guard' or 'custodian'—in referring to the earth at the centre of the universe is one of the least comprehensible details in the *Timaeus*; not surprisingly, it has attracted a considerable amount of attention from ancient and modern commentators. Yet it also bears a suspiciously close resemblance to the word *phylakē*, 'guard-post' or 'custody', used in the Pythagorean scheme to describe the fire at the centre of the universe.[15] The suspicion that this resemblance is more than an accident is confirmed when we turn from the *Timaeus* to the end of its sister dialogue, the *Critias*. There, as we have seen, the centre of the universe is described as 'the most honourable home of the gods'; but no less interesting is the detail which Plato adds next.

And he gathered all the gods together in their most honourable home which stands steady at the centre of the universe *and watches over* (καθορᾶι) *everything that belongs to the world of becoming.*[16]

This idea of the centre of the universe as a watch-post is particularly striking: usually in Greek the idea of the gods 'watching over' the world of mortals is associated with a view from high up in the heavens, not from the centre of the universe. On the other hand, the idea is strongly reminiscent of the passage in the *Timaeus* about the earth at the centre of the universe acting as the 'watcher' or 'custodian' of night and

---

[14] Calcid. 166.9; Procl. *Eucl.* 90.14–23, *Tim.* iii. 144.3, 8.
[15] In his commentary on the *Timaeus* passage Taylor (240) actually refers to the Pythagorean idea of the centre of the universe as being *Dios phylakē*, 'Zeus' watch-post', but without making the obvious connection with Plato's own use of the word *phylax*, 'watcher'.
[16] *Crit.* 121c2–4.

day—especially as there are other details which bind the two passages together.[17] This makes it difficult to avoid the conclusion that both passages are more or less experimental attempts at explaining how the central point in the universe can be described as a 'watcher' or 'watch-post'. But, as we saw earlier (Ch. 13), the *Critias* passage is inextricably linked up with the Platonic reinterpretation of Philolaus' cosmology. In short, the similarities between the idea of a 'custodian' or 'watcher' in the *Timaeus*, the idea of 'watching over' in the *Critias*, and the idea of a 'guard-post' in the Philolaic system are too close to be a coincidence.

We can also go a step further. One other strange feature of the *Critias* passage is its unhelpfulness about what the home of the gods at the centre of the universe actually is. It has been suggested that here in the *Critias* we have the Philolaic scheme of a central fire; but chronological considerations—along with the fact that in the *Critias*' twin dialogue, the *Timaeus*, the 'venerable' place at the centre of the cosmos is firmly occupied by our earth—make this difficult to accept.[18] And yet the immediate context of the *Critias* passage, along with the reference to the centre as 'home of the gods', creates equal difficulties for the view that Plato definitely intended this centre to be the earth.[19] The almost inevitable solution to this dilemma, made all the more inevitable by what we have already seen of the Pythagorean and Academic background to the passage in the *Critias*, is that Plato was working with mythological and cosmological themes which he had not fully succeeded in integrating into the framework of his own ideas. Whether this was the case or not, however, it would seem certain that the references in the *Timaeus* and *Critias* to the centre of the universe as custodian or watch-post betray the kind of efforts made in the early Academy at explaining the Pythagorean idea of the centre as Zeus' defence-tower or prison.

What this means is that the Platonizing interpretation of

---

[17] The connection between the emphasis on the earth as 'first and most venerable' in the *Timaeus* and the idea of the centre as 'most honourable' is well brought out by Simplicius, *Cael.* 515.7–13. Cf. also Procl. *Tim.* iii. 143.22–6; Cherniss (1944), 564.

[18] Cherniss, loc. cit., objecting to Frank's hypothesis, 207, 217–18.

[19] Frank 217.

A History of Errors

'Zeus' prison' as meaning 'Zeus' watch-post' which we find in Calcidius and Proclus in fact appears to go back to Plato himself. Here we cannot exclude the further possibility that it goes back even before Plato to his Pythagorean contact and associate, Archytas. We have already seen evidence of the role played by Archytas' school in reinterpreting—often quite radically—Pythagorean mythological ideas which were current in the generation of Philolaus; Archytas' role as immediate source for the Pythagorean ideas embodied by Plato in the *Timaeus* is more than a matter of speculation; and it is quite possible that the idea of the centre of a circle or sphere as its most 'honourable' place is related to Archytas' exaltation of the geometric properties of circle and sphere.[20] However, to push this fundamental error in interpretation back even to before Plato only serves to emphasize how total it was soon to become. With the endless speculation in the Greek philosophical schools about watching and watch-posts, protecting or being protected, the essential had been lost. By the time of Plato and Aristotle the doors of understanding were closed. By the time of Calcidius, Proclus, and Simplicius they would seem to have been well and truly locked.

And yet they were not. The basic understanding of the Philolaic scheme did survive—although in a form and in a line of tradition which modern scholarship appears to have completely overlooked.

The most obvious key to appreciating the dynamic of the Philolaic system lies, as we have seen, in the fact that its central fire corresponds to the traditional Tartarus of Greek mythology: the Tartarus into which Zeus hurls his enemies. We can add that no doubt for Philolaus, as certainly for Pythagoreans both before and after his time, Tartarus had become not just a prison for the Titans but also the place reserved for the souls of the human damned.[21] When early Christian Fathers came across the passage in the *Phaedo* about the punishment of souls

---

[20] For Archytas and his school see Chs. 8–9, 11, 12 with n. 42. Circle and sphere: Archytas A23a (cf. A14), Burkert 331–2 (cf. 68–9).

[21] Arist. *Analytica posteriora* 94^{b}32–4 (DK i. 462.38–9) with KRS 236–8, Schibli 119 n. 35; Ael. *VH* 4.17 (DK i. 463.6–7) with Zeller, i/1. 561 and n. 5, Burkert 134 n. 81, 141–7, 168–9; Pl. *Phd.* 111d2–112a5, 112d2–3, 113a5–114b6 with Chs. 7–12 above.

in Tartarus, they were hard put to explain how pagans could have anticipated the Christian doctrine of hell-fire.[22]

However, it was precisely this mythological and eschatological aspect of the Philolaic scheme which had already been lost by the time of Plato and Aristotle. The expression *Dios phylakē* was no longer understood as meaning 'Zeus' prison'; so, by the time that Calcidius came to translate it into Latin in the fourth century AD, the best he could do was understand it in the light of Plato's *Timaeus* as meaning *Iovis custos*, 'Zeus' watcher'. But the correct interpretation was, eventually, to return.

In the section of his *Summa theologica* on the punishment of the damned, St Thomas Aquinas raises the issue of the location of hell. As one piece of ancient evidence relevant to the issue, he turns to the passage about Pythagorean cosmology in Aristotle's *On the Heavens*.

Pythagoras located the place of punishment (*locum poenarum*) in a sphere of fire which he said existed at the very centre of the whole universe; and he called this region 'Zeus' prison' (*carcerem Iovis*), as is apparent from Aristotle. However, it is more in keeping with Scripture just to say that it is situated underneath the earth.[23]

Here Aquinas, unlike Calcidius, has not only translated the crucial expression correctly but has also grasped its eschatological connotation.

It might at first seem tempting to attribute this double achievement to Aquinas himself. In fact his own commentary to *On the Heavens* survives: in the passage in question he again translates the Greek name quite correctly as *carcer Iovis*, 'Zeus' prison'.[24] However, Aquinas based this commentary not directly on the Greek text of Aristotle but on the Latin version of it by William of Moerbeke; and indeed, William's version already contains the translation *carcer Iovis*, 'prison of Zeus'.[25]

We can go back still further, beyond William of Moerbeke. The first person we come to is the strange figure of Albertus

---

[22] Cf. esp. Arnobius, *Adversus nationes* 2.14; K. Lehrs, *Populäre Aufsätze aus dem Alterthum*² (Leipzig, 1875), 309–10.

[23] *Summa theologica*, Part 3, Supplement, Question 97, Article 7.

[24] *In Aristotelis libros de caelo et mundo, de generatione et corruptione, meteorologicorum expositio*, ed. R. M. Spiazzi (Turin, 1952), 242b, §§ 484–5.

[25] Ibid. 240b, § 342 (*Iovis carcerem nominant hanc habentem regionem ignem*).

Magnus: Aquinas' teacher, and the man largely responsible for initiating him into Aristotle as well as into Arabic philosophy and alchemical literature.[26] Albertus' own commentary to *On the Heavens* also survives, and the relevant passage reads as follows:

> And the Pythagoreans . . . called the centre of the world a prison (*carcerem*) which encloses equally on all sides.[27] And they said that this is the place of the punishments of hell (*locum poenarum inferni*), where those who need to be kept under guard as they suffer punishment down in the realm of gloom are held captive, as Pythagoras explains in his precepts. That is where they are held, because that is the place of fire where those who have been condemned by Zeus are made to burn.[28]

It is important to appreciate what Albertus has done here. He has managed to condense into a few lines the essence of the mythological ideas that lie at the heart of the Philolaic scheme of a central fire. All the veils of mystification and misunderstanding thrown over the subject by Plato and Platonists, Aristotle and Aristotelians, have been drawn back at once. Apart from getting the crucial word 'prison' correct, his account of it as the place where Zeus throws those whom he condemns to be punished is as accurate as if he had Zeus' speech in the *Iliad* right in front of him.

No less significant is one other detail in Albertus' overall discussion of the Pythagorean central fire. After repeating the little that Aristotle himself has to say about it at this particular point in *On the Heavens*, he deliberately digresses from the text of Aristotle so as to throw light on the Pythagorean idea by pointing to the phenomenon of volcanoes and hot springs. The way in which he refers to his fuller discussion of these phenomena in another work of his leaves no doubt that the idea of appealing to volcanoes to clarify this point in the text of

---

[26] On the relationship between Albertus and Aquinas see M.-L. von Franz, *Aurora consurgens* (London, 1966), 411–21, 429, with further refs.

[27] *Pythagorici . . . etiam medium mundi carcerem aequaliter undique claudentem vocaverunt*. For Albertus' 'equally' see Kingsley (1994e), 198 with n. 9.

[28] *. . . et hunc locum poenarum inferni esse dixerunt, in quo custodiuntur qui custodiendi sunt sub orbe tristi poenas luentes, ut inquit Pythagoras in legibus suis, eo quod ibi est locus ignis, in quo crematur hi qui a Iove sunt condemnati*: Albertus Magnus, *De caelo et mundo* 2.4.1 = *Opera omnia*, v/1, ed. P. Hossfeld (Münster/Westfalen, 1971), 180.59–65.

Aristotle was his own original contribution. That other work of his to which he refers still survives; and there, in the relevant section, he talks specifically about the volcanic phenomena on Sicily.[29] As we have seen, volcanic phenomena—and, in particular, the volcanic phenomena of Sicily and southern Italy—are precisely where we do need to look for the origins of the Pythagorean idea of a central fire. Albertus plainly used his intuition; and his intuition was correct.

This, too, deserves some comment. Veering off at right angles from the text of Aristotle to talk about hot springs and volcanoes, digressing from philosophical generalities to focus on specifics, was a technique of interpretation which Albertus consciously used and cultivated. As he explains elsewhere, it was a method of 'supplementing' (*supplere*) what Aristotle himself had skimmed over or simply omitted; and it gave him the opportunity to round out Aristotle's essentially abstract and generalizing philosophical treatment through his own more down-to-earth approach and far greater interest in particulars—an approach and interest which he clearly inherited not from Aristotle, but from the Arabs.[30] Stated in simple terms, it seems to have needed a certain approach and mentality—a type of interpretative and encyclopaedic skill which was brought to its height by a man such as Ibn Sīnā—to explain key aspects of early Greek philosophy which later Greek philosophers were unable to understand, and for that matter not even interested in understanding.

With Albertus Magnus' striking insight, and originality, we may seem to have gone back as far in time as we can. But that,

[29] Ibid. 2.4.1 (180.46–53 Hossfeld), referring to the *Liber de causis proprietatum elementorum et planetarum* 2.2.2–3 = *Opera omnia*, v/2, ed. Hossfeld (Münster/Westfalen, 1980), 96.81–99.63. On Albertus' often startling originality as a commentator on Aristotle see L. Thorndike, *A History of Magic and Experimental Science*, ii (London, 1923), 531–2. For the general nature of the *Liber de causis* cf. ibid. 535, 569, 577–8; and for the origins of Albertus' information about volcanoes and hot springs at this particular point in the work, Vodraska 38–43.
[30] Cf. Albertus Magnus, *Physica* 1.1.1 = *Opera omnia*, iv/1, ed. Hossfeld (Münster/Westfalen, 1987), 1.23–30, with Peters's comments (104–5 and n. 31) on Albertus' indebtedness here to Arab tradition—and especially to Ibn Sīnā. On Albertus' self-conscious departure from Aristotle in his 'desire for concrete, specific, detailed, accurate knowledge concerning everything in nature' see also Thorndike, op. cit. ii. 535–42.

too, is not the case. Albertus based his own commentary to Aristotle's *On the Heavens* mainly on the first translation of the work into Latin: the one produced, in the second half of the twelfth century, by William of Moerbeke's predecessor Gerard of Cremona. In Gerard's version of the crucial passage in Aristotle, the Pythagorean term *phylakē* is rendered as 'prison' (*carcer*); and, what is more, this translation is immediately followed by the explanatory statement, 'that is, the place of punishment'.[31]

Here, in a nutshell, we have the main ideas which were later to be developed in more detail by Albertus Magnus and then by Aquinas. The word originally used by Pythagoreans in the sense of 'prison' has been translated into Latin using the word which means just that; and the explanation of this prison as being a 'place of punishment' is already present. As for the Latin term that Gerard uses here (*locus poenae*), it was—just like the Greek equivalent from which it derives—a standard Christian expression for hell or Gehenna, which in Judaeo-Christian literature corresponds exactly to Tartarus.[32] In other words Gerard has not only hit on the fact that the central fire according to Philolaus is a prison, but has also succeeded in putting his finger on its broad mythical and eschatological connotations; and he has managed to do so in what is not even meant to be a commentary on Aristotle, but is supposed to be a simple translation. This deserves some explaining.

Gerard made his translation from the Arabic, not from the original Greek; and the Arabic version of Aristotle that he used was itself based on an earlier Syriac translation, now lost. As I have shown elsewhere, however, his explanation of the Pythagorean prison as 'place of punishment' has no earlier authority and is due simply to a misreading, by Gerard himself, of his Arabic text. His error was understandable enough: for Christians down to and beyond the Middle Ages, 'prison' was itself a synonym for hell—which was of course conceived of as full of fire. But it was an error none the less.[33]

---

[31] Ed. I. Opelt in Hossfeld's edition of Albertus Magnus' *De caelo et mundo*, 180.79 (*scilicet locum poenae*).
[32] For the expression see Kingsley (1994e), n. 5; for the relation between Tartarus and Gehenna, Dieterich 200–2 and (1891), 36–7.
[33] Kingsley (1994e), 196–202.

This mistake of Gerard's accounts for the way that Albertus Magnus and Thomas Aquinas came to treat the Pythagorean central fire as the place for punishment of souls after death; and yet it still leaves us with the fact that Gerard, Albertus, William of Moerbeke, and Aquinas all translated the Pythagorean *phylakē* correctly as 'prison'. This time Gerard was not the innovator. On the contrary, he simply translated into Latin the unambiguous word for 'prison' which he found in his Arabic text (*ḥabs*); and that, in turn, was no doubt just a translation of the equally unambiguous word for 'prison' in the earlier Syriac (*ḥbušya*). As for the original translator from Greek into Syriac, he could easily have chosen another word for *phylakē*: one which would have harmonized exactly with the sense of 'guard-post' offered by Aristotle himself in his own discussion of the term. Yet he plainly did not. Instead of paying attention to the context in Aristotle, he paid attention to the sense of 'prison' almost invariably attached to the word *phylakē* in early Christian literature—and, in particular, to the idea of hell as a fiery prison which not only influenced Gerard of Cremona but also happened to play a pronounced role in the literature of early Syriac Christianity.[34]

Gerard of Cremona made a mistake with his 'place of punishment': a mistake which proved fundamentally correct. The Syriac translator made no mistake at all in his translation of the crucial word for 'prison'—unless, that is, we consider it a mistake that he totally disregarded Aristotle's own misinterpretation of the word. In both cases, what Gerard and the Syriac translator did amounts to the same thing. It needed a translator or translators capable of departing radically from the learned Greek philosophical tradition, and of veering off at right angles from Aristotle himself, to get back to what the Philolaic system was really about.

However inclined we may be to dismiss these details as insignificant, there are some undeniable implications here for our broader understanding of the role played by Christian, Arab, and medieval writers in the transmission of Greek philosophy. Normally it is assumed that Arab and medieval writers may occasionally have preserved Greek material lost to

[34] Ibid. 202–4.

us elsewhere, but that these gains need to be offset against the misunderstandings and trivializations arising as a result of classical philosophy being poured into the alien mould of different interests, cultures, and languages. This approach—this simple, linear view of cultural corruption—is quite adequate if 'Greek philosophy' is equated either with Aristotelian philosophy or with Aristotle's interpretation of his predecessors (which amounts to the same thing). But when one begins to perceive the extent of the damage inflicted on the understanding and interpretation of the Presocratics by Aristotle and his contemporaries, then things become much more complex. Instead of Greek philosophy assuming the appearance of a dead-straight line which was bound to be refracted and distorted in the medium of Christian, Muslim, or medieval culture, another more rounded picture starts to emerge: a picture suggesting that the course of Greek philosophy is itself a kind of curve, and that it needed a further deviation to return—via the detour of Platonic and Aristotelian misunderstandings—closer to the original Presocratic point of view.

In our case, both Gerard and the Syriac translator were well equipped to produce an adjustment of this kind. First, it needed someone with a different approach and cast of mind—ideally someone with a different language and from a different culture—to cut through the almost endless spinning of abstract speculations by Aristotelians and Neoplatonists about the meaning of the Pythagorean central fire, the almost endless refinements in misinterpreting the expression 'Zeus' prison'. Second, as Christians, it was only natural for them to bring the theological and ultimately mythological dimension back into whatever they touched. With Plato this dimension had been purified; with Aristotle it had been removed altogether not just from the Philolaic cosmology but from the natural world as a whole; while with the late Neoplatonists there had been such a proliferation of levels of existence that it was impossible to say anything straightforward about anything at all. As Christians, both Gerard and the Syriac translator were quite capable of living with the hard edge of something that Plato and Aristotle had already become too ideologically refined to accept: the discomforting, uncompromisingly mythical vision of the centre of the universe as a prison. Paradoxically, Gerard's vision of the cosmos in the twelfth century was closer to Philolaus' than was Aristotle's. The labels 'Pythagorean' or 'Christian' were irrelevant; all that mattered was the mythology.

In this sense it is clearly no coincidence that Gerard of Cremona or the Syriac translator of *On the Heavens* managed to introduce modifications into the text of Aristotle which ended up making it reflect more accurately what Philolaus had once intended. Just how close the similarities in this case between Christian and Pythagorean mythical perspective actually were, emerges from two points in particular. First, the New Testament 'prison' as place of punishment was identified in the New Testament itself not only with Gehenna but, specifically, with Tartarus: just as was the case for Philolaus.[35] Second, this New Testament idea of a cosmic prison was considered *primarily* the place where the fallen angels are bound and *secondarily* a place of punishment for the souls of the damned, with no clear or necessary distinction between the two. Again, this is an exact parallel to how Tartarus was no doubt understood by Philolaus: as primarily the place where the Titans are bound and secondarily the place of punishment for the souls of the damned, with no sure or necessary distinction.[36] These incidental similarities are meaningful enough in themselves, but behind them lies a more fundamental piece of common ground. Both Philolaus on the one hand, and the Syriac translator and Gerard on the other, were ultimately indebted to the same mythological tradition. The Philolaic idea of Tartarus as a prison, *phylakē*, goes back via earlier Greek sources—in particular the episode of the binding and imprisonment of the Titans in the Hesiodic *Theogony*—to Babylonian mythological tradition of the second and early first millennium BC: to the theme of the 'bound' or 'imprisoned gods' (*ilū kamûtu*, *ilū ṣabtūtu*) who were thrown by Marduk into the 'prison' (*kišukku*) of the underworld.[37] As for the New Testament idea of hell or

---

[35] Cf. 2 Peter 2: 4 ταρταρώσας with Reicke 52–3, 67–8, 116.
[36] Ibid. 52–70, 90–1, and *passim*; above, with n. 21.
[37] T. G. Pinches, *Proceedings of the Society of Biblical Archaeology*, 30 (1908), 57–60, 77–83; *Enūma eliš* 4.111–127, 7.27 = R. Labat, *Le Poème babylonien de la création* (Paris, 1935), 130–3 with 132 n. 127, 164–5; E. F. Weidner, *Handbuch der babylonischen Astronomie*, i (Leipzig, 1915), 19.24; G. Meier, *Archiv für Orientforschung*, 14 (1941–4), 146.126; W. G. Lambert, ibid. 19 (1959–60), 117.23–30; *Chicago Assyrian Dictionary*, K 127b-128a and 465a, § 45a; cf. also Livingstone (1986), 135, 151–5, and, for the underworld as a

Tartarus as a prison, *phylakē*, this goes back via earlier Jewish sources—in particular the account of the binding and imprisonment of the fallen angels in the *Book of Enoch*—to the same Babylonian tradition.[38]

There is also another similarity here between the Pythagorean on the one hand and the Christian on the other. As we have seen, in Greek philosophical tradition the discomforting but mythically resonant idea of the centre of the universe as a prison was soon tidied up, made more respectable, through mistranslating the crucial word *phylakē* as 'watch-post': a step that appears already to have been taken by Plato and the first generation of the Academy. Exactly the same thing happened quite independently in Christian tradition, when Calvin took the single most important reference in the New Testament to hell as being a prison, *phylakē*, and mistranslated the word as 'watch-post'. Bo Reicke appropriately condemned Calvin for his 'vain spiritualizing',[39] and there is a certain sense in which it is profitable to see what can be learned from viewing Plato and the early Academy as the Protestants of Pythagorean tradition.

It is tempting to say that chance or coincidence played the decisive part in the way that first the Syriac translator of *On the Heavens* and then Gerard of Cremona so accurately interpreted the Pythagorean central fire; but they played a part only superficially. Ultimately, it was contact with the same mythological tradition on which Philolaus himself had depended that was to ensure the Syriac translator translated the crucial Pythagorean term just as he did—Aristotle notwithstanding. There was naturally no need for him to be aware of his religious and mythological conditioning; all that matters is that the conditioning worked. And of course his correct translation was to

---

prison where gods are bound, T. Frymer-Kensky's comments, *JAOS* 103 (1983), 131–9. For the oriental, and specifically Babylonian, background to the Titan episode in the *Theogony* see Weidner, op. cit. 19–20; H. G. Güterbock, *American Journal of Archaeology*, 52 (1948), 130–2; F. M. Cornford, *The Unwritten Philosophy* (Cambridge, 1950), 105–6, 112–14; F. Dornseiff, *Antike und alter Orient*[2] (Leipzig, 1959), 44–7, 63–5; M. L. West, *Gnomon*, 35 (1963), 11.

[38] Reicke 236–40, referring to the Pinches and *Enūma eliš* texts, locc. citt. For the relation between the NT idea of hell as a prison, *phylakē*, and *1 Enoch*, cf. Reicke 52–3, 66–70, 116–17, and *passim*; for the Babylonian cosmological background to *1 Enoch* see most recently S. Dalley, *JRAS*[3] 1 (1991), 13–16.

[39] Reicke 66 n. 2.

prove fundamental in creating the later tradition of understanding and interpretation. Without it, Gerard could never have expanded the reference to a prison into the additional mention of it as 'place of punishment'; Albertus Magnus would never have been able to piece the various fragments of the picture together; and Thomas Aquinas would never have had the opportunity to refer to the Philolaic system in his *Summa theologica* as an example of pre-Christian eschatology.

This tradition of understanding the Philolaic central fire did not stop with Aquinas. On the contrary, the sheer authority of the *Summa* made sure that over time it would reach a wider audience than ever. To mention one example: in 1714, at London, a book was published by Tobias Swinden called *An Enquiry into the Nature and Place of Hell*. In the light of our earlier analysis of the Philolaic scheme, some of his comments are quite remarkable. So, in considering the possible locations of hell, he points out:

If you run over all the Poets . . . you shall find that with one general Consent they placed *Hell*, not only κάτω *below*, but ὑπὸ χθονὸς quite *under the Earth*. And the Gods of *Hell*, according to their Fashion of having Gods for all Places, were stiled Ὑποχθόνιοι *subterraneous*, and *positi sub Terrâ Numina Mundi*, i.e. *The Deities of the World placed under the Earth*.

This is exactly the course of reasoning that Philolaus evidently pursued over two thousand years before Swinden ever put pen to paper. Swinden immediately goes on to cite as demonstration of his point the passage in the *Iliad* where Zeus talks of Tartarus as being as far below Hades as the earth is below the heavens, and the similar statement in Hesiod's *Theogony*: the very same passages on which Philolaus himself appears to have based his whole system.[40]

Finally, Swinden comes to Aquinas. First he makes the most of the fact that the Catholic Church, and notably Aquinas in his *Summa theologica*, admitted the severe difficulties involved in locating hell inside the earth: difficulties which made them concede that nobody can tell where hell really is 'excepting by

[40] *Enquiry*, 129–31; above, Ch. 13.

A History of Errors 213

express Revelation from the Spirit of *God*'. But then he continues:

And yet in the same place where *Aquinas* saith this, he telleth us that *Pythagoras* placed the Seat of Punishment in the Sphere of Fire, and that he also placed that in the middle, not of the Earth, but of the Universe, or whole World, and for this he quoteth *Aristotle de Coelo*.

Indeed *Aristotle* telleth us there, not of *Pythagoras* himself, but of certain *Pythagorick*, or *Italian Philosophers* that placed the Sphere of Fire there, and that they also called it *Jupiter's* Prison, οἳ Διὸς Φυλακὴν ὀνομάζουσιν τό τ' αὐτὴν ἔχον τὴν χώραν Πῦρ.

So fair an Hint as this might, methinks, have given some free philosophizing Christian, an occasion to take this Matter into his Consideration . . .[41]

It is, to say the least, ironic that all the essential elements for an understanding of the Philolaic scheme are present in a book written just before the dawn of modern scholarly endeavour: ironic, and yet not unexpected. Modern scholarship would hardly dream of learning anything fundamental about early Pythagorean cosmology from an eighteenth-century divine, or for that matter from Thomas Aquinas or Albertus Magnus, but prefers instead to drink without precaution from a tainted stream.

[41] *Enquiry*, 154–5.

III
MAGIC
*

15
The Magus

*

EMPEDOCLES' equation of Hades and fire has turned out to be a key capable of opening many doors. Ideas, themes, and philosophical tendencies which on the surface might seem unrelated or even divergent prove to be linked with each other at a deeper level; apparently unconnected doctrines turn out to derive from a common mythological source. And in tracing these links we repeatedly find ourselves drawn back into the orbit of ancient Pythagoreanism. Often in the modern literature it is emphasized how little we know about Pythagorean ideas before Plato's time. But on closer analysis one of the major reasons for this lack of knowledge proves to be, very simply, that we have not been looking in the right direction: into the literary, mythological, and geographical substratum underlying the supposedly 'philosophical' ideas.

However, the fact remains that the ground covered so far in clarifying Empedocles' idea of a fiery underworld has in a fundamental sense been limited. This is because the course which has been followed has, by and large, been to approach the idea from an impersonal or cosmological point of view. And yet the whole question of fire in the underworld also has a more personal, more immediately human dimension—a dimension plainly hinted at in the ancient literature about Empedocles but which, predictably, modern scholarship has not wanted to take seriously because it would seem to take us into the realms of fantasy and legend. We have already come far enough, however, in uncovering the mythological background to Empedocles' ideas for there to be no stopping half-way.

First of all we need to fetch some skeletons out of their cupboards and face up to a side of Empedocles which, ever since the days of Plato and Aristotle, has for the most part been assiduously ignored. There are some lines of Empedocles which Diels discreetly tucked away at the very end of his edition of the cosmological poem—in spite of the fact that they almost certainly belong right at the beginning.[1] There they will have provided a crucial indication of how Empedocles wanted the poem as a whole to be understood, and of what the person to whom he addresses the poem could expect to gain as a result of taking its contents to heart. The lines—fragment 111 in Diels's collection—are important enough to deserve translating in full.

> And all the remedies that exist as defence against sufferings and old age:
> These you will learn, because for you alone will I make all these things come true.
> And you'll stop the force of the tireless winds that chase over the earth
> And destroy the fields with their gusts and blasts;
> But then again, if you so wish, you'll stir up winds as requital.
> Out of a black rainstorm you'll create a timely drought
> For men, and out of a summer drought you'll create
> Tree-nurturing floods that will stream through the ether.[2]
> And you will fetch back from Hades the life-force of a man who has died.

The modern reactions to this passage are, for the most part, amusing. Diels was quick to admit that its content can be summed up in one word: magic—or as a later writer was to say,

---

[1] The formula 'you will learn, because . . .' (πεύσῃ ἐπεί . . ., B111.2) plainly echoes the same formula in B2.8-9 (ἐπεί . . . πεύσεαι), and should be placed in the same context. Diels's evident wish to make the fragment form part of an epilogue to the cosmological poem is also inconsistent with the basic fact that in Empedocles' time the convention of a poetic epilogue would seem not yet to have existed in the sense Diels required (van Groningen 70-82, 217-18; van der Ben 87 n. 71).

[2] Literally 'that will flow in the *aither*', τά τ' αἰθέρι ναιήσονται. The expression is unusual, and has given rise to problems both in antiquity (see the MS variants in Bollack, ii. 9) and more recently; but it is very simply explained as a humorous variant on the standard epic αἰθέρι ναίειν, 'to dwell in the *aither*' (*Il.* 2.412, Hes. *Op.* 18, etc., always at the line-end). On Empedocles' penchant for deliberate variations of this kind see Ch. 4. Bollack's alteration of ναιήσονται to ναιετάουσι (iii. 24-5) is not only unnecessary grammatically (cf. DK i. 342.13, apparatus), but also removes the graphic effect of the future tense: the water is not up in the *aither* all the time but *will* build up at the command of the magician, who is able to combine different elements just as he is able to separate them out. For the dative αἰθέρι without a preposition cf. e.g. D. B. Monro, *A Grammar of the Homeric Dialect*² (Oxford, 1891), 139; and for the identical corruption of αἰθέρι in some of the MSS to ἐν θέρει, Aesch. *Ag.* 6.

'pure magic'.³ But relegating it to the end of Empedocles' poem was not enough. For Diels any mention of magic in a work supposedly devoted to philosophy and science was unacceptable, so he chose to interpret the passage in a different way. According to him, the fragment as a whole 'suggests nothing more than what science promises to give its adepts today: information about the laws of nature which will enable one to become their master'.⁴ In other words, what Empedocles says is not to be taken literally; once allegorized, or interpreted metaphorically, it is no longer embarrassing. This type of interpretation has found its followers. So, for example, Empedocles' assurance that he will pass on his knowledge to 'you alone'—the 'you' in the Greek is singular and evidently refers to Pausanias, Empedocles' disciple—has recently been taken to indicate that here he is 'assuming a very limited audience capable of appreciating a complex philosophical argument'.⁵ The fact that the fragment itself neither contains nor implies anything even remotely resembling 'a complex philosophical argument', and that what it does say could hardly be less complex, less philosophical, and less argumentative, is obviously considered insignificant.

Faced with the same challenge of making Empedocles conform to our image of what an ancient philosopher should be, van Groningen found a simpler solution. Claiming that nowhere else in the surviving fragments of Empedocles do we find any comparable promises, he decided that the only 'sane' conclusion was to reject the entire passage as a forgery.⁶

As helping to characterize a chapter in the history of modern scholarship, these attempts at dealing with the Empedocles fragment are not only amusing but also revealing. More than anything else, they are reminiscent of the attempts made by Jewish and Christian scholars at explaining away embarrassing passages in the Bible by either bowdlerizing or allegorizing them. Efforts of this kind tell us a great deal about the prejudices and preoccupations of the scholars, but nothing at all about the original texts in question.

---

³ Diels 138; van Groningen (1956), 49.
⁴ Diels 140.
⁵ Wright 262. For Pausanias cf. Emp. B1; below, Ch. 16 n. 8.
⁶ van Groningen (1956), 47–61, esp. 49.

In fact it is quite clear that any attempt at denying the fragment its literal, obvious meaning is wrong. Elsewhere Empedocles does make equally outrageous claims—although with them, too, van Groningen attempted unsuccessfully to deny the fact. Empedocles claims he is immortal, that he has transcended the human condition, and is able through his mystical or occult powers to free men and women from their mortal sufferings; in these other passages there can be no doubt that he means what he says quite literally.[7] The external evidence is equally instructive. The Hippocratic corpus vividly reminds us of the need to understand Empedocles' promises literally when, in the name of the new medical rationalism, it lashes out at those 'magicians and purifiers ... who claim to know how to create storms and fine weather, rain and drought, and all the rest of their nonsense'.[8] There is also the specific report that one of Empedocles' pupils, Gorgias, 'had himself witnessed Empedocles performing magic'.[9] It is very possible that by Empedocles' 'magic' Gorgias himself had been referring primarily to his teacher's uncanny rhetorical powers and mastery of the spoken word;[10] but even if that were true, this reference of his is inseparable from Empedocles' own claims to magical powers and would no doubt have gained extra force for just that reason. What we are faced with is a

---

[7] Cf. esp. B112, also B113–14; and B129 with A. A. Long's comments, *CQ* 16 (1966), 259 n. 1. The 'well-healing utterance' (εὐηκέα βάξιν, B112.11) which the sick flock to him to hear is plainly no rational diagnosis but, as the context implies, an oracular insight into the state of the individual which Empedocles derives from prophetic inspiration: for the oracular connotations of the word βάξις cf. e.g. Aesch. *Pr.* 658–69; Soph. *Electra* 638, *Tr.* 87; Apollonius Rhodius 1.8, 2.767; Eus. *PE* 5.9.5. In other words, Empedocles was not approached by people *either* for prophetic oracles (μαντοσυνέων, B112.10) *or* for healing: his healing was itself prophetic and obtained by magical means, making him the perfect example of a *iatromantis* or 'inspired healer'. As for Empedocles' claim at the start of B112 that he is 'an immortal god ... honoured among all', van Groningen tried to take this away from him as well (op. cit. 51–2); but here too, he is wrong. Cf. Wilamowitz (1935), 479; Zuntz 189–91; S. Panagiotou, *Mnemosyne*[4], 36 (1983), 276–85.

[8] *De morbo sacro* 1.10, 29, cf. 31; for the relevance to Empedocles, Burkert 292–3 with n. 78, Lloyd (1979), 37.

[9] Satyrus *ap.* D.L. 8.59 (ὡς αὐτὸς παρείη τῶι Ἐμπεδοκλεῖ γοητεύοντι). Diels tried to argue the evidence away, but unsuccessfully: cf. Rostagni, i. 156 with n. 4; Burkert (1962), 48 and n. 60; Wright 10. For Gorgias' relation to Empedocles see Burnet (1930), 201 n. 2; Guthrie, ii. 135 and iii. 198, 269–70; G. B. Kerferd, *Siculorum Gymnasium*, 38 (1985), 595–605.

[10] Cf. Gorgias, DK 82 B11, §§ 10, 14; below, Ch. 16 n. 53.

tradition of Empedocles as a magician which does not begin either with Gorgias or with some later writer, but which begins with Empedocles himself. The circumstantial evidence merely serves to make any attempt at disposing of fragment 111 altogether indefensible. As for van Groningen's attempts to dismiss the fragment as un-Empedoclean on grounds of style, the result, again, is total failure.[11]

That leaves us with Empedocles' promise to his disciple to make his magical powers available to 'you alone' (μούνωι σοί). This, quite obviously, has nothing to do with assumptions about only 'a very limited audience' being capable of understanding his complex philosophical arguments. On the other hand, it has everything to do with the established tradition throughout the ancient world of transmitting esoteric and magical powers on a one-to-one basis from spiritual (as well as physical) 'father' to 'son'.[12] The closest parallels to Empedocles' assurance that he will 'make all these things come true for you alone'—and they are the only real parallels in the whole surviving body of Greek and Latin literature—occur in the world of Graeco-Egyptian mysticism: the magical papyri (especially the famous Paris papyrus), the Hermetica, and alchemical literature.[13]

---

[11] Diels's statement that 'every single word of the fragment establishes its authenticity' (140 n. 2) is exaggerated, but basically sound. For πεύσηι ἐπεί in B111.7 see above, n. 1. The emphatic ἀνθρώποις, 'for men', at the start of a line (B111.7) is typically Empedoclean: cf. B8.2, 17.26, 23.10, 112.8, 114.3, Bidez 165–6 and, for the implied message that the speaker is looking down on mortals from a superior position, Empedocles' explicit statement in B113. Van Groningen's dismissal of the fragment as un-Empedoclean because it contains 'only' two *hapax legomena* (op. cit. 60) is absurd; for refutation of his other stylistic arguments see Bollack, iii. 20–1, also Flashar 549 with C. J. de Vogel's accompanying remarks, ibid. 551.

[12] Dieterich (1891), 162–3, (1911), 11, (1923), 146–55; Norden 290–1; Reitzenstein (1927), 40–1 with 40 n. 1; Festugière (1967), 147–50 and (1950), 332–54; Grese 66–7, 86; Meuli, ii. 1124; Burkert 179–80 and (1982), 19 with n. 96; Faraone 116–17. Cf. also Bidez and Cumont, i. 267 s.v. 'Hérédité'; Kranz 67 (referring to B111); Dodds 303 n. 49; Burkert (1983), 118–19.

[13] With Emp. B111.2 (μούνωι σοὶ ἐγὼ κρανέω τάδε πάντα) cf. e.g. PGM IV.254–6, 'let this matter be transmitted to you alone (εἰς σὲ μόνον), and guarded by you without being imparted to anyone else', 476–85, I.192–4, XII.93–4, XXXVI.293–4 (σὲ μόνον . . . καὶ ὑπὸ μηδενός . . . εἰ μὴ ὑπὸ σοῦ μόνου); Delatte (1927), 413.11–12, 416.23, 417.20–1, 479.3–8 ('. . . keep this tradition to yourself, because through it I have been able to perform marvels in the cosmos'); BR ii. 30.7–8, 34.19–20 (μηδενὶ μεταδιδόναι εἰ μὴ μόνον . . .); the Latin *Asclepius* 1 (*praeter Hammona nullum vocassis alium* . . .). For Empedocles' own period the closest parallel is Soph. OC 1522–32.

Similarly, when viewed as a whole, fragment 111 displays a distinctive physiognomy. The magical content of the fragment is a clear indication as to how we are to understand the 'remedies', *pharmaka*, mentioned in the first line; but it also warns us against giving the word too specific a meaning. A primary reference is, no doubt, to magical remedies extracted from plants.[14] However, the word *pharmakon* also had the broader connotations of 'spell' or 'charm'; and here in Empedocles it is best understood as referring not just to the remedies themselves but also, implicitly, to the incantations (ἐπωιδαί) which were recited during the gathering of the plants and during their preparation.[15] We find the best examples of such spells in the magical papyri—and particularly in the famous Paris papyrus.[16] This is the second time the papyrus provides the closest parallels to what we find in Empedocles; and as we will see later, this is no coincidence.

Empedocles' claim that his remedies can act as a 'defence' or 'safeguard' against 'sufferings and old age' (κακῶν καὶ γήραος ἄλκαρ) is, in terms of the traditional Greek religion we are most accustomed to, scandalous. Apart from death itself, nothing was a more certain reminder of the sharp dividing-line between human and god than the unavoidability of old age. Old age was a deliberate imposition on mortals by the gods so as to keep them in their place; only a god had the power to postpone or take it away.[17] Even in the choice of his wording (γήραος

---

[14] For this basic sense of the word *pharmakon*—especially in an overtly magical context—the most helpful discussions are by A. Abt, *Die Apologie des Apuleius von Madaura und die antike Zauberei* (Giessen, 1908), 112–15, F. Pfister, *RE* xix. 1446–7, and Lloyd (1983), 119–33, esp. 120. *Pharmaka* in B111 as plants: Bollack, iii. 23. It is worth noting that Diocles, who is known to have been influenced by Empedocles in a number of other respects, wrote a famous 'Ριζοτομικόν or 'root-cutting' manual: cf. Lloyd, op. cit. 120 and, for the cultural background, Raven 169–72.

[15] A. Delatte, *Herbarius*² (Liège, 1938), 90–115, 157–8. It was also normal to use incantations during the actual treatment: cf. e.g. Pl. *Charmides* 155e–157c, R. Kotansky in *MagH* 108–10.

[16] Cf. esp. *PGM* iv.286–95; Delatte, op. cit. 93, 97–101, J. Scarborough in *MagH* 138, 156–7.

[17] So, typically, *Il.* 4.320–1, 9.445–6, *Od.* 13.59–60. It was one thing to imagine people who live free from 'sickness and baneful old age' in the dream-world of the Hyperboreans (Pi. *Pyth.* 10.29–50); but the most that a Greek could expect for himself was something to help make the austerities of old age a little easier to bear, and even that he was bound to consider a gift from the gods (Pl. *Laws* 666b: Dionysus). For Pindar's realm of the Hyperboreans cf. G. M. Bongard-Levin and E. A. Grantovskij, *De la Scythie à l'Inde* (Paris, 1981), 40–63, with Pl. *Phd.* 110e–111c.

ἄλκαρ, 'defence against old age') Empedocles made it clear that he was issuing a flagrant challenge to the standard Greek view, embodied in the Homeric *Hymn to Apollo*, of humanity as 'senseless and helpless, incapable of finding a remedy for death and a defence against old age' (γήραος ἄλκαρ, 192–3). In turning the words of the hymn on their head Empedocles was affronting not only literary tradition but also—to the extent that the Homeric poems were viewed as the major repository of Greek values and wisdom[18]—the most fundamental of religious attitudes and assumptions. Essentially there is little to choose between his implied message here and his declaration elsewhere that he was no longer a human but a god.[19] This is not to say that Greek myth of Empedocles' time did not happen to know of exactly what he mentions: a 'plant that defends against old age' (φάρμακον γήρως ἀμυντήριον). It did—in the form of legends emanating from the eastern, Asiatic edge of the Greek world.[20]

Next, Empedocles claims to be able to control the wind and rain. Strictly this is not surprising: in the sphere of magic, being able to influence the weather was a natural extension of influencing the human body.[21] And yet command of the winds

---

[18] Ch. 13 with n. 62.
[19] B112.4–5; above, n. 7.
[20] Cf. esp. Soph. frs. 362–6 (Phrygia); Xanthus, *FGrH* 765 F3 (Lydia); also Ibycus, fr. 342 Davies with M. Davies, *MH* 44 (1987), 68–73 and Burkert (1992), 123–4; Aeschylus' Κρῆσσαι (frs. 116–20 Radt) and Sophocles' Μάντεις (frs. 389a–400) with G. Weicker, *RE* vii. 1408–12, 1415–16 and R. F. Willetts, *Klio*, 37 (1959), 23–6 (Crete). The first of these Sophocles stories has obvious affinities with the episode about the plant of rejuvenation in the Gilgamesh legend, which had been widely disseminated for centuries throughout western Asia and Anatolia: S. Dalley, *Myths from Mesopotamia* (Oxford, 1989), 45–7, 118–19. For Medea see the *Nostoi*, fr. 7 Bernabé with Aesch. fr. 246a, E. Maass, *Neue Jahrbücher*, 31 (1913), 628–32, J. Griffin, *JHS* 97 (1977), 42. For the Iranian mythology regarding plants of rejuvenation and immortality see Kingsley (1994*d*), 193 with n. 26. As for the magical use of herbs in actual rituals of mystical death, resurrection, and immortalization, cf. the so-called Mithras Liturgy in the Paris magical papyrus (*PGM* iv.475–82, 770–813), Psellus' commentary on the *Chaldaean Oracles* 1132a (169.11–12 des Places; Lewy 178 n. 4), and Martianus Capella 2.110, 2.141, with P. Boyancé, *Revue archéologique*, 30 (1929), 211–19 and *MEFRA* 52 (1935), 95–6.
[21] Cf. e.g. the Hippocratic *De morbo sacro* 1.10–31 (by implication); *Codex Theodosianus* 9.16.3 = *Codex Iustinianus* 9.18.4. Herbalism and weather magic were intimately connected, as we can also see from the legends surrounding names such as Epimenides—Pfeiffer 97; Lloyd (1983), 120 n. 18—or Medea (Fiedler 9). J. Scarborough, *MagH* 142, forces the evidence in trying to draw the same connection out of Homer.

and rain was traditionally in the hands of Zeus, which explains why the bid to exert influence over the weather of an extremely limited and restricted kind could be embodied in institutionalized religious worship and ritual.[22] However, almost as if to shatter any idea that what he had in mind was simply to pray to Zeus and invoke his assistance, Empedocles throws in the assurance that Pausanias will be able to change the weather 'if you so wish' (ἢν ἐθέλῃσθα, B111.5). The phrase has the general sense of 'if you feel inclined'; and to that we can add that it was no doubt chosen deliberately by Empedocles because of its frequent use in Homer and Hesiod when referring to the special divine powers of gods and goddesses.[23] The casualness of the phrase, not to mention its theological overtones, could hardly convey more clearly that Empedocles was not counting on obtaining the approval of some higher divinity; and it explains very well why this kind of magic was always bound to bring down on itself the charge of impiety.[24]

The closest parallel we have in classical literature to Empedocles' description of having free power to influence the weather is a passage in Diodorus Siculus about the Telchines, the legendary smiths who were able to produce clouds, rainstorms, hail, and snow 'whenever they chose' (ὅτε βούλοιντο).[25] The closeness of the parallel is significant. The Telchines had a kind of divine status, like Empedocles. In addition to their power of control over the weather they supposedly also had knowledge about the magical uses of plants.[26] And their mythical home was the islands of Rhodes and Crete: two islands which, on the one hand, had extremely close links with oriental craft-lore, metallurgical practices, and mythology while, on the other hand, it was settlers from these two islands who founded and populated Empedocles' home town of Acragas—bringing their

---

[22] Cook, iii/1. 103–12, 284–338; Kingsley (1994d), 193 and n. 24.
[23] West (1966), 163, with refs. Bollack notes that 'the formula constantly refers to the good will of the gods' (iii. 24), but misses the implication.
[24] Cf. the Hippocratic *De morbo sacro* 1.28–30 (ἀσεβές ... δυσσεβεῖν), echoed centuries later by the Christian condemnation of magicians who claim to be able to bring down rain and hail 'wherever they choose' (ἔνθα βούλονται) because this is to show contempt for the will of God (ps.-Justin, *Quaestiones et responsiones* 31, *PG* vi. 1277c–d).
[25] Diod. 5.55.3 = Zeno of Rhodes, *FGrH* 523 F1; Fiedler 9, Cook, iii/1. 296.
[26] Eust. *Il.* 771.60 (ii. 789.7–8 van der Valk); for the combination, see n. 21.

The Magus

religious and cultural traditions with them.[27] It is important in this context to note the extent to which, both before and during Empedocles' time, mythological ideas or religious practices incorporated in the Pythagorean and related lineages of the Greek West were a harmonizing synthesis of local, native traditions with other living traditions transported to their new homes by Greek immigrants from the Asiatic coastal regions and Anatolia.[28] This same general process also helps, for example, to explain the evidence for transfer of religious traditions from Anatolia to Empedocles' home town of Acragas in the particular case of the cult of Persephone and Demeter.[29] But to return to Diodorus Siculus: he also supplies the word routinely used when describing the man who lays claim—like the mythical Telchines—to have power over the wind and rain. He is like a 'magus', μάγος.[30]

Finally, we have the fetching of a dead man's life-force back from Hades. The precise wording chosen here by Empedocles quite clearly indicates that what he had in mind was not just some kind of necromantic invocation but an actual descent to the underworld to fetch back the dead man's vital soul—in

---

[27] For the Telchines, Crete, Rhodes, and the oriental background cf. Strabo 6.2.7 with S. Marinatos, *Kadmos*, 1 (1962), 93; also Burkert (1992), 9–11, 15–23, 26–7, 54–5, 62–3. For Rhodians, Cretans, and the settlement of Acragas see A. Holm, *Geschichte Siciliens im Alterthum*, i (Leipzig, 1870), 134–5, 138–9; Freeman, i. 398–9, 431; J. A. de Waele, *Acragas graeca*, i (The Hague, 1971), 5, 100–1, 186–7, 191; N. Demand, *GRBS* 16 (1975), 348–51; also P. E. Arias, *Problemi sui Siculi e sugli Etruschi* (Catania, 1943), 8, 18–20.

[28] For Pythagoras and early Pythagoreanism cf. Burkert 111–13 and (1969), esp. 23–6. It will be noted that the cult of Apollo Oulios at Velia was 'exclusively oriental', and specifically Anatolian: Fabbri and Trotta 69–75 and *passim*, esp. 70–1, 75; D. Morelli, *I culti in Rodi* (Pisa, 1959), 106, 110. As for the title 'lord of the lair', φώλαρχος, applied at Velia to the successive leaders of its organization of 'doctors' or healers, it is clearly related to the practice of incubation—or inducing dreams—for healing purposes: cf. e.g. S. Musitelli, *PP* 35 (1980), 241–55. In fact the closest parallel for the use of the term 'lair', φωλεόν, in precisely this context is a description by Strabo of healing through incubation as practised in north Caria, not far inland to the south-east from Phocaea; and Herodotus emphasizes that the Phocaeans took their cult objects and religious practices with them when they sailed west to settle, eventually, in Velia. Cf. Strabo 14.1.44 (a reference which I owe to Walter Burkert); Hdt. 1.164–8 with Fabbri and Trotta 71; and for the affinities of the Velia finds with Pythagoreanism, Ch. 20 n. 22. On the origins of the Sibyl at Cumae cf. Parke 87–94; and for the oriental origins of Etruscan divination, Burkert (1983), 117 with n. 20.

[29] Burkert (1979), 131.

[30] Diod. 5.55.3; cf. the Hippocratic *De morbo sacro* 1.10 (μάγοι), Fiedler 17–23, Lloyd (1979), 16.

exactly the same way as does a shaman.[31] It is not difficult to understand why such a promise by Empedocles, in the last line of a fragment tucked away by Diels at the very end of the cosmological poem, should have been put out of mind to the extent that it has. The idea of trying to bring someone back from the dead was, in the framework of normal Greek morality, almost unthinkable. Even for someone half-divine like Asclepius, the attempt was bound to end in disaster. To be sure, Orpheus would seem originally to have had the power to fetch the dead back to life; but exposure to the influence of Greek moralizing appears to have guaranteed that his success would be suppressed and converted into a tragic story of failure. And yet behind the suppression lies what was suppressed: in terms not only of formal and structural analogies but also of historical contacts, there can be no separating the Thracian Orpheus from central-Asiatic shamanic tradition. To ignore these connections, or try to banish the word 'shaman' from discussions of Greek religion on the grounds that phenomena in the Greek world must only be explained using Greek words, is simply to perpetuate the myth that ancient Greeks were magically sealed off from external influences—and is to carry the suppression of those foreign influences further than the Greeks ever did themselves.[32]

In view of these points it is certainly no accident that the closest parallel from ancient literature to Empedocles' image of a person capable of descending to and returning from the underworld at will is the account by Lucian of the practices of a Zoroastrian magus at Babylon.[33] Not only were these Persian Magi the people who provided the Greeks with their word *magos* or 'magus' in the first place: we also know that their own religious and magical traditions are inextricably linked with the traditions of north-Asiatic shamanism, and a major problem in

[31] B111.9 with Ch. 4 and nn. 15–18; Kingsley (1994*d*), 188 and n. 7.
[32] For refs. to the points in this paragraph see ibid. 187–97. Regarding Orpheus' failure cf. also above, Ch. 11 with n. 19; and for his links with shamanism, F. Thordarson in *The Nart Epic and Caucasology* (Maykop, 1994), 347–8.
[33] *Menippus* 6 (Bidez and Cumont, ii. 40), on a meeting with 'one of the Magi, disciples and successors of Zoroaster, who I heard are able—through certain spells and rituals—to open the gates of Hades and take down safely whomever they want (ὃν ἂν βούλωνται) and then bring him back up again'. Cf. Kingsley (1994*d*), 193 and n. 25 (on the Hellenizing element in Lucian's account and its oriental background), and below, Ch. 19 with n. 10 (on Lucian as source for ancient magic and religion).

The Magus

understanding the influence of shamanic traditions on the Greeks has been due to failure to appreciate the role played by Iranians as intermediaries in the process of transmission.[34] From the closeness of the parallel in Lucian to Empedocles' own words in fragment 111 we can understand why later writers felt it natural to make Empedocles a pupil of the Magi.[35] But even more significant is the fact that this tradition of linking him with the East almost certainly goes back to his own lifetime, because the very first reference to him in the surviving body of Greek literature—by his contemporary, Xanthus of Lydia—appears to have presented him in the context of a discussion of the Persian Magi.[36] In other words even before his death there would appear to have been people attesting, indirectly if not directly, to Empedocles' magical powers.

*

In a remarkable paper published recently, Ava Chitwood has attempted to dispose of the evidence of fragment 111 not by interpreting the passage allegorically, or by rejecting it as a forgery, but simply by making it disappear. Her method deserves some brief comment.

It is now well known that ancient biographers often created the life stories of classical poets out of nothing: that, with no genuine facts to work from, they took fictional events described by the poets in their writings and transformed the details into incidents supposedly experienced by the poets themselves. The Greek flair for inventing history is nowhere more apparent.[37] By appealing to this phenomenon, Chitwood argued that all the supposed incidents in the life story of Empedocles as presented in such graphic detail by his biographers simply

---

[34] For these points see Kingsley (1994*d*), 191–4 and *passim*.

[35] Plin. *HN* 30.2.8–9 (Bidez and Cumont, ii. 10) and Philostr. *VA* 1.2 (Emp. A14); cf. also Apul. *Apology* 27 (Bidez and Cumont, ii. 268).

[36] D.L. 8.63 = Arist. fr. 66 = Xanthus, *FGrH* 765 F33. On the literary and cultural background to the reference cf. Momigliano 28–38; Grottanelli 649–68, esp. 661. For its genuineness, and its origin in the section of Xanthus' work devoted to the Persian Magi, see Kingsley (1995*c*).

[37] Cf. Lefkowitz, *passim*; Kingsley (1990), 264; J. Mansfeld, *Prolegomena* (Leiden, 1994), 180.

derive from passages in his own poetry innocent of any autobiographical significance.[38] But apart from the tenuousness of her arguments in general, she has failed to note the most fundamental point of all: that some of the poetry from which, according to her, ancient biographers unscrupulously fabricated stories about Empedocles' life is explicitly autobiographical to begin with. What the biographers did with the material later on is a separate issue. We do not have to be told that Diogenes Laertius 'says that Satyrus believed that Empedocles laid claim to the powers mentioned in fragment 111';[39] we know that Empedocles laid claim to the powers in question from the fragment itself. The modern stereotype of the intellectual certainly makes it tempting to try to drive a wedge between mere ideas and the living of them: to convert Empedocles into the image of an armchair philosopher by claiming that,

in the end, we know very little about the historical Empedocles and a great deal about the Empedocles of legend. The historical Empedocles might not, perhaps, have been very interesting, and it is not important that he should have been. Empedocles' importance lies in his philosophy, taken as philosophy and not as biography. On the other hand, the legendary Empedocles is such an appealing figure that we would regret his not having been invented.[40]

But distinguishing between the historical or 'important' and the legendary or 'appealing' is not so simple. The legend of Empedocles was created in the first instance not by later biographers. It was created, as we have seen, by Empedocles himself; and as Burkert has pointed out,

how could he have created it, how could he have called himself a god, if he was not able actually to perform, or at least to pretend to perform extraordinary and amazing feats? His saga must have its roots not in literature, but in reality.[41]

The implications of understanding fragment 111 literally are considerable. Once the magical side to his activities has been granted, as it must, the possibility of being able to draw a neat dividing line between Empedocles the magician and Empedocles the philosopher immediately disappears. With fragment

[38] *AJP* 107 (1986), 175–91.
[39] Ibid. 184.
[40] Ibid. 191.
[41] Burkert 153–4. Cf. Bidez 134–6; Weinreich, ii. 179–83; Dodds 145.

111 placed where it almost certainly belongs, at the beginning of the cosmological poem, we are given a clear indication of the way in which we are meant to understand the poem as a whole. Learning about the elements, about the history and constitution of man, about the nature of plants, animals, metals, physical phenomena, astronomy, and the structure of the universe is not some end in itself: not the satisfaction of some theoretical desire for knowledge. On the contrary, the promise of magical powers—mastery of the elements, mastery of life and death—plainly implies that through grasping and absorbing the teachings contained in the poem, and letting them grow inside him,[42] the disciple will eventually gain the ability not just to understand the powers of nature but also to control them. That, after all, was the original role of magic; that is why, before the realm of magic was torn in two by the mutually hostile claims of religion and science, to accumulate knowledge and information about the world—and be able to synthesize and use it—was precisely the task of the magician. In the words of Marcel Mauss,

Magic is linked to science in the same way as it is linked to technology. It is not only a practical art, it is also a storehouse of ideas. It attaches great importance to knowledge—one of its mainsprings. In fact as far as magic is concerned, knowledge is power.... It quickly set up a kind of index of plants, metals, phenomena, beings and life in general, and became an early store of information for the astronomical, physical and natural sciences. It is a fact that certain branches of magic, such as astrology and alchemy, were called applied physics in Greece. That is why magicians received the name of *physikoi* and that the word *physikos* was a synonym for magic.[43]

That Empedocles' concern with nature and cosmology was magical in just this sense, and far from simply theoretical or disinterested, has sometimes been realized.[44] It has also been pointed out that a passage about Empedocles in the Hippocratic writings implies intelligent Greeks were quite aware that his knowledge of cosmology and nature was meant to have a

[42] B110 (below, nn. 47–8); cf. also B17.14.
[43] *A General Theory of Magic*, trans. R. Brain (London, 1972), 143 = *Sociologie et anthropologie*³ (Paris, 1966), 136. See also Delcourt's comments, 133.
[44] Cf. esp. W. Nestle, *Philologus*, 65 (1906), 548; R. D. Ranade and R. N. Kaul in S. Radhakrishnan (ed.), *History of Philosophy Eastern and Western*, ii (London, 1953), 40.

practical application, especially in the sphere of healing.[45] It has been suggested as well that, in this concern with practical application and the mastery of natural forces, Empedocles had more in common with the modern scientist than with our familiar image of the abstract philosopher.[46] And yet there is a world of difference between the modern practical scientist, attempting to manipulate various forms of matter, and Empedocles—for whom, as he explains in another fragment, there is nothing that is not vibrantly and knowingly alive. For him everything—even the words spoken by a man of understanding—has an existence, intelligence, and consciousness of its own.[47] He is quite explicit as to how he intends his words, and specifically the words in his cosmological poem, to be received. They are not to be argued with: they are to be treasured. They are to be tended and guarded, otherwise they will fly away. They contain a mystery. More specifically, the terminology Empedocles uses refers in a deliberately allusive way to the standard vocabulary employed in the mysteries of Persephone and Demeter. If we follow up the allusion we see that he is referring to the stage when—after the pupil to whom the poem is addressed has gone through the preliminary stage of purification (*katharmos*), and has then absorbed the knowledge transmitted to him by his teacher (*paradosis*)—'there is nothing left to learn'. The teachings take over; the words spoken by the teacher take root and grow inside the disciple, transforming not just his vision but his being, and providing him with everything he will need in future.[48]

[45] *De veteri medicina* 20.1–2 (51.6–15 Heiberg = Emp. A71); Wright 263. For συνεπάγη (51.12 Heiberg) cf. Emp. B15.4, 75.1, 86.1, 107.1, Plut. *De amicorum multitudine* 95a–b (DK i. 326.5, 8); for ἡ γραφικὴ τέχνη (51.13–14), Emp. B23, J. Jouanna, *Hippocrate: De l'ancienne médecine* (Paris, 1990), 209.
[46] Guthrie, ii. 133–4.
[47] B110; for the identity of the σφε in B110.1 cf. Wright 258–9, and compare Plutarch's comments on B5 at *Qu. conv.* 728e (τὰ δόγματα).
[48] On the need to treasure, tend, and guard Empedocles' words cf. B110, also B5 with Plut. ad loc. Mystery terminology: Emp. B110.2 (εὐμενέως καθαρῇσιν ἐποπτεύσῃς μελέτῃσιν), correctly noted e.g. by Delatte (1934), 26 and n. 2, Traglia 163. For *epopteia* and its central place in the grades of the mysteries—*katharmos, paradosis, epopteia*, crowning, divine bliss—the chief texts are Theon of Smyrna 14.18–15.7 Hiller and Clem. *Str.* 5.71.1; cf. Dieterich (1911), 118–19, Farnell, iii. 131–2, Wilson 7–10, Burkert (1987), 69–70, 91–2, 153–4, and also *PGM* iv.504 with MT iii. 233, 235–6. On the mystery idea of *paradosis* or transmission cf. Burkert (1987), 153 n. 14; for its relevance to Empedocles' cosmological poem, ibid. 69–70 with n. 17, and B111.2

This 'initiation vocabulary'[49] is no more a metaphor or allegory than the promises in fragment 111. And just as in the case of fragment 111, to ignore it is as pointless as trying to explain it away: it forms a fundamental part of the framework and context in which Empedocles' poetry needs to be approached. Once again, if we are to understand Empedocles at all we need to understand him on his own terms; otherwise the whole enterprise of interpretation will be nothing but a collection of anachronisms and empty words. Here, the logic in Empedocles' choice of ideas and vocabulary should be clear. He considered his words capable of acting, of having an effect; as someone who considered himself divine it is perfectly understandable that he attributed to them the magical properties of the divine word which typically has a life of its own, is 'a living thing, a natural reality that sprouts and grows'.[50] In the first instance, as he carefully explains, these words of his had the power to affect and transform the person who absorbs them. Later, it was only inevitable that the same power could be used by the transformed person to affect and change the world around him.

In short, it is not enough to accept that Empedocles may have had a 'magical' side to his character: to suppose that although first and foremost a philosopher he may, also, have had an interest in magic—like someone who takes up stamp-collecting or conjuring as a spare-time hobby. Those who try to understand Empedocles as a phenomenon in his own right invariably end up describing him as an individual who managed to combine in himself the contradictory roles of magician and philosopher, wonder-worker and thinker.[51] But the contradiction is in ourselves, not in Empedocles. To

---

with n. 13 above. For the phrase 'there is nothing left to learn' see Clement, loc. cit., who also describes how—after the initial purifications and the intermediate stage of learning—ahead of the pupil lies nothing but direct 'understanding of nature and all things'; Burkert (1987), 69. On the natural growth of Empedocles' words inside the prepared disciple, and their transformation so as to provide him with everything he will need, cf. B110.3–5 (Ch. 19 n. 35).

[49] Wright 259.
[50] Detienne (1973), 53–5; cf. Kingsley (1990), 264 n. 120.
[51] So e.g. Guthrie, ii. 123, 125–6. Weinreich, ii. 181–3 and Dodds 145–6 come closest to presenting a fair picture of the situation, but still underestimate the extent to which the 'functions' of magician and philosopher were actually intertwined.

suppose that a magician is someone incapable of devising coherent theories about the structure and working of the universe is in the last resort sheer prejudice, denied by all the evidence. To assume that Empedocles must have shared our questionable distinctions between rational and irrational forms of inquiry or understanding is historically implausible, and is not necessarily to compliment him. Most importantly, there is no point in acknowledging that Empedocles was a magician unless we are prepared to see that the poetry of his which has come down to us was an integral part of his magical activity. Taken out of context, this poetry can be interpreted in any number of ways; but placed in its magical matrix, the evidence starts to fit together in a way fundamentally independent of modern prejudices and preconceptions. And, as we will now see, we can also observe this same process at work in the case of the ancient legends about Empedocles' death.

16
From Sicily to Egypt
*

THE ancient Greeks invented a number of different stories about Empedocles' death. They had him drown, hang himself: it is easy to forget how snide Greek writers could be about the philosophers in whom they claimed to show an interest. However, there is one particular version of his death which stands out from the rest. For one thing, it was by far the most famous of them all; with the inevitable malicious comments and innuendoes it survived down to the end of antiquity. But it also attracted a different kind of attention: already in the first century or two after Empedocles' lifetime it had become a focus of intense criticism from people who otherwise could hardly agree on anything. For the realist it was too fantastic and romantic. For the romantic it seemed too trivial and ignoble.

According to this story Empedocles died by throwing himself into Mount Etna. He had, apparently, hoped to deceive everyone into believing he was a god by making any trace of his body disappear. But, so the story goes, one of the bronze sandals he used to wear gave the trick away. The sandal was ejected by the volcano and discovered, exposing the hoax so that eventually it became clear (ὕστερον γνωσθῆναι) what he had really done.[1] For obvious reasons this story has been dismissed by both modern and ancient writers as nothing but a 'farce'.[2] But even so, it deserves looking at a little more closely.

The first point worth noting about the story is its age: of all the versions of Empedocles' death that circulated in antiquity it is clearly the oldest.

If we want to trace it back in time the best person to start

---

[1] D.L. 8.69 (Hippobotus). Cf. esp. Strabo 6.2.8 and 10; Luc. *DM* 6 (20).4; *Suda*, s.v. Ἐμπεδοκλῆς Μέτωνος.
[2] Bidez 40.

from is Timaeus of Tauromenium, who produced his history of Sicily in the early third century BC. Always the rationalist, Timaeus had little patience or sympathy for stories about jumping into volcanoes and sandals of bronze; in the absence of any more credible account of Empedocles' death he considered it his duty to find one. Taking his cue from the fact that no evidence for any tomb or memorial seemed to survive, he suggested that Empedocles may have died outside of Sicily and came to the conclusion that he ended his days, unknown, in the Peloponnese.[3] The conclusion sounds eminently reasonable, and has been welcomed by scholarship either as the most probable account of Empedocles' death or even as established fact.[4] It could be correct. But, as Bidez pointed out a century ago, Timaeus' theorizing is sheer guesswork—motivated, as we will soon see, by a strong element of polemic—and historically is worthless.[5] He knew no more about the matter than we do.

In dismissing the fantastic stories about Empedocles' death, along with the credulous people who disseminated them, Timaeus mentions one guilty party in particular: the writer Heraclides of Pontus.[6] With Heraclides—a pupil and friend of Plato, and Empedocles' junior by roughly a hundred years— we are brought back to the mid-fourth century BC. And already by the time of Heraclides the situation had become complex: with him we are faced not just with one version of Empedocles' death but with two. Heraclides himself appears to have written a fictional dialogue specifically about Empedocles. In it he described how a celebration banquet was held in a Sicilian field after Empedocles had performed one of his miracles; and how in the middle of the night, after everyone else had fallen asleep, Empedocles was suddenly snatched up to heaven accompanied by a flashing of celestial lights.[7] But in the same dialogue Heraclides also refers to another version of Empedocles' death: the version we have already looked at. He makes Pausanias— evidently a real disciple of Empedocles whom Heraclides had adopted as a dramatic character in his work of fiction—

---

[3] Timaeus *ap.* D.L. 8.67, 71–2 (... τετελεύτηκεν οὖν ἐν Πελοποννήσωι) = *FGrH* 566 F6.   [4] So e.g. Freeman, ii. 354; Guthrie, ii. 131 with n. 1.
[5] Bidez 48, 174.
[6] D.L. 8.71–2 = Timaeus, *FGrH* 566 F6 = Heraclid. fr. 84.
[7] D.L. 8.67–8 = Heraclid. fr. 83.

expressly contradict and deny the legend which had Empedocles die by leaping into Etna. It would seem that the legend already had the connotations of hoax and fraud from which Heraclides felt Empedocles ought to be protected.[8] In short, Heraclides proposes one version of Empedocles' death, but not without feeling the need to refute another.

On the relationship between these two accounts of Empedocles' death that we find in Heraclides, Bidez adopted a strange but adamant position. According to him the story of Empedocles' ascent to heaven and evident apotheosis was older than the story about the leap into Etna.[9] And yet the evidence that we have tends to point in precisely the opposite direction. On the one hand, the trouble Heraclides took to contradict the legend of the leap into Etna shows plainly that—as Bidez himself admits[10]—it must have been well established before his time. On the other hand, many of the details in the alternative version which Heraclides favours reveal the tell-tale signs of Heraclides' own creative workmanship: of his own skill at inventing dramatic settings for the philosophical dialogues that he liked to put into the mouths of famous people from the past.[11] In other words, if forced to choose, one would have to conclude that the story of the leap into Etna is the older of the two versions. However, as we will see later, trying to discriminate between death by leaping into Etna and death as apotheosis is to risk creating a false dichotomy. Heraclides would in fact seem to have been responsible not so much for inventing the idea of a celestial ascent as for reworking and bowdlerizing an earlier version which described the leap plus the apotheosis together. The trouble is that in cutting out the embarrassing

---

[8] D.L. 8.69 = fr. 85. Cf. R. D. Hicks's Loeb edition of Diogenes (Cambridge, Mass., 1925), ii. 384 n. 1; Wehrli 89, ad loc.; Bolton 154; Gottschalk 16–17. For Pausanias see D.L. 8.60–1; Wehrli, loc. cit.; Gottschalk 16 n. 9.

[9] Bidez 35–9, followed by Dodds 167 n. 65.

[10] Bidez 39.

[11] For the Platonic influence in the setting of the dialogue at a banquet cf. Bidez himself, 35 n. 3, Gottschalk 17; for 'the idea of a dialogue describing the last day on earth of a great philosopher', Gottschalk, loc. cit.; for the celestial lights ($\phi \tilde{\omega} s$ οὐράνιον καὶ λαμπάδων φέγγος, D.L. 8.68), Bidez 35 n. 4, Heraclid. fr. 93 (καταλαμφθῆναι ὑπὸ τοῦ φωτὸς τοῦ περιθέοντος κύκλωι τοὺς θεούς), also frs. 98–100 and 116 (ὑπὸ μεταρσίου φωτὸς καταυγαζόμενον); and for the voice greeting Empedocles from heaven, Bolton 165–6. On Heraclides' artistry as a creator of fictional settings for his philosophical dramas see Kingsley (1990), 263–4.

details of the jump into Etna he managed to destroy the dynamic of the legend as a whole.

Bidez also passed another premature judgement on the story of Empedocles' leap into Etna. Prepared as he was to admit that the legend was older than Heraclides, he chose to dismiss the detail about the bronze sandal as a later addition: an addition invented after the time of both Heraclides and Timaeus.[12] But here too, his conclusion is directly contradicted by the evidence. According to our only source of information on the matter, Heraclides made his fictional Pausanias reject not just the story of the leap into Etna as a whole but—very specifically—the detail about the bronze sandal being ejected by the volcano and exposing Empedocles' fraud; there is no reason whatever to ascribe this to a misunderstanding on the part of our source.[13] And there are, as it happens, other more general reasons as well for assuming that the mention of the sandal belonged to the legend in its early stages. The motif of the sage who dies a miraculous death by vanishing into thin air but who leaves a tell-tale item or items of his dress behind is a common feature of folklore.[14] This would tend to suggest that the peculiar mention of the bronze sandal is more likely to have been an integral part of the original legend than simply a later addition by some malicious biographer or satirist. And it also suggests something else: that originally the detail of the sandal was meant to act as a 'sign' which, far from contradicting the miraculous nature of Empedocles' death, would in fact confirm

---

[12] Bidez 8 n. 1, 70–2; so e.g. Wright 16 ('later elaboration').

[13] D.L. 8.69 (... πρὸς τοῦθ' ὁ Παυσανίας ἀντέλεγε) = Heraclid. fr. 85; for the sequence of the passage cf. F. Leo, *Die griechisch-römische Biographie* (Leipzig, 1901), 77–9 and Hicks, loc. cit. (above, n. 8).

[14] For one example from the other side of the world see J. Blofeld, *The Secret and the Sublime* (London, 1973), 147: 'A more classic Taoist concept is that of achieving the divine state either by fashioning a spirit-body or else by transmogrification. History, unfortunately, affords but few indisputable examples of the latter. However, we do know that the divine Ko Hung vanished from human ken, thoughtfully leaving a pile of discarded garments lest it be supposed he had wandered off into the mountains and lost his way home.' Cf. also the case of Elijah and his mantle in 2 Kings 2: for later comments on the significance of his leaving the mantle see Abraham Maimonides, *The High Ways to Perfection*, ed. S. Rosenblatt, ii (Baltimore, 1938), 264–7, 318–19, and for the parallel between Elijah and Empedocles in this and other respects, Grottanelli 651–62 (esp. 653, 661), Culianu 35.

it.¹⁵ As for the malicious twist to the story which makes the discovery of the sandal expose Empedocles as a fraud, rather than reveal his genuineness, this must almost certainly have been introduced at some later stage—either deliberately or through a basic misunderstanding of the mythical motif. In other words, here we would appear to have a perfect example of one of the most pervasive features of ancient Greek mythology: the tendency to rationalize the obscure, with the common result of giving motifs, symbols, or ritual practices a meaning the very opposite of the significance they would seem originally to have had.¹⁶ The anatomy of this process also helps to give us a rough indication as to the antiquity of the legend about the bronze sandal and the leap into Etna. Because the malicious version seems already to have been the one that Heraclides knew and reacted to, we must suppose that the original story goes back at least to the early fourth century BC—quite probably to the late fifth century—and so brings us very close to Empedocles' own lifetime.

There is one other, highly significant, aspect to the detail about the bronze sandal. Apart from being put aside as a 'late elaboration' of an originally much simpler story, it has also been described as an 'obvious fiction'.¹⁷ As for being a late elaboration, we have seen that this is contradicted not only by the folklore evidence but also by Diogenes Laertius' statement to the contrary. As for being an 'obvious fiction', this too is hardly an accurate description. There could in fact be few things less obvious than why anyone should invent a story of someone walking around in sandals made of bronze. Various suggestions have been made as to why Empedocles should have worn bronze sandals;¹⁸ none of them comes near to the

---

¹⁵ Correctly Grottanelli 661; Culianu 35.
¹⁶ Cf. e.g. Parmentier 31–61, and n. 22 below (the sound of bronze as apotropaic); West 143–5, 154–5 (conversion of the beneficent initiator into a demon); and for the rationalization, Frazer (1911), 311 with nn. 3–4, 312 with n. 1, Fiedler 35, Dodds 163 n. 44 *ad init.*, Burkert 146, 155–9, J. F. Kindstrand, *Anacharsis* (Uppsala, 1981), 19–23, 74–5.
¹⁷ Wright 16.
¹⁸ For van Groningen, (1956), 48 n. 2, bronze was an obvious alternative to leather which Empedocles—with his horror of animal sacrifice—would supposedly have objected to wearing. For S. Campailla bronze 'is a symbol of eternification': 'Empedocles walks on bronze, already erecting for himself in his lifetime a kind of pedestal for

point or is even remotely convincing. If the desired intention was to make Empedocles boast his immortality by openly imitating the ways of the gods, then the obvious step would have been to portray him wearing golden sandals—not sandals of bronze.[19] As a piece of fiction the detail makes no sense; but factually it makes no sense either. Walking around in bronze sandals would be not only extremely uncomfortable but, in practice, virtually impossible. The biographers are careful to smooth over and rationalize the oddity of the incident about the one bronze sandal being thrown up from the volcano at Empedocles' death by stating that it was his normal custom to go around in sandals of bronze all the time.[20] And yet if Empedocles was to wear bronze sandals at all it can only have been for a very limited, and special, occasion—such as a particular ritual.

In fact the solution to the whole problem of Empedocles' bronze sandal was provided over a century ago by Dieterich— even though Dieterich himself made no mention of the detail in the Empedocles legend and seems not to have noticed its relevance to what he says. In one of his earliest works he gathered together various texts, and chiefly an important passage from the Paris magical papyrus, which show that in the ancient world a bronze sandal was a symbol connected specifically with underworld ritual and magic. What is more, the evidence he gathered shows that—in the strange world of mythical and magical symbolism to which the motif of the bronze sandal belongs—it was invariably a matter of *one* sandal of bronze, and one alone. This one bronze sandal was the chief 'sign' or 'symbol' of Hecate who, as the 'controller of Tartarus' and mediator between this world and the next, grants the magician access to the underworld.[21]

---

eternity', *Studi offerti a Francesco della Corte* (Urbino, 1987), v. 668. Similarly West (1966), 156 considered them simply 'a symbol of his divinity'.

[19] *Il.* 24.340–1, *Od.* 1.96–7, 11.604, Hes. *Th.* 12, 454, 952, Virg. *Aen.* 4.239–41, etc.; Delcourt 88–9, West, loc. cit.

[20] So e.g. D.L. 8.69 (χαλκᾶς [sc. κρηπῖδας] εἴθιστο ὑποδεῖσθαι), 8.73, Strabo 6.2.8 (... τῶν ἐμβάδων τὴν ἑτέραν ἃς ἐφόρει χαλκᾶς), Suda, s.v. Ἐμπεδοκλῆς Μέτωνος (ἔχων ... ἀμύκλας ἐν τοῖς ποσὶ χαλκᾶς ... ἐπῄιει τὰς πόλεις).

[21] *De hymnis Orphicis* (Marburg, 1891), 42–4 = (1911), 101–2. The most relevant passages—either already cited by Dieterich and subsequently re-edited, or discovered after his time—are *PGM* IV.2292–4 (τοῦτο γάρ σου σύμβολον τὸ σάνδαλόν σου ἔκρυψα,

From Sicily to Egypt 239

The correspondence with the story of Empedocles' leap into Etna is unmistakable. It becomes even more remarkable when we bear in mind the number of times that the Greek magical papyri, and especially the Paris papyrus, have thrown light not just on the legends surrounding Empedocles but on his actual life and poetry. And this is not all. We have already seen that Empedocles' own equation of Hades and fire has its roots in the volcanic regions and phenomena of Sicily—and in the mythology associated with Mount Etna in particular. Here, under the guise of a legend which takes us back very close to Empedocles' lifetime, we have a direct linking of underworld symbolism (the bronze sandal) with fire (Mount Etna): in other words, Empedocles' supposedly 'philosophical' equation of underworld with fire is presented in precisely the kind of mythological and geographical form that we have already come to expect. What the motif of the bronze sandal has succeeded in doing is to plunge us into a web of correspondences linking the magical papyri themselves not only with the legend of Empedocles' death but also with Empedocles' own life and his own ideas. The full significance of this web, as well as its contours, will begin to emerge as we start to uncover the historical and geographical background to those later magical texts.

Bronze for the Greeks had a special affinity with the dead,

καὶ κλεῖδα κρατῶ. ἤνοιξα ταρταρούχου κλεῖθρα Κερβέρου . . ., cf. *GMPT* 79 n. 286), 2333–40 (εἶτα κἀγώ σοι σημεῖον ἐρῶ· χάλκεον τὸ σάνδαλον τῆς ταρταρούχου, στέμμα, κλείς . . .) = E. Heitsch, *Die griechischen Dichterfragmente der römischen Kaiserzeit*[2] (Göttingen, 1963), 189.48–50, 190.89–96 = *PGM*[2] ii. 251.48–50, 253.89–96; Marcellus Empiricus, *De medicamentis* 15.89, ed. Heim 501 § 117 (ταρταρούχου χάλκεον τὸ σάνδαλον); the Oxyrhynchus lead plate published by Wortmann (1968a), 62.57–8, by Jordan (250.57–8), and in *Suppl. Mag.* i. 196.57–8 = 203.1–2 (κλείς, κηρύκειον, τῆς ταρταρούχου χάλκεον τὸ σάνδαλον); Porph. fr. 359.62–70 Smith = *Περὶ ἀγαλμάτων*, 14*–15* Bidez (on Hecate's 3 aspects: τῆς δ' αὖ πανσελήνου ἡ χαλκοσάνδαλος σύμβολον). At *PGM*[2] LXX.10–11 the supplement τῆς ταρταρούχου ⟨χάλκεον τὸ σάνδαλον⟩ is demanded not only by these parallels—cf. Wortmann (1968b), 156 § 4—but also by the metre (Jordan 255–8). For the motif of the single sandal in ritual, magic, and taboo cf. Frazer (1911), 310–13; Kerényi 360; Gernet 225–6; and note the demonic connotations of wearing only one sandal in Near Eastern religious traditions: S. Shaked in P. Gignoux (ed.), *Recurrent Patterns in Iranian Religions* (Studia Iranica, Cahier 11; Paris, 1992), 151–2. Compare also the slightly different—but clearly related—magical tradition from Sicily mentioned by Weinreich, *Antike Heilungswunder* (Giessen, 1909), 166 and n. 4. On the general significance of sandals in magic and underworld ritual see Eitrem 367–8 and his *Hermes und die Toten* (Christiania, 1909), 44–5; and on Hecate as intermediary between this world and the other, Graf (1974), 73, S. I. Johnston, *Hekate Soteira* (Atlanta, 1990), 11–75.

the demonic, and the underworld. This was true in particular for the early Pythagoreans; in connection with the mysteries of Persephone and Demeter; and in the context of oracles and prophecy.[22] As for the bronze sandal, it points directly to Hecate and—through her—to the mysteries of Persephone.[23] That brings us, once again, onto familiar ground: as we have seen repeatedly in looking into the background to Empedocles' equation of Hades and fire, the craters and volcanic phenomena of Sicily—including Etna itself—were not just viewed as entrances to the underworld but were associated specifically with the mythology and the mysteries of Persephone. The extreme oddity of the feature of the one bronze sandal would, in itself, be quite enough to establish a link between the Empedocles legend and the magical papyri; these subsidiary connections merely serve to confirm it.

That leaves the question as to how we are to explain such links and correspondences. The first step to answering this question involves making some basic observations about the magical papyri themselves. All of them were composed, and subsequently discovered, in Egypt; they date for the most part from the first five centuries AD, although there are exceptions

[22] Pythagoreans: Porph. *VP* 41 = Arist. fr. 196; Parmentier 51-2, 57-61; Burkert 170-1, 376-7; and for the idea cf. G. Contenau, *La Magie chez les Assyriens et les Babyloniens* (Paris, 1947), 114-15. Persephone and Demeter: Apollodorus, *FGrH* 244 F110b, with the further evidence cited by A. B. Cook, *JHS* 22 (1902), 15; Parmentier 39, 52-3. Oracles and prophecy: Cook, op. cit. 5-28; Parmentier 43-50. As Parmentier has shown (31-61), the function of bronze in relation to the underworld was not only— or even primarily—apotropaic; for the basic affinity between bronze and the world of the dead see also Apollodorus, loc. cit. (οἰκεῖος τοῖς κατοιχομένοις), Eur. *Helen* 1346 (χαλκοῦ δ' αὐδὰν χθονίαν), and Macr. *Sat.* 5.19.9-11. For Hecate and bronze cf. e.g. Soph. frs. 534-5, Virg. *Aen.* 4.509-14, Macr. *Sat.* 5.19.9-10 (use of bronze for cutting and holding plants under Hecate's auspices) with Martinez 2 n. 6; Theocr. 2.10-16, 35-6 with Parmentier 40-2 (magical invocation of Hecate through sounding bronze); and the Orphic *Argonautica* 965-77 (Hecate and the Furies approach when invoked by sounding of bronze) with F. Vian's note on 965, Dieterich (1911), 102 n. 1, and A. D. Nock, *JHS* 46 (1926), 50-3 for the parallels between the scene and rituals in the Paris magical papyrus. Also relevant to Hecate's bronze sandal is Empusa, an underworld bogey so closely associated with Hecate as to be identified with her, who is portrayed as having a fiery face and a leg of bronze: Aristophanes, *Frogs* 293-4; *Suda*, s.v. Ἔμπουσα; Rohde 590-3; Dieterich (1911), 102.

[23] Wortmann (1968*a*), 78, (1968*b*), 155-9. On Hecate and the Persephone mysteries cf. also Maass 176-80; Richardson 155-6, 295. Both in classical times and, especially, in the later magical literature the two goddesses tend to merge: Roscher 119-20, Wilamowitz (1931-2), i. 170, 173, Richardson 295.

From Sicily to Egypt 241

which are significantly older.[24] As for the later ones, even they plainly represent no more than the end result of a long process of textual transmission. This is especially true in the case of that great miscellany of spells, evidently produced during the late fourth century AD in the region of Thebes in Upper Egypt, which is known as the Paris magical papyrus: here the scribe went to considerable trouble to preserve the variant readings he had collated from older magical texts, and in so doing he provides brief glimpses into the history of this continuous tradition.[25] But particularly important for our purposes is the fact that over the course of the centuries the writers of these texts incorporated and absorbed into their magical material fragments of earlier Greek poetry, hymns, and liturgical formulae used in the mysteries.[26] Interestingly, one of the most revealing passages of all in this respect is the section of the Paris papyrus which—itself in the form of a hymn—refers to Hecate and the bronze sandal: long ago Dieterich gave strong reasons for concluding that this mention of the bronze sandal derives, like many other details in the magical papyri, from Orphic literary tradition.[27]

Tracing the geographical and historical background to some of these older strata of Greek material in the magical papyri is not difficult. Strangely enough, however, the implications of being able to do so have hardly if at all been appreciated. There are, to begin with, important analogies between some of the more specific practices described in the papyri and the magical

[24] Festugière (1981), 281–3 n.; *PGM* ii. 177; Betz in *GMPT*, p. xli.
[25] Nock, i. 177–80; see also Kuster 13–14; E. Heitsch, *Philologus*, 103 (1959), 221–2; Martinez 7 n. 31. For the dating of the Paris papyrus itself cf. Kuster 12; Nock, i. 180; E. N. Lane, *The Second Century*, 4 (1984), 25–7. For comments on its place of origin see Fowden 168–73.
[26] Cf. Dieterich (1923), 213–18, 256–8; Nilsson, iii. 130–1; Nock, i. 180–1; Heitsch, op. cit. 218, 222; Betz in *GMPT*, pp. xliv-xlv.
[27] *PGM* IV.2334–5 with Dieterich (1911), 101–8, who refers in particular to the σῶσον refrain in the verse fragment preserved by Marcellus Empiricus and compares it with the standard refrains in the Orphic hymns (2.14 σῷζ'... σώτειρα, 9.12, 14.8, 12, and *passim*); add now the Gurob papyrus (3rd c. BC) = *OF* fr. 31 I.5 and 22 (σώισομ με), and cf. R. Kotansky in *MagH* 120–2. See in general Dieterich (1891), 132–3, Kuster 14–15, 52–5, 79–81, Kern, *OF* pp. 193–4, 312, 319–20, 341–2, Nock, i. 187, West 255–6. As within Orphic tradition itself (Ch. 10 with n. 40), it will naturally have been a matter more of adapting, reshaping, and restructuring earlier materials than of simply reproducing them: cf., with regard to the bronze sandal motif, Wortmann (1968*b*), 156 with n. 4, and in general Dilthey 382–3; Dieterich (1911), 13–14, 101; Kuster 14–15; Nilsson, loc. cit.; also Heitsch, op. cit. 220.

practices described by various Latin writers—for example Lucan, who was writing at Rome in the mid-first century AD.[28] But we can go back further, as well. A major component of what we find in the papyri is, as one would expect, Egyptian: a basic stock of native magical tradition onto which other ideas and practices—primarily Greek and Jewish[29]—were grafted. The components of this synthesis which are most recognizably Greek in origin are the elements of Hecate magic; and the closest and most remarkable analogies here are to be found in Theocritus. By origin from Syracuse, Theocritus left Sicily in the early third century BC and—like many other Sicilians of the time who, disturbed by political conditions on the island, emigrated to Alexandria—he ended up in Egypt. This would make it possible to suppose that the details of magical ritual recorded in his second *Idyll* simply reflect his experiences and encounters in Egypt. But that is not the case. As was already noted in antiquity, behind these accounts of Hecate magic which Theocritus' poetry shares in common with the Graeco-Egyptian papyri stands another figure to whom Theocritus was in this respect indebted. That was the Sicilian writer Sophron, who lived and worked at Syracuse during the fifth century BC.[30] In other words, when we trace the Hecate magic of the Egyptian papyri back in time, we arrive at the very same country and the very same century in which Empedocles himself lived—and in which the legend about his death probably started to evolve.

No doubt there is a sense here in which we can speak in general terms of Mediterranean magical traditions: the Mediterranean then, as now, was a small world.[31] But there is also a more specific sense in which we can talk of magical traditions passing down to Egypt from the Greek West, and particularly from Sicily. In fact there are small but telling

---

[28] Nock, i. 185–7; for Lucan cf. e.g. MT ii. 87–8.
[29] For the Jewish dimension cf. e.g. M. Gaster, *JRAS* (1901), 109–17; Nilsson, iii. 134–8; Smith (1973), 180–1 and 220–48 with refs.; D. R. Jordan, *HTR* 84 (1991), 343–6; and also Wasserstrom 160–6.
[30] Schol. Theocr. 269.13–270.7, 271.22–272.2 Wendel, with C. Wendel, *Überlieferung und Entstehung der Theokrit-Scholien* (Berlin, 1920), 92 and Nock, i. 184–5; cf. Dilthey 388–90, Gow, ii. 33–6. For the Hecate magic in Sophron and the magical papyri see also T. Kraus, *Hekate* (Heidelberg, 1960), 90, 161.
[31] Ch. 12 with n. 16; *LSS* 59.

From Sicily to Egypt 243

details which, quite apart from the Sophron connection, point in their own right to a Sicilian origin for Hecate and Persephone material embodied in the magical papyri. For instance there is the divine name Pasikrateia: it only occurs in this form in the Paris papyrus and in the great Gaggera sanctuary of Demeter, Hecate, and Persephone at Selinus, up the south coast of Sicily from Acragas. Here, at Selinus, it appears to have been a local cult title for Persephone.[32] The importance of Sicily as a place with its own, very vital blend of religious and magical traditions—no doubt infused, especially at Selinus, with oriental elements received from the Phoenicians[33]—should not be underestimated. It is plainly significant that the oldest surviving examples in the ancient Greek world of a certain type of magical spell come from Sicily and date from the sixth or fifth centuries BC; and it is no coincidence that the discovery of a number of them either in, or right next to, the Gaggera sanctuary at Selinus points to a very close link between the type of 'binding' magic in question and the local cult of Demeter, Hecate, and Persephone.[34] What is more, the goddess invoked at the start of the fullest and most important of these magical texts from the Gaggera is almost certainly none other than the Persephone–Pasikrateia whom we find invoked in the Paris papyrus.[35] In other words the circle linking Sicilian magic, Selinus, and the magic of the Egyptian papyri is complete. And to this evidence for contacts between Sicily and Egypt at the level of cult and magic we must add the evidence noted earlier on: evidence of Pythagorean and Orphic ideas datable to the fifth century BC being carried from southern Italy or Sicily down to Alexandria and, even further, down into the

[32] Πασικράτεια: *PGM* IV.2774, G. Kaibel, *Inscriptiones graecae*, xiv (Berlin, 1890), § 268 = Calder 15; cf. *PGM* IV.2745–7, A. Holm, *RhM* 27 (1872), 368, Farnell, iii. 137, Calder 32, Zuntz 103–5. The Gaggera sanctuary was evidently a significant site for the cult of Hecate as well as Persephone, with Hecate performing the role of gate-keeper to the main sanctuary: Zuntz 98–9; T. Kraus, op. cit. 90; Wortmann (1968*b*), 158; H. Sarian, *LIMC* vi/1. 986–7.

[33] Pace, i. 416; Sfameni Gasparro 1–19, 50–2 (Selinus), and *passim*; *LSS* 140.

[34] Dubois 39–54; Faraone in *MagH* 16; *LSS* 125–8. For the material see esp. E. Gàbrici, *Monumenti Antichi*, 32 (1927), 384–91; Jordan, *GRBS* 26 (1985), 172–80. Any Athenian influence on the earliest examples can be excluded: K. Forbes, *Philologus*, 100 (1955), 244.

[35] S. Ferri, *Notizie degli Scavi di Antichità*, 5–6 (1944–5), 168–71 (τὰν ἁγνὰν θεὸν); W. M. Calder III, *Philologus*, 107 (1963), 167–8; Dubois 50–1. For Persephone in Sicily as ἁγνή see also Richardson 181, 222, 265.

depths of Upper Egypt. These ideas related to notions about the other world, in particular to the idea of a fiery underworld; and they may also have included elements of cult initiation involving a ritual of immortalization, and ascent to heaven, as a result of throwing oneself into a *krater*.[36]

As to how and when the transmission of such mythological, mystical, and magical ideas occurred, there is no lack of options. We know for example—again from Theocritus but not from him alone—that by the third century BC a good number of Syracusans had, just like him, left their home for Egypt and become either temporary or permanent residents at Alexandria.[37] But we must also bear in mind that the compilers of the Egyptian papyri appear still to have been at the receiving end of magical traditions imported from Rome as late as the fourth century AD.[38] It would be perfectly understandable if such traditions were preserved intact in southern Italy over a period of many centuries, and in fact we see exactly the same thing happening in very similar cases.[39] Needless to say, this raises some major questions about the transmission of ancient magical, mystical, and religious ideas which we will have to come back to later when considering the history of Pythagoreanism.

These pointers to what can be described as a Sicily–Egypt connection obviously have an immediate bearing on the issue we are concerned with: that is, the striking similarity between

[36] Above, Chs. 5, 10 n. 26, 11 with nn. 9–11, 13 with n. 68.

[37] Theocr. 15; Gow, ii. 262, 267, 277, 290; Fraser, i. 65, 101, 309, 672. On the question of trade connections between Sicily and Egypt cf. ibid. 151, 153, 159. For the opposite movement of religious ideas and practice from Egypt to Sicily see Pace, i. 416–17, iii. 673–86; Loicq-Berger 59–60; Sfameni Gasparro 7–18 and *passim*; Kingsley (1994c), § 1.

[38] E. N. Lane, *The Second Century*, 4 (1984), 25–7.

[39] On 'the curious case of the fourth-century-BC gold tablet from Petelia that was apparently found enclosed in a tubular necklace dating to the second or third century AD, either carefully handed down from one generation to the next or disinterred at a later date and reused,' see R. Kotansky in *MagH* 115; Zuntz 355–6. For more certain evidence of continuity of cult tradition in southern Italy and Sicily cf. ibid. 334–5 ('. . . witnesses to a tradition strong enough to last into the age of the Soldier Emperors . . .'); Jordan 256 (a 4th-c. BC lead tablet from Selinus containing an even older text which recurs, around half a millennium later, in Latin transliteration on a silver tablet 'said to come from Rome'); Ch. 20 with n. 22. In general, for the continuous use of certain types of magical text in Sicily and southern Italy from the 6th c. BC down to the 4th c. AD see A. Audollent, *Defixionum tabellae* (Paris, 1904), 179–286, 556; Jordan, *GRBS* 26 (1985), 172–82.

From Sicily to Egypt 245

references to Hecate's bronze sandal in magical texts from southern Egypt and the details about a bronze sandal in a legend of Empedocles' death which dates back to the fifth or early fourth century BC. Given the existence not just of general affinities but also of actual links between Sicilian and Graeco-Egyptian magic, nothing could be less surprising; in broad terms, all the evidence fits remarkably well into place. However, by a stroke of good fortune we can also be more specific about the link between Sicily and Egypt in this particular case.

Reference to a bronze sandal as magical symbol of Hecate occurs not only in the great Paris papyrus but also in two other documents which, at this particular point in their texts, are very closely interrelated. These are, first, another Graeco-Egyptian magical papyrus (*PGM* LXX); and secondly a lead tablet, found at Oxyrhynchus in central Egypt, dating from around the third century AD, and only published in the 1960s.[40] The initial editor of this lead tablet assumed the text it contained was in prose, but recently it has been shown that two brief passages are in fact fragments of mystical poetry.[41] The first of these passages is the one that mentions the bronze sandal; it is separated from the second by only three brief lines of text. What is more, the parallel text in the magical papyrus does not distinguish formally between the two verse passages but combines words from both of them in a continuous sequence. These factors tend to suggest that the two verse extracts are probably related and in origin are likely to have derived from 'the same worship'.[42] This conclusion, in turn, is supported by the fact that both of the verse extracts sing the praises of the same three goddesses: Hecate, Demeter, and Persephone. And there is one other point worth noting. Naturally it is impossible to know just how much of the adjoining prose sections either in the papyrus or on the tablet are likely to derive from the same source; but it is difficult not to notice that the papyrus contains,

---

[40] *PGM* LXX.10–11; Wortmann (1968*a*), 62.57–8 = *Suppl. Mag.* i. 196.57–8, 203.1–2.
[41] For what follows see Jordan 250.57–61, 250.64–251.73, 252–3, 255–8. When considering the lead tablet it is important to bear in mind not only the parallels with *PGM* LXX but also the overall similarities in content to magical material in the Paris papyrus (ibid. 246–7, 253–5). For the occurrence of related magical texts on papyrus and on metal cf. Dieterich (1911), 44–7, 108.
[42] Jordan 258.

mixed up in the same sentence in which it quotes from the second of the two verse fragments found on the lead tablet, a reference to initiation in an underground chamber. The wording of the papyrus—'I have been initiated and I have descended into the chamber'—points to the mysteries of Demeter and Persephone, to the natural and artificial chambers used in the mysteries of the goddesses throughout Sicily and southern Italy, and to precisely the type of initiation into their mysteries which we know was practised in the West by early Pythagoreans.[43]

While it would seem safe to derive both of the verse passages on the lead tablet from the same 'chthonic cult',[44] the origin of the second verse fragment can in fact be isolated with precision. The greater part of it happens to be an almost exact reproduction of a block of verse preserved on another lead tablet which was apparently found at Selinus in Sicily and is datable to the fourth century BC. To be more exact, this Sicilian tablet already contains errors in the transcribing of the text which point to a previous history of transmission and 'suggest a model at least as early as the fifth century'.[45] This fact has major implications of its own, as we will discover later; but as we have already seen, what applies to the second verse passage almost certainly applies to the first as well. In other words we are brought back to Sicily, and to the fifth century BC, not just for the general elements of Hecate lore in the Egyptian magical texts but also—much more specifically—for the symbolism of the single bronze sandal. On the one hand, the evidence of the lead tablets helps to place on an even firmer footing the correspondence between the detail of the bronze sandal in the Empedocles legend and the symbolism of the bronze sandal in the magical papyri. On the other hand, any doubts as to the safety of tracing not only the second but also the first verse fragment on the Oxyrhynchus lead plate back to Sicily are effectively overruled by the fact that we find precisely the item

---

[43] τετέλεσμαι καὶ εἰς μέγαρον κατέβην, *PGM* LXX.13–14. For the word μέγαρον cf. Burkert 155–9 with n. 195 (parallels overlooked by Betz, 292–3). For the Pythagorean, Sicilian, and Italian evidence see further Chs. 6 (with n. 11), 9–10. The editors of *PGM*², followed by Betz (288, 292–5), claimed to find a reference in *PGM* LXX.13–14 to mysteries of the Idaean Dactyls, but this must be excluded (Jordan 257–8).

[44] Jordan 245.
[45] Ibid. 256.

From Sicily to Egypt 247

which it mentions—the single bronze sandal—featured in a Sicilian legend of the fifth or early fourth century BC.

There is no need to emphasize the significance of this Sicily–Egypt connection for our understanding of the magical traditions preserved in the Egyptian papyri. However, it should also be clear by now how valuable the Egyptian papyri and related magical documents can be for understanding the Empedoclean legends which we find circulating in Sicily very shortly after his death. The detail about the bronze sandal is an obvious case in point, but it is by no means the only example of a connection between the magical papyri and the legends surrounding Empedocles. A typical example it may be, but unique it certainly is not.

For instance, there is the picturesque story about how one day at a banquet Empedocles prevented a man from murdering someone in an uncontrolled outburst of anger ($\theta\upsilon\mu\acute{o}s$) by, at the critical moment, reciting a single verse from Homer. The verse acted like a charm, and the man was transformed in the nick of time. In essence, what Empedocles is made to do here corresponds exactly to the magical practice of reciting a single verse from Homer as a 'charm for restraining anger' ($\theta\upsilon\mu o\kappa\acute{a}\tau o\chi o\nu$): a practice referred to repeatedly in the Paris magical papyrus.[46] Our earliest written source for the Empedocles story is Iamblichus; this makes it tempting to dismiss it as a late invention dating at least from after the time of Plato and Aristotle.[47] But that would not be correct. The story in Iamblichus forms an integral part of a discussion about the early Pythagorean use of music and song for magical purposes to counteract passion—and particularly anger ($\theta\upsilon\mu\acute{o}s$)—which can be traced back, via Aristoxenus, to the pre-Platonic period;[48] it has been demonstrated that this incantatory use of poetry for harmonizing the emotions was a common magical practice among the Pythagoreans of Empedocles' time;[49] it has

---

[46] Iam. *VP* 113 = Emp. A15, plus the later versions of the story cited by Sturz, 65–6 (add Fic. *Op.* i. 651); *PGM* IV.467–8, 831–2. The connection between the Empedocles story and the passages in the Paris papyrus has been seen by Heim (517), by Boyancé (126–9), and by J. J. Winkler in *MagH* 236 n. 22. Regarding the antiquity of the $\theta\upsilon\mu o\kappa\acute{a}\tau o\chi o\nu$ as a type of magical spell cf. S. West's comments, *ZPE* 101 (1994), 11–12.

[47] So Bidez 94.   [48] Iam. *VP* 110–14; Burkert 212, 376.

[49] Boyancé 93–131; Laín Entralgo 74–82 and *passim*.

also been shown that this magical use of the spoken word for curative purposes meshes perfectly with what we can see of Empedocles' healing methods from his own poetry;[50] and, contrary to what is usually supposed, the tendency was for later biographers of men such as Pythagoras and Empedocles not to create but to tone down and rationalize, or even remove altogether, this kind of miraculous tale.[51] As in the case of the leap into Etna and the detail of the bronze sandal, the story cannot be dismissed as a late invention but can and must be taken as a reliable indication of the type of ideas and practices known to Pythagorean circles in the West during the fifth century BC. In other words, in the 'charm for restraining anger' by reciting a line from Homer we have a characteristically ancient Pythagorean practice reappearing, almost a millennium later, in the Egyptian magical papyri.[52]

And yet that, too, is only a part of the whole story. As we keep discovering, it is not just a matter of correspondences between the Empedocles legends and the magical papyri. It is also a question of links between the papyri, the Empedocles legends, and Empedocles himself. Inseparable from the Empedocles of legend, able to charm and manipulate emotions with his words, is the Empedocles who used magical incantations—the same type of incantations we find mentioned or referred to in the magical papyri—and who claimed that his own poetry could have a magical effect.[53] As for the legend about his death, we saw earlier in this chapter that it clarifies his own equation of Hades and fire by presenting it in precisely the context we had already come to expect: the context of volcanic phenomena and Mount Etna, underworld mythology and the

[50] Ibid. 82–98. See also Ch. 15, with n. 48, on the transformative power of the spoken word, and F. Pfister, *RE* Suppl. iv. 341, on Empedocles and ἐπῳδαί.
[51] Burkert 145–6; Ch. 19 with n. 14.
[52] See Heim's comments, 515, on the relations in this case between later magical practice and earlier Pythagorean tradition. Nock's remarks on the use of Homer in the magical papyri (i. 180 n. 27) are somewhat invalidated by his failure to note the Pythagorean precedents.
[53] Ch. 15 with nn. 14–15, 48; above, n. 50. With the ability of the legendary Empedocles to manipulate people's emotions we must also compare the remarks on the magical effectiveness of the spoken word made by Gorgias, who was in a number of senses a successor of Empedocles. Cf. Gorgias, DK 82 B11 §§ 8–10, 14 with Ch. 15 n. 9; J. de Romilly, *Magic and Rhetoric in Ancient Greece* (Cambridge, Mass., 1975), 3–22; Morton Smith's remarks on the continuing link between magic and rhetoric in the magical papyri, *Classical World* (1977), 396; Faraone, *TAPA* 119 (1989), 155 and n. 17.

mysteries of Persephone. Strictly speaking there is no need to depend on such indirect inferences alone, because enough evidence survives in the fragmentary remains of Empedocles' poetry to indicate his own familiarity with the darker side of the Persephone mysteries.[54] But behind all these details looms one basic fact. This is the historical reality of Empedocles as a magician—not just any magician but a magician who claimed to be able to use his magical powers to descend to the underworld.[55] In the light of this picture of his involvements and interests, it is almost superfluous to add that the passage from Lucian which provides the closest parallel to this claim of his explicitly states that such a descent to the underworld would be bound to involve encounter with, and invocation of, Hecate and the Furies as well as Persephone.[56]

In short, in confronting us with the ritual aspect of the story about Empedocles' death, the magical papyri have ended up confronting us with an Empedocles who is implicated in his own legend. The background to the story is also the background to the poet and to his poetry. What that means for our general understanding of him—and for our understanding of his equation of Hades and fire in particular—will become clearer in due course. But first of all we need to take a closer look at the story about Empedocles' death on Etna.

[54] For his reference to Baubo (B153) cf. Dieterich (1911), 127 and n. 2; Guthrie (1952), 135–7; Richardson 215–16; Graf (1974), 165–71. For the links between Baubo and Hecate see Dilthey 392–5; Rohde 590–3; Dieterich (1911), 127–8; Ganschinietz (1913), 67; U. Pestalozza, *Nuovi saggi di religione mediterranea* (Florence, 1964), 54–5. She is often mentioned by name in the magical papyri: *PGM* II.33, IV.1257, 2202, 2715 (cf. 2958), VII.692, 886, XIII.924, Hopfner, ii/2. 607, 610, MT i. 198, and twice in the lead plate which also contains the reference to the bronze sandal (Jordan 249.7–8, 250.46 = *Suppl. Mag.* i. 194.7–8, 195.46). For the relation of the Furies and Empusa—both closely linked with Hecate (above, n. 22)—to Empedocles' poetry see van Essen 63–4; G. Lanata, *Medicina magica e religione popolare in Grecia* (Rome, 1967), 34–5.
[55] B111.9; above, Ch. 4 with n. 17, Ch. 15.
[56] Luc. *Menippus* 6–11; H. D. Broadhead, *The Persae of Aeschylus* (Cambridge, 1960), 304–5; Ch. 15 with n. 33; above, n. 54.

17
The Hero

*

WHEN Empedocles' biographers presented him as always going around wearing sandals of bronze, they were generalizing falsely from some much more specific data. As so often, they were falling in with the tendency among Greek writers to rationalize and so misinterpret the details of magic, ritual, and myth.[1] Again, when they turned the detail about the discovery of the single bronze sandal—symbol of Hecate, magical sign of the ability to descend to the underworld—into a proof that Empedocles was a fraud they either did not understand the ritual symbolism or, for the sake of their parody, deliberately chose to ignore it.

In these respects the magical symbolism of the bronze sandal has been like an Ariadne's thread, allowing us to move behind the superficial parodies and distortions to something far more substantial. But this still leaves us with an obvious question: the question as to the original nature and significance of the other details in the legend before the veils of misunderstanding were drawn over it. Answering the question is not a major problem. On the contrary, there is an inevitability in how—once we perceive that behind the legend lie elements of ritual and that the leap into Etna represents, mythologically, a descent into the underworld—the question starts to answer itself. In the process it will become possible to view the details considered so far in a wider perspective. As we have already seen, Empedocles' poetry provides a context for understanding aspects of the legend about his death, and vice versa; but it has been the failure to provide a broader context for them both which has caused so many misconceptions in the past.

The first point to consider is the fact that, according to the

[1] Ch. 16 with nn. 16, 20.

The Hero

legend about Etna and the bronze sandal, Empedocles' jump into the volcano resulted in his death. In a ritual context—and, as we have seen, that is what we are concerned with—death, and especially death in the form of a descent to the underworld, is hardly ever just death. Almost invariably it is no more than the preliminary stage in a dynamic process of dying to be reborn at a different level, with a new identity.[2] As far as the Empedocles legend is concerned, it is clearly no coincidence that the ritual sequence of descent into the underworld, death, and regeneration is known to have been practised by the earliest Pythagoreans in the West. Nor is it a coincidence that—just as with the Empedocles legend—we find this Pythagorean practice too being repeatedly parodied, rationalized, and misunderstood by Greek writers who 'are either blind to, or deliberately distort, the religious significance of the facts that they report'.[3]

More specifically, dying to be reborn often had the connotation for Greeks of dying to one's mortality in order to be reborn as immortal or divine. This was true in particular for the early Pythagoreans; in the mystical circles of southern Italy and Sicily which were closely associated with early Pythagoreanism; and, to judge from the hints in the Homeric *Hymn to Demeter*, it would also appear to have played some form of role in mysteries associated with Persephone.[4] Significantly, the theme of ritual death and immortalization recurs in the Paris magical papyrus, where it culminates in an ascent into the heavens.[5] Behind the various ideas here—ritual death and descent, immortalization and ascent—it is not difficult to detect an underlying schema of descent into the underworld as prelude to a celestial ascent. This apparent illogic is the logic of

---

[2] Nock, i. 101–2; A. Brelich, *Paides e parthenoi*, i (Rome, 1969), 33–5; Seaford 261–2; Graf 100; Brashear 21–9.

[3] Eliade, *Zalmoxis* (Chicago, 1972), 24–7 (esp. 26). Cf. Hdt. 4.94–6; Hermippus, fr. 20 Wehrli, Tert. *An.* 28.2–4, Schol. Soph. *Electra* 62, *Suda*, s.v. ῞Ηδη; Burkert 155–60, Wehrli, *Hermippos der Kallimacheer* (Basle, 1974), 56–7.

[4] Cf. in general Dieterich (1923), 134–79. For early Pythagoreanism see Ap. *Mir.* 6 and Iam. *VP* 140 = Arist. fr. 191 (153.22–154.1, 155.5–7 Rose) with Burkert 159–60 and below, Ch. 19; for the gold plates (and their relation to Pythagoreanism), this chapter with the refs. in nn. 24, 27; for the *Hymn to Demeter*, Richardson 232–4.

[5] *PGM* IV.475–732 with Festugière (1950), 304 n. 1, 307 n. 6, and Fowden 82–4. For one of the more obvious features shared between this section of the Paris papyrus and Pythagorean legend see West (1971), 214 n. 1; Burkert 165 n. 249.

myth. One dies to be reborn; one descends into the depths in order to ascend.[6] Then there is the fact that by leaping into Etna Empedocles jumps into fire. For the Greeks the function of fire was above all purificatory, particularly in relation to the underworld and the dead. This fundamental idea was so strong that it managed to survive intact in spite of the cruder Christian dramatizations of hell-fire, suffering, and punishment, and it is still very much in evidence in contemporary Greek folklore and ritual.[7] Also, by Empedocles' time we find throughout the classical world the idea that in one way or another fire is able not just to purify but to immortalize as well by stripping away everything transient and corrupt. We find it in Sicilian folklore; in magical ritual; and, once again, referred to in the mysteries associated with Persephone.[8] Of particular significance in this context is the mythology about Heracles, which was extremely influential in Sicily. There it was closely linked to local ritual, had become implicated (as elsewhere) in the Persephone mysteries, and showed a noticeable tendency to gravitate towards sites of

[6] Cf. Ganschinietz's fundamental comments, 2365–6; L. Koenen, *ZPE* 9 (1972), 195; J. Kroll, *Gott und Hölle* (Leipzig, 1932), 3–4, 11 n. 6, 64–5, 313, 457 and n. 2. As the framework for Christian eschatology, cf. esp. Ephesians 4: 9–10 and the Apostles' Creed; as the essential 'initiatory schema' in shamanism, Eliade 235–6, 447 ('no one can reach heaven without having first gone down to hell'). Awareness of this basic and evidently widespread mythological dialectic goes a long way to resolving the apparent problem of whether specific visionary journeys were meant to be descents or ascents; the over-simplistic classifications adopted by e.g. Clark (33 with n. 49) and Culianu (5) inevitably falsify more than they explain. Particularly worth noting in other-world journeys is the importance of arriving at a place which simultaneously gives access to the lowest depths of the cosmos and to the greatest heights of heaven, plus the fact that this place is invariably arrived at through an initial descent or 'death'. Cf. e.g. Parm. B1.9–10, 13, and 26–7 with, respectively, Pellikaan-Engel 51–3, 59–61 (descent to the underworld), Ch. 2 n. 14 (the gates as *aitheriai*, reaching up to heaven: compare Pi. *Pyth.* 1.19 on Etna as a 'pillar reaching up to heaven', κίων οὐρανία), and A. H. Coxon, *The Fragments of Parmenides* (Assen, 1986), 10, 16, 167 with Ch. 5 n. 15 (Parmenides arrives by the road of death); Pl. *Rep.* 614b–621b; Plut. *De gen.* 589f–592f (oracle of Trophonius); Apul. *Met.* 11.23; and for general comments on the mythology of a cosmic 'axis', Eliade 259–87.
[7] Diels, *Sibyllinische Blätter* (Berlin, 1890), 47–8, Dieterich 197–201 and (1911), 476–7; C. Schmidt, *Gespräche Jesu mit seinen Jüngern* (Leipzig, 1919), 526–9; Danforth 122–31 and *passim*.
[8] Sicily: E. Maass, *Neue Jahrbücher*, 31 (1913), 633; Cook, i. 785–6. Magic: Dieterich 197 n. 7; Maass, op. cit. 629–32; C. Picard, *RHR* 107 (1933), 153. Mysteries of Persephone and Demeter: W. R. Halliday, *CR* 25 (1911), 10–11; Richardson 231–4. See in general Rohde 21, 49 n. 41, 302, 334 n. 127; Frazer (1919), 179–81; Edsman, *passim*; Richardson 240; below, nn. 16–17, 24.

The Hero

volcanic activity considered as points of access to the underworld.[9] What is more, there is clear evidence that already by the fifth century BC Heracles had become a figure of major importance for Pythagoreans in the West, as an archetype of the spiritual 'hero'; and there can be little doubt that the theme of his immortalization through fire—a theme of oriental, and specifically Phoenician, origin—along with his subsequent ascent into the heavens to join the company of the gods had become a part of this tradition at least by the time of Empedocles.[10] Against this background of mythological ideas there is little to be surprised at in the resemblance which has often been noted between Empedocles' well-known lines about the festivities awaiting the purified and divinized soul among the gods in heaven after death, and the famous passage in the *Odyssey* about Heracles' festivities among the gods after his apotheosis.[11]

The idea of fire as purifying—and immortalizing—became so deeply imbedded in the consciousness of the Greeks that even at their most facetious they were unable to escape it. So we find Diogenes Laertius composing two malicious epigrams about Empedocles' death which he considered worth placing

---

[9] Bidez 38 and n. 4 (noting the relevance to the legend of Empedocles' death); Croon 38–46; Jourdain-Annequin 273–300, esp. 275–82, 286–7, 290–4. For Heracles in Sicily cf. also Ciaceri 90–6, 275–85; Gruppe 990–3; G. Capovilla, *Eracle in Sicilia e Magna Grecia* (Milan, 1925), 3–24; Burkert (1979), 83–5.

[10] Detienne 19–53; cf. also Boyancé 242–7 and in *REA* 44 (1942), 191–216, H. Koller, *Symbolon*, 7 (1971), 33–52. The main texts are: Diod. 12.9.2–6 on Milo of Croton, plainly based on historical data (Bayet 16–23, Detienne 20–1; for Milo cf. Burkert 115, 153 n. 184, 293 n. 82, and below with n. 90), Iam. *VP* 40 and 50 (Detienne 21–4; both passages contain 'local tradition of south Italy' and probably derive, via Apollonius of Tyana, from Timaeus: cf. Burkert 100 with nn. 10 and 12, 104 n. 37, also Rostagni, i. 53–4 and, for Apollonius, below with n. 90), Iam. *VP* 152 and 155 (cf. Pi. fr. 133.3–5; Boyancé 243–4; Detienne 43–5), 222 (also via Apollonius, Burkert 100 and n. 13), Herodorus, *FGrH* 31 F1 with Porph. *VP* 34–5, F4 and 14, F19 with Aulus Gellius 1.1 = Plut. *Lives*, fr. 7 Sandbach (cf. Jacoby, *FGrH* ad locc. and Detienne 24–32), Iam. *Pr.* 21 (139.24–140.7 des Places; Detienne 33). Regarding Heracles and Pythagoreanism see also Graf's comments (1974), 146–9. For the oriental background to Heracles' death and resurrection by fire see G. R. Levy, *JHS* 54 (1934), 44–53, B. C. Brundage, *JNES* 17 (1958), 230–1, Jourdain-Annequin 127, 506–11; and in general cf. Burkert (1979), 78–83 and (1992), 87, 112–13, 124, Dalley 61–6.

[11] Emp. B146.3 (... ἔνθεν ἀναβλαστοῦσι θεοὶ τιμῆισι φέριστοι), 147 (ἀθανάτοις ἄλλοισιν ὁμέστιοι, αὐτοτράπεζοι); *Od.* 11.602–3 (αὐτὸς δὲ μετ' ἀθανάτοισι θεοῖσιν τέρπεται ἐν θαλίηις). Cf. e.g. Detienne 47–9, Wright 292 (note also Zuntz 268), and, for the imagery, Vermaseren 59 and n. 162.

as an appropriately witty summary at the end of his section on Empedocles. The first of them runs:

> There came a time, Empedocles, when you purified your body
>   with living flame
> and drank immortal fire from the mixing-bowls of the craters.
> I won't go so far as to say that you threw yourself into the
>   stream of Etna willingly,
> but that you wanted to disappear and fell in without wanting to.

The second starts:

> True, there's a story that Empedocles died when he
> fell from a chariot and broke his right thigh;
> but if he leapt into the craters of fire and in them drank Life
> ($\pi\acute{\iota}\epsilon\ \tau\grave{o}\ \zeta\hat{\eta}\nu$)...[12]

These verses were already situated in their proper context a century ago by Dieterich and Bidez. As Bidez pointed out, both of the epigrams contain a reference to the 'highly characteristic theory regarding the divine nature of fire: it is fire that gives both life and immortality.'[13] And we can press the matter further. Diogenes Laertius was not an inventive man; he was a compiler of ideas, not a creator of them. The story he mentions at the start of his second epigram about Empedocles dying by falling from a chariot was not his own, as we know from Diogenes himself.[14] Similarly the idea of Empedocles finding purification, life, and immortality in the flames of Etna was not his own invention either: he simply cites it either to criticize or mock it. So, we are left with the important conclusion that there were others before him who saw Empedocles' leap into Etna as an act intended to purify and immortalize. With this conclusion one more piece in the jigsaw falls into place: alongside the symbolism of the bronze sandal we are now presented with another aspect of the original legend before the distorting parodies and rationalizations started to set in; and, as could have been predicted from the way that the detail about the bronze sandal was manipulated and turned on its head, we can

---

[12] D.L. 8.75; Bidez 90–1. On Empedocles' 'wanting to disappear' ($\lambda\alpha\theta\epsilon\hat{\iota}\nu\ \dot{\epsilon}\theta\acute{\epsilon}\lambda\omega\nu$) see Ch. 16 with n. 1; Luc. *Peregrinus' End* 1. For the idea of 'drinking life' cf. Dieterich (1923), 172.   [13] Bidez 91.

[14] D.L. 8.73 (Neanthes). Cf. Bidez 64 and, for the detail about the thigh injury, 65 n. 2.

The Hero 255

see that all that was needed to produce the distortion was just a slight twist. According to the hostile rationalizations Empedocles jumped into Etna because, fraud that he was, he wanted to create the illusion that he had miraculously disappeared and become immortal; according to the original version he jumped in so as to become immortal in fact.[15]

Dieterich focused on another aspect of Diogenes' epigrams. He pointed out that in the first one we have a perfect example of the familiar idea that fire—whether in the form of a pyre or in the form of lightning—not only releases and immortalizes the soul but also allows it to make its way, fully purified, straight to the immortal gods.[16] So we come back again to Empedocles' own words about the fate of the purified soul after death, and to the mythology about Heracles—purified by fire of everything in him that is 'not Zeus'—rising upwards to Zeus' domain.[17] It goes almost without saying that in this mythological context nothing could be more natural than for Empedocles, already conscious of his divinity while alive but not yet separated from his physical body, to rise up to the gods via Mount Etna: the volcano which classical poets never tired of describing as 'reaching into the heavens', as 'shooting its lightnings of flame up to heaven', as 'belching into the *aither* and licking the stars'.[18]

---

[15] Cf. Horace's comments, *Ars poetica* 464–6, with A. M. Etman, Τὸ πρόβλημα τῆς ἀποθεώσεως τοῦ Ἡρακλέους ἐν ταῖς Τραχινίαις τοῦ Σοφοκλέους καὶ ἐν τῶι *Hercules Oetaeus τοῦ Σενέκα* (Athens, 1974), 58–9.

[16] Dieterich 197, citing the standard phrases such as 'purifying his limbs with fire he was off, to the immortals' (πυρὶ γυῖα καθήρας ὤιχετ' ἐς ἀθανάτους) or 'up to the gods the divine part of him flew, sifted thoroughly by the fire' (τὸ θεῖον ἀνέπτατο ἐς τοὺς θεοὺς διευκρινηθὲν ὑπὸ τοῦ πυρός) from later literature and inscriptions. On the scope and antiquity of the idea cf. Rohde 334 n. 127, 581–2, and below, nn. 17, 24.

[17] Empedocles: above, n. 11. Heracles: Theocr. 24.82–3, Ov. *Met.* 9.250–5, 262–5, Luc. *Hermotimus* 7, Wilamowitz (1909), 79–81, Edsman 234–7, Jourdain-Annequin 506–11. Cf. also Luc. *Alex.* 40 on the soul of Pythagoras returning to Zeus' realm after being struck by lightning (πάλιν ἐς Διὸς εἶσι Διὸς βληθεῖσα κεραυνῶι). For the antiquity of the Pythagoras mythology in this passage see Burkert 142, 159–60; for Pythagoras, lightning symbolism, initiation, and purification in the context of Crete, Porph. *VP* 17 (Κρήτης δ' ἐπιβὰς ... ἐκαθάρθη τῆι κεραυνίαι λίθωι) with S. Marinatos, *Kadmos*, 1 (1962), 93; P. Faure, *Fonctions des cavernes crétoises* (Paris, 1964), 110–19, 127–31, 249–50; Burkert 151–2, 376–7.

[18] Pi. *Pyth.* 1.19 (κίων οὐρανία: cf. Hdt. 4.184 and above, n. 6); Lucr. 1.722–5 (... *ad caelumque ferat flammai fulgura*); Virg. *Aen.* 3.572–6 (*prorumpit ad aethera ... et sidera lambit*). See also Lucr. 6.644–5, 669–70 with Monro ad loc.; *Aetna* 608–10, Sen. *Herc. Oet.* 285–6, Claud. *Rapt.* 1.164–5; Ch. 6 n. 29; Ch. 18.

So neatly, so closely do the various pieces of evidence fit together that it is as though behind the distortions and parodies there is an alphabet waiting to be read. And the evidence of Diogenes' epigrams shows that there were others in antiquity who knew how to write and read this alphabet. The alphabet, as we have seen, belongs to the language of ritual. One dies to be reborn, one descends in order to ascend, one burns away one's mortality before ascending to the gods. As for Heraclides of Pontus' theatrical apotheosis story, together with his polemical rejection of the details about the leap into Etna, he clearly took what he wanted but disregarded the rest; he accepted the ascent but ignored the descent. In doing so he destroyed the mythical dialectic and lost the legend's real dynamic. There is nothing surprising in that. On the contrary, as someone very close to Plato and intimately involved in the early days of the Academy, in this case Heraclides simply manifests again the same tendency we already encountered in the case of Philolaus' central fire: that is, the tendency of the early Academy to eradicate the subtleties or ambiguities of earlier mythology—and, more specifically, mythology received from Pythagorean sources—by turning a blind eye to its darker side and purging it of its dynamic complexity.

There is one other source of information which we have so far not examined: a source which in modern scholarship has been appealed to more than any other single text, or group of texts, for its relevance to Empedocles' religious ideas. This is the inscribed gold plates, intended to accompany the dead person into the underworld, which have been found in graves across much of the Greek-speaking world: in Crete, in Thessaly, but above all in southern Italy. The earliest examples are from around 400 BC and, just as with the text on the lead plate from Selinus, must be assumed to be based on an archetype dating from the fifth century at the very latest.[19]

Two points need making about these gold plates. The first concerns the issue of heroization. As we have already seen, the legend of Empedocles' death resonates with heroic undertones and invites comparison with the mythology of Heracles in

[19] Burkert in *OMG* 84; West (1975), 229.

The Hero

particular. The gold plates, too, ascribe a fundamental role to the process of heroization. In fact they present a possibility virtually unparalleled in Greek religion: the possibility for the initiate's experience to culminate, at death, in heroization and apotheosis.[20] The geographical spread of the evidence—and in particular its concentration at Thurii, one of the greatest centres for the cult of Heracles in southern Italy—leaves little doubt that this initiatory heroization schema was more or less consciously modelled on the figure of Heracles.[21] Also, to supplement the texts on the gold plates themselves we have the fragment of a poem by Pindar which closely mirrors the theological doctrine they contain, and which ends with the heroization of the liberated soul: in terms plainly reminiscent of Empedocles' own lines on the final incarnation of purified souls before their return to heaven, it describes those who live the lives of great kings, champions, or sages before ultimately being transformed into 'sacred heroes' (ἥρωες ἁγνοί). And we have Pindar's second *Olympian* as well: a poem which is preserved in full, was written for Theron of Acragas when Empedocles was still a young man, starts off in its opening sentence with a reference to Heracles, and maintains the general theme of the hero right through to its mystical passage on reincarnation and the final liberation of the soul.[22]

There is one other aspect to the issue of heroization in the gold plates which has not been given the emphasis it deserves. All three of the plates found in one of the burial mounds at Thurii, on the western edge of the Gulf of Tarentum, present the individual with whom each of them was buried as dying from being struck by lightning.[23] The continuing debate as to whether the people involved died as a result of actually being

---

[20] The evidence relating to heroization is both textual and circumstantial. For the textual evidence see Zuntz 358–9.11 (compare also 302–3.7, 304–5.7, 318); Janko 97 (where, however, the sharp distinction drawn between heroic and divine status is unjustified: cf. van Essen 58, Zuntz 345), 99; S. G. Cole, *EA* 4 (1984), 43–4 (Petelia, Thurii, Hipponium). For the circumstantial evidence cf. Zuntz 289–90, 344–5; Burkert, *Gnomon*, 46 (1974), 326 and in *OMG* 93 (Thurii). See also Boyancé's general remarks, 242–7.

[21] Bayet 417–18. Cf. also H. Grégoire's comments, *Bulletin de la Classe des Lettres de l'Académie Royale de Belgique*, 41 (1955), 445–52; Detienne 46 n. 1.

[22] Pi. fr. 133 with Boyancé 247 (cf. Emp. B146); *Ol.* 2 with B. L. Gildersleeve, *Pindar: The Olympian and Pythian Odes* (London, 1893), 142 and Nilsson's comments, ii. 665–6.

[23] Zuntz 300–1.4, 302–3.5, 304–5.5.

struck by lightning, or whether the lightning was part of a symbolic ritual instead, has directed attention away from one fundamental point. For the Greeks death by lightning, which was seen as the purest form of fire, was a standard way of attaining heroic status. Considering the importance of the theme of heroization with regard to the gold plates—and that includes the ones from Thurii—this is rather obviously what is being alluded to here; and the obvious becomes even more so when we compare the surviving evidence for people who died from lightning becoming the centre of a hero cult at neighbouring Tarentum.[24] In other words these references in the gold plates to death by lightning are inseparable from other mythological allusions of the same kind, such as the legend of Heracles' being taken up to heaven by lightning after stepping onto the funeral pyre or the stories of Pythagoras' death—or initiation—through an actual or ritual thunderbolt.[25] In short, through the gold plates we encounter not only the theme of the spiritual hero but also, more specifically, the theme of heroization as an immediate result of death—whether actual or symbolic—by fire.

There is also a second reason for considering the gold plates, a reason intimately related to the first. This is for what they have to tell us about immortalization. As has been noted time and time again, the startling declarations of divinity and immortality which we find on the plates—'happy and most blessed, you will be a god instead of a mortal' on one of them, 'from a man you have become a god' on another—provide the most direct and eloquent parallels to Empedocles' own famous assertion that he has become 'an immortal god, no longer

---

[24] For lightning, death by fire, and heroization see Rohde's fundamental comments, 100–1, 193 n. 68, 576 n. 154, 581–2 and, regarding the gold plates, 448 n. 54. The heroic implications of the death by lightning mentioned on the gold plates were clearly seen by Cumont, 330–1. For heroization and the gold plates from Thurii see the refs. above, n. 20; for the evidence from Tarentum, Clearchus, fr. 48 Wehrli. Examples of recent literature on the death by lightning in the Thurii plates and its literal or symbolic significance are: Zuntz 314–16; Burkert in *OMG* 93–5; Lloyd-Jones 274–5; Seaford, *HSCP* 90 (1986), 4–6. It is worth emphasizing that, for the Greeks, death by lightning as punishment for a crime was in no way incompatible with death by lightning as a prelude to heroization: cf. e.g. Clearchus, loc. cit., Rohde 581–2.

[25] Heracles: Rohde 581, Edsman 234–5. Pythagoras: above, n. 17, and cf. Wieten 113–14.

mortal'.[26] Just as significant, in the light of what was said earlier about the dynamic of death and rebirth implicit in the legend of Empedocles and Etna, is the fact that we can be more precise about the process of immortalization referred to on the gold plates. New discoveries plainly demonstrate what a number of scholars had already suspected: that it was considered the end result of a dialectical sequence of death followed by rebirth or regeneration 'into another condition'.[27] In other words, what the plates have to say on the theme of immortalization provides striking analogies not only to Empedocles' own poetry but also to the legend about his death; and it is justifiable to suppose that anything we are able to deduce about the background to these declarations in the gold plates is also likely to throw light on both the historical and the legendary Empedocles.

First of all, a few words need saying about the nature and identity of the gold plates. In the nineteenth century the habit gradually evolved of describing them as 'Orphic', and for a number of reasons they came to be associated by several scholars with the Bacchic mysteries of Dionysus.[28] But along with the age of scepticism towards all matters Orphic, inaugurated by Wilamowitz in particular, these trends became more or less a thing of the past. When Zuntz published his *Persephone* in 1971, he could seriously claim that these so-called Orphic plates were completely misnamed: that they were purely Pythagorean in nature and had no connection whatever with anything Bacchic or Orphic. However a new gold plate, found at Hipponium in 1969 and published in 1974, introduced a crucial new factor: it presents the same material found on a number of the other gold plates in an explicit context of

[26] Zuntz 300–1.8, 328–9.4 (cf. also 333, A5.4); Emp. B112.4 (Ch. 15 n. 7). See e.g. Guthrie (1952), 175; Gernet 17; Zuntz 252.

[27] *TGL* 10–11. This conclusion had already been arrived at by e.g. Dieterich (1923), 171–3, Wieten 102–19, on the basis of the gold plates known since the 19th century; the newly-found gold plates from Thessaly—'Now you have died and now you have come into being, thrice-blessed, on this very day' (*TGL*, loc. cit.)—vividly confirm it. Cf. also Graf 99–100. With the dramatic emphasis on 'now ... now you have come into being ... on this very day' (νῦν ... νῦν ἐγένου ... ἄματι τῶιδε) it is worth comparing Acts 13: 32–4 with Dieterich's comments (1923), 177. But also relevant is the standard insistence in magical texts on the desired action being accomplished 'now, now ... on this very day' (ἤδη ἤδη ... ἐν τῆι σήμερον: *Suppl. Mag.* i. 33.6–7, 212.9–10, cf. 24.12–14, 31.15–18, 39.6–40.11, and *passim*, *PGM* I.262, VII.471–3). On the relation of the gold plates to magic see Ch. 19.

[28] Cf. e.g. Wieten, *passim*, with refs. to earlier literature.

initiation into the Bacchic mysteries.[29] After the publication of this new text Zuntz tried to cover his tracks by claiming that the reference in its final line to initiates into the Bacchic mysteries (μύσται καὶ βάχχοι) was 'untypical' of the gold plates as a whole and simply an idiosyncratic, insignificant deviation from what he considered their essentially Pythagorean tenor and content.[30] And yet the evidence strikingly fails to bear his assessment out. The text on the plate from Hipponium is not only the fullest and longest of the series discovered so far, but also the earliest; and various factors point to the conclusion that its reference to Bacchic mysteries occurred in the archetype from which a number of the other gold plates derive as well.[31] Next, as though to bring the point home even more forcefully, two further gold plates—this time in the shape of ivy leaves— were discovered in 1985 in Thessaly. From the point of view of content, they bear a very close relation to the various gold plates found in southern Italy; and yet they shed a dramatic new light on them by highlighting not only Persephone—who was central to the earlier texts—but also Dionysus *Bacchios* as key figures in the liberation of the soul.[32] A Bacchic background for the gold plates, discussed and interpreted as Pythagorean by Zuntz, now appears guaranteed.

However, Zuntz was not entirely wrong. As he well appreciated, the gold plates are not to be distinguished from Pythagorean ideas and ideology; on the contrary, the evidence suggests that they represent a vital aspect of the religious landscape in the Greek West which early Pythagoreans rapidly assimilated and inherited.[33] But that appears to present us with a major problem: any direct point of contact between Pythagorean ideas and Bacchic mysteries would seem incompatible with the familiar concept of Pythagoreanism as a purely

---

[29] Cf. esp. G. Pugliese Carratelli, *PP* 29 (1974), 108–26, 139, and 31 (1976), 462–6; Burkert (1977), 2–4; Cole 223–38; Feyerabend 1–10.

[30] Zuntz (1976), 145–51.

[31] Janko 89–100, Feyerabend 1–10; cf. also Pugliese Carratelli, *PP* 29 (1974), 122, Lloyd-Jones 263–70, Burkert (1985), 293–6, Graf 93. On the rationale behind the process of abbreviation in the later gold plates see Janko 91, 97.

[32] *TGL* 3–16; Graf 89–91, 97–9. As both Tsantsanoglou and Parássoglou (*TGL* 5, 10) and Graf (95–7) point out, the Thessaly texts effectively undermine Zuntz's strict segregation of the gold tablets into two separate categories, A and B.

[33] Burkert 112–13; Kingsley (1994c), § 1. Cf. also Dieterich (1911), 91; Wieten, *passim*; Zuntz 321–2, 337–43, 379–85; Pugliese Carratelli, op. cit. 118–19, 122–3, 143.

Apollonian, upwardly mobile movement. An obvious solution to the problem is to find a bridge between the Bacchic element and the Pythagorean which links them while at the same time keeping them apart, and the perfect candidate for such a bridging factor is close to hand: Orphic literature. By reinstating the gold plates as 'Orphic' texts, all criteria appear to be fulfilled and all the problems bypassed.[34] The bridge certainly seems to hold strong at every level. In general terms, enough has already been said about the use and production by Pythagoreans of Orphic literature centring on the fate of the soul after death.[35] On the other hand, the links between Orphic texts and Bacchic mysteries are undeniable. It has often been noted that, in its relation to the mysteries of Dionysus, Orphic literature shows a concern not with the coarser revels and celebrations but with a more esoteric aspect of the mysteries, with a kind of mystery within the mysteries focusing on the fate of the soul after death; recent discoveries from throughout the Greek-speaking world have only served to emphasize this link between the Orphic and the Bacchic in relation to burial rites, death, and life after death.[36] And to return to the gold plates: although for a long time parallels have been observed between what they say and what later sources tell us about Orphic theology, the precise way in which the two new samples from Thessaly mention Dionysus together with Persephone ties the genre in even more directly to the little we know of Orphic sacred poetry.[37]

[34] Cf. Burkert in *OMG* 81–104, 184, Graf 87–102.
[35] Above, Chs. 10–12. Cf. also Lloyd-Jones 264–6.
[36] Cf. e.g. Graf 98 and, for observations on the increasing evidence for close ties between Orpheus, Orphic literature, Orphic rites, and Bacchic mysteries, Burkert (1977), 4, 6–8, (1982), 4–5, 8, (1985), 294–5, 297–8; West 16–18, 24–6, and in *ZPE* 45 (1982), 27–9; L. Zhmud', *Hermes*, 120 (1992), 162–3.
[37] Cf. already *TGL* 11–12; also Graf 90–1, and for earlier literature Boyancé 77–8 with n. 7. The most important texts to compare with the new gold plates—and in particular with the sentence 'Tell Persephone that Bacchios himself has liberated you'—are *OF* fr. 229, where Proclus mentions 'those who in the writings of Orpheus have become initiates of Dionysus and Kore' and then quotes the same Pythagorean doctrine of liberation from the 'cycle' which is cited on the longest of the gold-plate texts from Thurii (Zuntz 300–1.5, 321); and fr. 232, which presents Dionysus the Liberator as offering humans 'freedom from the crimes of their ancestors' in terms plainly reminiscent both of the Italian gold plates (Zuntz 302–3.4, 305.4) and of the doctrine alluded to earlier by Pindar (fr. 133; cf. Kern, *OF* p. 246 on fr. 232, Graf 9–10) and, later, by Plato (*Phaedrus* 244d–e, 265b; to Plato's 'madness' corresponds the

At the same time, however, the idea of making Orphic literature a bridge between the Bacchic and the Pythagorean is bound to prove misleading. From recent discussions of the matter, it is easy to come away with the impression that the Orphic factor is needed to create some kind of buffer between two fundamentally incompatible elements: the Pythagorean and the Bacchic.[38] And yet this idea of incompatibility is one that should never have been given too much credence. Apollo may have been important for Pythagoreans, but at a purely theoretical and general level we need to bear in mind that the Apollonian and Dionysiac elements of Greek life were intimately intertwined.[39] At a more specific level, there are telltale pieces of evidence pointing to a direct link between Pythagoreans and Bacchic mysteries. Philolaus is credited with a work called *Bacchae*: a piece of information which it is highly dangerous to attempt to argue away. The historical Archytas refers in his writings to details from Dionysiac ritual, which is hardly surprising when we consider the major religious importance of Dionysus at Tarentum—not to mention the fact that, as ruler of the city, Archytas will no doubt have been expected to preside in person at the performance of the local mysteries of Persephone and Dionysus.[40] But even more significant is the context in which we came across these details about Philolaus and Archytas in the first place: in the context of an Orphic poem apparently called the *Krater* or 'Mixing-Bowl'—an archetypally Bacchic symbol—and written by Zopyrus, a Pythagorean roughly contemporary with Philolaus and a generation older than Archytas.[41] Again, it is in that web of

Orphic 'frenzy', οἶστρος, of *OF* fr. 232.5). For the terminology of liberation in Plato's *Phaedo* and its relation to Dionysiac vocabulary, cf. *TGL* 12.

[38] So e.g. Cole 238, Graf 97 n. 26 (classification as Bacchic 'appears to exclude the Pythagoreans').

[39] Cf. Rohde's classic exposition, 287–91; Burkert (1985), 222–5; G. Casadio, *Religioni e Società*, 9 (1990), 135–8. It is worth noting how Heracles, in particular, managed to maintain the closest alliance with both Dionysus and Apollo.

[40] Ch. 12 with nn. 50 (Philolaus), 57 (Archytas); Feyerabend 19 with n. 66.

[41] Chs. 10–12; cf. also Feyerabend 21 and n. 77, with Ch. 12 n. 41, and the scene of Bacchic revelry in Plut. *De sera* 565e–566a, with Ch. 11 n. 17. It is against this background that we need also to assess the tradition which credited Arignote, the daughter of Pythagoras, with a work called *Bacchica*—'which is about the mysteries of Demeter'—and a further work called *Dionysiac Rites* (Τελεταὶ Διονύσου). Cf. the *Suda*, s.v. Ἀριγνώτη = Thesleff (1965), 51; Clem. *Str.* 4.121.4; Maass 163 n. 61; Thesleff

allusions to Orphic sacred poetry which shows through at various points in Plato's *Phaedo*—and which, as we have seen, was part of the legacy to Plato from the Pythagoreans—that we find the single verse which perfectly encapsulates the notion of a Bacchic esotericism. This is the verse, cited by Plato in a discussion about the fate of the soul after death, which refers to the narthex or giant fennel stalk, a famous attribute of Dionysus and also a ritual object apparently given to initiates into his mysteries:

> There are many who hold the narthex, but the *bacchoi* are few.[42]

The fact that in isolating this direct link between Pythagoreanism and what can be described as Bacchic esotericism we are drawn into the orbit of Orphic literary tradition is, naturally, significant. Even though the appearance of a fundamental conflict between the Bacchic and the Pythagorean may just be an illusion, and even though there may be no need for a third party to keep the two apart, the Orphic factor is certainly a point where the two terms meet. Here, at this meeting point of the Bacchic, the Orphic, and the Pythagorean, we encounter one piece of evidence which has come to appear increasingly meaningful in the light of the recently discovered gold plates. Burkert has pointed out that the plate from Hipponium in particular—with its combined mention of Bacchic initiation and of ideas also familiar to us from Pythagoreanism—not only clarifies but also helps to confirm the text of a disputed passage in Herodotus; the passage was very probably written in southern Italy during the second half of the fifth century BC, and so is closely related to the plate in both time and space.[43] According to the fuller version of this text, Herodotus states that certain 'so-called Orphic and Bacchic' burial rites are 'really Egyptian and Pythagorean'.[44] In other words Herodotus,

---

(1961), 11. Thesleff's assertion (ibid. 104) that the *Bacchica* attributed to Arignote appears to 'reveal an antiquarian interest in mysteries, but no actual engagement' is quite arbitrary.

[42] Pl. *Phd.* 69c = *OF* frs. 5, 235; West 159; Henrichs (1982), 228 n. 137; above, Ch. 10.

[43] Burkert (1985), 293–4. For Herodotus at Thurii cf. F. Jacoby, *RE* Suppl. ii. 205–47.

[44] Hdt. 2.81. A shorter variant of the text preserved by one MS family simply refers to Orphics and Pythagoreans; but apart from the way that the longer version, with its reference to Bacchic rites, is clarified and justified by the Hipponium find (cf. e.g.

living and writing at a time when the gold plates were actually being used, considered himself able to discern Egyptian origins and Pythagorean affinities in funereal practices which were referred to by his contemporaries as either Orphic or Bacchic. The question of Egyptian influence is one we will come back to later.[45] As for implying that practices which went by the name of Orphic were really Pythagorean in origin, he had good reasons for doing so—except that in this particular case he was probably reversing the direction of the indebtedness.[46] Otherwise, his summarizing of certain religious practices as Orphic, Bacchic, and Pythagorean provides a good working definition for the handful of gold plates that still survive.

In purely general terms, it would appear that the gold plates need to be understood in a Bacchic and Orphic context. But where their central theme of death, rebirth, and immortalization is concerned we can also be rather more specific. Immediately after the declarations on the plates from Thurii about becoming 'a god instead of a mortal', or 'from a man becoming a god', we find the notorious saying 'I am a kid who has rushed for the milk' (ἔριφος ἐς γάλ' ἔπετον).[47] From the sequence of the statements here it is quite clear that this imagery of a young goat rushing for milk is itself meant to be understood as referring to the process of immortalization.[48] The enigmatic nature of the saying has, understandably, given rise to heated debates about its exact meaning and significance. By the early twentieth century it had become common to interpret the milk

Burkert, loc. cit.), it was in fact always to be preferred to the shorter version on purely textual grounds. See Nock, *Conversion* (Oxford, 1933), 277 (Linforth's retort, 40, is unsatisfactory).

[45] Ch. 18 with n. 32.

[46] For Herodotus' good reasons cf. Nock in *Studies Presented to F. Ll. Griffith* (London, 1932), 248; Ch. 10 with the refs. in n. 10. For the inversion see Nock, loc. cit.; above, with n. 33.

[47] Zuntz 300–1.8–9, 328–9.4 (with ἔπετες for ἔπετον); cf. *TGL* 10, 13 for the variants on the plates from Thessaly. For πίπτειν ἐς = 'rush to' see *TGL* 13, confirming the explanation already provided by Vollgraff, 22–4, van Essen, 59; and also Zuntz 319.

[48] On this point there can be no doubt; cf. e.g. Festugière's comments (1972), 37 n. 10 *ad init.* The inference is further confirmed by the new plates from Thessaly: *TGL* 13.

The Hero

and the kid as allusions to Bacchic mysteries.[49] At a very obvious level this interpretation has the evidence on its side. One of Dionysus' ritual titles appears to have been 'the kid' (ἔριφος); the habit of associating or identifying him with a kid is specifically linked in our ancient sources with Metapontum, the famous centre of Pythagoreanism a little to the east of Thurii on the Gulf of Tarentum; and, last but by no means least, abundance of milk is a standard feature of Bacchic imagery.[50] However, another more recent line of interpretation—which developed alongside the tendency to either deny or, at the very least, minimize any trace of Bacchic influence on the gold plates—has been to dismiss the saying about the kid and the milk as no more than a proverb describing the attainment in abundance of something one has hoped for. In other words, the imagery of milk and kid has no significance and is simply to be understood metaphorically.[51]

This reduction of the saying to the status of a mere metaphor is unacceptable for two different categories of reason: one external to the text of the gold plates, the other internal. First of all there is the basic fact that the saying about the kid rushing for milk needs, as the context shows, to be understood in terms of immortalization: obviously of relevance here is the further fact that later in antiquity initiates were given milk to drink as part of a ritual of rebirth and divinization in the Attis mysteries, which in their origins and development are related to the

[49] Cf. e.g. Dieterich (1911), 95–7; Thomas 140; DK i. 17.13 n. The best overall discussion of the saying remains Wieten's, 40–5, 119–41; cf. also Griffiths 295–6. The significance of the new gold plates from Thessaly is well summarized in *TGL* 13 (in contrast to Graf's somewhat over-hasty conclusions, 93–4).

[50] Dionysus 'the kid': Hesychius, s.vv. Εἰραφιώτης (καὶ ἔριφος παρὰ Λάκωσιν) and Ἔριφος, Apollodorus, *Bibliotheca* 3.4.3 with Frazer ad loc., and cf. Cook, i. 674–5, Vollgraff 21 and n. 1, Farnell, v. 107–8, 130–1, 164–9, 232–7, 303. Dionysus ἐρίφιος at Metapontum: Apollodorus, *FGrH* 244 F132, Thomas 140–1, Wuilleumier 499. Emending Hesychius' double mention of ἔριφος to ἐρίφιος simply on the basis of this Apollodorus passage is unjustified, particularly in the light of the mythological and ritual evidence for Dionysus' transformation into, or identification with, a kid or goat (Apollodorus, *Bibliotheca* 3.4.3 and Cook, Vollgraff, and Farnell, locc. citt.; correctly Kerényi 245 and n. 171). For the milk imagery see below, nn. 55, 69.

[51] R. S. Conway, *Bulletin of the John Rylands Library*, 17 (1933), 76; Nilsson, *Geschichte der griechischen Religion*, ii (Munich, 1950), 225; Zuntz 323–7; Lloyd-Jones 276. The parallel from Aelian recently adduced by Graf to support his own interpretation of the saying as a metaphor for 'going back to the beginning' is both inadequate and unilluminating (Ael. *VH* 8.8 ἐν γάλαξιν εἶναι = 'to be at the breast', cf. e.g. Pl. *Tim.* 81c, *Laws* 887d; Graf 95).

earlier mysteries of Dionysus.[52] To this evidence we have to add a passage in one of the magical papyri which refers, too, to milk in the context of a ritual of divinization; and also a strange passage in the *Katasterismoi* attributed to Eratosthenes, which brings together the themes of suckling milk, of Heracles as prototype for all 'sons of Zeus', and of the attainment of celestial immortality.[53]

Secondly, there is the internal evidence of the gold plates themselves. To begin with, on principle there is little point in attempting to deny the occurrence of Bacchic imagery here when it seems clear—thanks to the new discoveries in Hipponium and Thessaly—that the gold plates as a whole need to be understood in a Bacchic context; on the contrary, what we can now infer about this context simply provides extra reinforcement for an interpretation of the statement about the kid and the milk which was always more or less self-evident. In fact, even before the recent finds, the case for seeing the kid and the milk as Bacchic allusions was somewhat understated. We only have to consider, for example, that Dionysus is not just associated in classical sources with kids, but is also associated with them specifically in the context of suckling milk.[54] Then there is the matter of the new gold plates from Thessaly. Abundance of milk is a characteristic feature of Bacchic imagery, but very typically it is mentioned alongside a reference to wine: on the overtly Bacchic plates from Thessaly we find a bull and a ram—rather than a kid—rushing for milk, immediately followed by a reference to wine.[55] And yet the

---

[52] Sallustius, *De deis* 4 (γάλακτος τροφὴ ὥσπερ ἀναγεννωμένων...). Cf. Nock (1926), pp. liv–lv with nn. 70–1, 73, and P. Borgeaud in Graf 100. Graf's own dismissal of the Sallustius passage as 'isolated in the material relating to ancient mysteries' (100–1) flies in the face of Nock's observations on the passage (loc. cit.), of the evidence—cf. e.g. Dieterich (1911), 497; Burkert (1979), 102, 104—for fundamental links between Bacchic and Attis mysteries, and of the evidence supplied by the magical papyri (below, n. 53).

[53] *PGM* 1.20–1 with Dieterich (1923), 171–2, Harrison 596–7, Eitrem 101–2; Eratosthenes, *Katasterismoi* 44 with Detienne 46 n. 1. On the authorship of the *Katasterismoi* cf. R. M. Bentham, 'The Fragments of Eratosthenes', Ph.D. thesis (London, 1948), pp. xxviii–xxix; for its reference to 'sons of Zeus' (οὐ γὰρ ἐξῆν τοῖς Διὸς υἱοῖς τῆς οὐρανίου τιμῆς μετασχεῖν εἰ μή τις αὐτῶν θηλάσει τὸν τῆς Ἥρας μαστόν) see above, with n. 17, and Zuntz 333, A5.2 (Διὸς τέκος) with Olivieri 19.

[54] *Etymologicum Magnum*, s.v. Εἰραφιώτης; ibid., s.v. ἔρεψα = *SH* § 1045. Cf. Vollgraff 26 n. 3.

[55] *TGL* 10 (a6, b5). For Bacchic milk and wine cf. Eur. *Bacchae* 142, 706–10; Horace, *Odes* 2.19.10–11; Philostr. *VA* 6.11 and *Imagines* 1.14 (οἶνον ἐκ πηγῶν γάλα τε οἷον ἀπὸ

Thessaly plates help also in another respect, thanks to the explicit evidence they provide for the process of immortalization mentioned on other gold plates being the end result of a sequence of death and rebirth: on them, the refrain about rushing for milk comes just after their initial description of the initiate as 'coming into being' or 'being born'.[56] It is difficult not to be struck by the obvious connection between this theme of birth or rebirth on the one hand, and the imagery of milk on the other; it becomes altogether impossible to ignore the connection when we bear in mind the later parallel from the mysteries of Attis about initiates being given milk 'as though they were being reborn' (ὥσπερ ἀναγεννωμένων).[57]

There is one more factor here which also seems to need taking into account. In the longest example of the series of gold plates recovered from Thurii, the assertions about becoming a god instead of a mortal and being a kid who has rushed for milk come just after the statement: 'I have made straight for the breast of Her Mistress, queen of the underworld.'[58] The most varied interpretations, and even translations, have been offered for this line; but the parallels cited by Zuntz from Homeric literature put its meaning beyond any doubt. The individual in question makes straight for the breasts of Persephone, queen of the underworld, just like an infant to the breast of its nurse or mother.[59] Ultimately, only prejudice and preconception can

μαζῶν); Nonnus 45.306–10; C. Bonner, *TAPA* 41 (1911), 185. For Dionysus as bull see Farnell, v. 284–5, 302, 303; *TGL* 13 and n. 18; Graf 94 n. 20. For the ram in a Bacchic context cf. *OF* fr. 31 1.10 (West 171), also Kerényi 379.

[56] *TGL* 10 (a1, b1); above, with n. 27.
[57] Above, n. 52.
[58] δεσποίνας δ' ὑπὸ κόλπον ἔδυν χθονίας βασιλείας, Zuntz 300–1.7.
[59] Zuntz 319 ('like an infant which finds safety and rest ὑπὸ κόλπωι of its mother or nurse, thus he has found his haven with Persephone. . . . Persephone's ward "rushes", or "dives", trustfully for safety to his goddess, as . . . a child to its mother or nurse'). Zuntz's parallels—esp. *Il.* 6.400, 467–8, 483, 8.271–2 (δύσκεν for ἔδυν on the plate), *Dem.* 187 (ὑπὸ κόλπωι), 231, 285–6; add Homeric *Hymn* 26.3–4—exclude any interpretation of the words on the gold plate as describing the initiate passing 'under the lap' or out of the womb of Persephone and being reborn: so e.g. Burkert (1985), 295 with n. 24. In comparing Pl. *Rep.* 620e–621b Burkert seems to mean by rebirth physical reincarnation, but this interpretation is anyway unacceptable: the initiate has just been described as escaping from the 'cycle', which is no doubt the cycle of reincarnation. Cf. Zuntz 300–1.5, 320–2, and e.g. G. Casadio in P. Borgeaud (ed.), *Orphisme et Orphée* (Geneva, 1991), 135. When considering the initiatory διὰ κόλπου in the mysteries of Sabazius, it is important to note Arnobius' clear distinction (*Adversus nationes* 5.21) between *in sinum* and *ab inferioribus partibus atque imis*.

justify failing to see in this and the other statements on the gold plates the use of a consistent, coherent, and starkly simple imagery: a new birth, making straight for the maternal breast, rushing for milk.[60] What lends an extra dimension of significance to this imagery is the fact not only that the kid was an animal associated and identified with Dionysus, but also that according to Orphic theology the mother of Dionysus—from whom he was brutally separated at the start of a sequence of events which was to result in the creation of humanity—was Persephone.[61]

Zuntz was well aware, even without the explicit evidence of the Thessaly plates about a new birth, of this thread of imagery and of its more immediate implications. But he would have none of it:

The speaker is standing before the chthonian Goddess. Is he, the *renatus*, rushing to suck the milk of immortality from her *lactea ubera*? This idea, though quite proper with Egyptian devotees of Isis, makes him shudder who has the slightest notion of Persephone, the goddess of the dead.[62]

However, there were aspects of Persephone which Zuntz deliberately ignored. In particular he dismissed any serious consideration of the Persephone presented in Orphic mythology, which he intermittently showered with various forms of horrified abuse.[63] But, as we have already seen in passing, the theology of the gold plates has unmistakable affinities with

[60] The point is well noted in the *editio princeps* of the Thessaly plates, *TGL* 13 with n. 16. As with the saying about the kid and the milk, the attempt has been made to reduce the statement about rushing for Persephone's breast to nothing but a metaphor and mere banality: cf. Festugière (1972), 37 n. 10 ('... l'expression est tout à fait banale ...'). But making for the breast of the queen of the underworld is not, by any stretch of the imagination, a metaphor for 'sinking into the bosom of the underworld'; 'the bosom of Persephone' is not on the same level as the expression 'the bosom of the sea'; and even in *Il.* 6.135-7 there is a clear idea of Thetis maternally gathering the vulnerable and frightened Dionysus to her protective breast (cf. also schol. on 135 Διώνυσος δὲ φοβηθείς, ii. 154.14–15 Erbse).

[61] *OF* frs. 58, 195, 198, 210, 303; Orphic *Hymn* 30.6–7 (not 44.3–9); West 74, 95, 252; Burkert (1985), 200, 297–9.

[62] Zuntz 324. The main instigator of the explanation attacked by Zuntz was Dieterich (1911), 96–7 (here Zuntz's *lactea ubera*); cf. also G. E. Rizzo, *Dionysos Mystes* (Naples, 1914), 77, V. Macchioro, *Zagreus*¹ (Bari, 1920), 85–6, Vollgraff 24–6, 52, van Essen 59.

[63] 80 n. 3, 81 n. 3 ('mythographical eccentricities', 'hotch-potch of mythological curiosities'), 86 n. 3, 338, 398.

ideas presented in Orphic poetry;[64] and there are times when, 'shuddering' apart, it is best to let the general trend of the evidence speak for itself. This applies to the imagery of the milk, to the Bacchic and Orphic framework of the gold plates, and to the association of Dionysus with a kid. Lastly, it may even apply to the capacity of a Bacchic initiate to imitate or identify with Dionysus while remaining, in certain fundamental respects, distinct from him. In an attempt to forestall any interpretation of the saying about the kid and the milk which would involve suggesting some mystical bond of identity between the initiate and Dionysus, Zuntz claimed that to aspire for union with a god may have been acceptable in Egypt but 'it is beyond the horizon of Greek religion': that 'it is axiomatic that no Greek cult of any kind ever aimed to achieve identity of god and worshipper, alive or dead', and that 'no man or woman was ever thought to have been transformed into Dionysos'.[65] In fact the evidence does point to imitation of, impersonation of, or identification with Dionysus in the context of Bacchic mysteries during Hellenistic and later antiquity, and this is almost certain to have its roots in the mysteries of an earlier period.[66]

There is also another point worth noting here in passing. This is the fact that—not at all surprisingly, when we consider the very close affinities between the god Dionysus and Heracles the hero—the idea of imitating or identifying with Dionysus in later times often tended to go hand in hand with the idea of imitating Heracles. Imitation of Heracles is a theme which we will soon come back to.

While still on the subject of milk and Bacchic imagery, there is one other text that needs mentioning. This is the fragment of verse preserved on the two lead plates referred to earlier: the one from Selinus, in Sicily, and the other from Oxyrhynchus in Egypt. It will be remembered that the Oxyrhynchus plate

[64] Above, with n. 37.
[65] Zuntz 325-6.
[66] Rohde 258, 262; Nock (1926), p. liv n. 71; Dodds (1960), 82-3; Kerényi 350-7; Henrichs (1982), 157-8; S. G. Cole, *EA* 4 (1984), 46. The distinction which Boyancé attempted to draw, *REA* 68 (1966), 52, between imitation of Dionysus and identification with him might seem both convenient and attractive; however, in practice it is impossible to maintain.

Chapter 17

contains two fragments of mystical poetry which evidently derive from the same chthonic cult; in the case of the first fragment a Sicilian original can be inferred, while the greater part of the second in fact occurs as well on the much earlier lead plate from Selinus.[67] The first of these two verse fragments from Oxyrhynchus alludes in passing to the one bronze sandal of Hecate, holder of Tartarus (τῆς Ταρταρούχου χάλκεον τὸ σάνδαλον), and ends with a plea for salvation to Persephone, daughter of Demeter (σῶσόν με, σωσίκοσμε Δήμητρος κόρη). The poetry of the second fragment—as also preserved, in only slightly better form, on the lead plate from Selinus—is exquisitely evocative. Before going on to mention Hecate and the dark terrors of the underworld it appears to sketch a scene—'under the shadowy mountains in the darkly gleaming land'—in which a goat, sacred animal of Demeter, is tugged along out of Persephone's garden at twilight by a child to be delivered from the burden of 'the tirelessly flowing stream of its fresh milk'.[68] The imagery here—the pastoral scene, the abundance of milk, the reference to the 'child' leading the goat, and even the mention of Demeter and Persephone—is, taken as a whole, plainly Bacchic in nature.[69] But also worth noting is the fact that, in the verse fragment as it continues on the Oxyrhynchus tablet, we find Aphrodite's name directly juxtaposed with Persephone's in a way which naturally implies the two goddesses were to at least some extent being

---

[67] Above, Ch. 16; Jordan 250.57–61, 250.64–251.73, 256–7.

[68] For a reconstruction of the Oxyrhynchus text with the help of the version from Selinus, cf. Jordan 256. I base the end of my paraphrase on the reading βριθομένη[ν] in the Selinus text instead of the θεσόμενον (probably to be construed as θησάμενον: cf. Suppl. Mag. i. 204) from Oxyrhynchus. My special thanks to David Jordan for providing me with more detailed information about the Selinus text than is presented in his published paper.

[69] For the milk cf. e.g. Eur. Bacchae 142, 700, 708–10 and Hypsipyle, fr. 57 von Arnim, Pl. Ion 534a, Orphic Hymn 53 (heading plus text), Antoninus Liberalis 10, Nonnus 45.298–303, 309–10, H. Usener, RhM 57 (1902), 177–8, Vollgraff 21–6, Dodds (1960), 142, and above, n. 55; for the goat, Henrichs (1982), 157–8 and above, n. 50; for the combination of Dionysus, Persephone, and Demeter, Thomas 141–54, Richardson 319–20, Burkert (1985), 222, 294, Lloyd-Jones 263, Feyerabend 2, 5 with n. 20. For the mountains cf. also Eur. Bacchae 72–87, 116, 165, 726, 977–86 and passim; Wuilleumier 498; Henrichs, ZPE 4 (1969), 225–34 and HSCP 82 (1978), 148–9; Burkert (1985), 291, 293. With the mention of the child (π[αῖς) on the Oxyrhynchus tablet is safely reconstructed from the Selinuntine version) tugging at the goat we have another very probable connection with Dionysus: cf. Henrichs (1982), 149–50.

identified.[70] Even though this juxtaposition is cited outside its original context,[71] it is clearly significant. Zuntz devoted the central, and climactic, part of his book to demonstrating that the merging of attributes of Persephone and Aphrodite occurred at one cult centre in particular: at Locri, on the southern tip of Italy. He also drew attention to the fact that, paradoxically, the fullest evidence we possess for this merging of their attributes at Locri is to be found not in Italy itself but in a place where artefacts from Locri were exported *en masse* and subsequently imitated and reproduced. That place was none other than the Gaggera sanctuary at Selinus: the same place where the older of the two lead plates appears to have been discovered.[72] In this respect the two lead plates from Selinus and Oxyrhynchus strengthen Zuntz's conclusions. And yet they also show that he was wrong in another respect: in trying to minimize the importance, or even the existence, of the cult of Dionysus in Sicily during the classical period.[73] Here the evidence of the plates simply emphasizes the force and relevance of two points which were already known: the role played by Dionysus at Locri in the context of the cult of Persephone, and the importance of Dionysus—alongside Persephone and Hecate—in the local cult at Selinus itself.[74]

It is impossible to read this verse fragment on the lead plates and not be reminded by its imagery of the gold plates from Thurii, with their saying about the kid and the milk. In both cases we have the abundance of milk, the goat or kid, the Bacchic imagery; and in both cases the scene is set against a background of 'Persephone's garden' or 'Persephone's sacred

---

[70] Jordan 251.72, 257 J = *Suppl. Mag.* i. 204.10 (κεστῶι ἀγαλλομένη Ἀφροδίτη, Περσεφονείη). The fact that this as well as the immediately surrounding lines continues in the same metre as the fragment preserved on the text from Selinus is—in itself—no guarantee that it too derives from the original stratum of the poem; and the way that the Selinus text continues with other hexameters certainly indicates that, if it did, it was extracted from another part of the original poem. But as we will see, there are good reasons to suppose it does indeed have a Selinuntine origin.

[71] For a restoration of the following line of verse cf. C. A. Faraone, *ZPE* 100 (1994), 81.

[72] Zuntz 173–7, esp. 176.

[73] Ibid. 86 n. 3, 94 n. 1. Cf. in contrast Ciaceri 221–6, Cole 234–5, and Lloyd-Jones's comments, 263.

[74] Dionysus at Locri: Thomas 149; Cole 235. Dionysus at Selinus: Paus. 6.19.10; Ciaceri 226; K. Ziegler, *RE* iiA. 1272, 1307.

meadows and groves'.[75] In both cases this background sets the scene in the underworld—as is made even clearer on the lead plates by the mention in the following verses of Hecate and her torches (and, on the Oxyrhynchus one, of Night and Erebus as well).[76] There is an important inference to be drawn here. The verse fragment on the lead plate from Selinus is reproduced in the second of the two verse fragments on the plate from Oxyrhynchus: this contains the Bacchic, underworld imagery which points back to the gold plates and the theme of ritual immortalization. But there is also the other verse fragment on the Oxyrhynchus plate: a fragment which evidently derives from the same source as the second one, which can be inferred to derive from Selinus as well, and which mentions the single bronze sandal as magical symbol of Hecate. That this particular combination of details in the two verse fragments—bronze sandal in the first, imagery associated with ritual immortalization in the second—is no accident becomes clear when we remember how fifth-century legend presents Empedocles as wearing one bronze sandal precisely at the moment of his bid for immortality.[77] The gold plates from southern Italy provide, as we have seen, major analogies both to the legend of Empedocles' death and to the fragments of his own poetry: analogies which, among other things, throw an interesting light on the reputable story that Empedocles himself visited Thurii.[78] But the lead plates from Selinus and Oxyrhynchus bring us closer still to an understanding of the legend, which is hardly surprising when we consider that they also bring us even closer to him geographically.

In view of these points of contact between Selinus and the Empedocles legend, there is also little to be surprised at in the fact that one version of the legend explicitly links him with Selinus. According to this version—preserved for us by Diogenes Laertius and ascribed by Diogenes to an otherwise

[75] Jordan 250.65, 256 B; Zuntz 328–9.6.
[76] Jordan 251.69–71, 256 F-H. Cf. e.g. Il. 9.569–72, Od. 11.36–47, Dem. 334–49 with Richardson on 335; OF fr. 1; ps.-Pl. Axiochus 371e (descents of Heracles and Dionysus) with J. P. Hershbell ad loc.; PGM IV.1402–4, 1432–4, 1460–4, 2334–8, 2854–61.
[77] Ch. 16 with the refs. in n. 1.
[78] Glaucus of Rhegium ap. D.L. 8.52 = Apollodorus, FGrH 244 F32a; Guthrie, ii. 131–2.

The Hero

unknown author, Diodorus of Ephesus—Empedocles performed what was considered a miracle by saving the inhabitants of Selinus from mass sterility and death. The local river had been afflicting them with a plague (λοιμός); but Empedocles ingeniously diverted two neighbouring streams from their course, channelled them into the river to purify its water, and put an end to the disaster. The story then goes on to describe how he suddenly manifested himself (ἐπιφανῆναι) to the people of Selinus as they were celebrating down by the river. Everyone stood up and worshipped him as a god, and so as to confirm their impression of him he went and threw himself into Mount Etna.[79]

From the sixteenth century until very recently, it has been traditional to assume a historical basis for this story by finding a reference to it in two famous coin series from Selinus which were minted around 450 BC. The coins feature the two local rivers, the Selinus and the Hypsas, and in the light of the Diogenes passage they have been interpreted as celebrating the town's miraculous deliverance from a plague.[80] However, since the 1930s it has been made clear that this interpretation of the coins is not only unjustified but wrong: there are no grounds whatever for reading into them any reference at all to a plague or to a miraculous cure.[81] What is more, elementary

---

[79] D.L. 8.70. On the phenomenon and terminology of divine manifestation cf. Smith 181, with the refs. in n. 52; for manifestations of Heracles (see below), Ister, *FGrH* 334 F53.

[80] Cf. e.g. G. F. Hill, *Coins of Ancient Sicily* (London, 1903), 83–6; B. V. Head, *Historia numorum*² (Oxford, 1911), 167–8. The interpretation is still reproduced in the British Museum *Guide to the Principal Coins of the Greeks*², based on the work of Head (London, 1959, 29); Guthrie, ii. 133; Wright 12. For further bibliography cf. L. Lacroix, *Monnaies et colonisation dans l'Occident grec* (Brussels, 1965), 27 n. 1, 29 nn. 1–2. On the dating of the coins see C. M. Kraay, *Archaic and Classical Greek Coins* (London, 1976), 219–20; for reproductions of them, ibid., plate 46 §§ 787–8.

[81] See most recently Lacroix, op. cit. 13, 26–42, 115–24, and M. C. Caltabiano, *LIMC* v/1. 610; also J. Longrigg in *Tria Lustra: A Festschrift in Honour of John Pinsent* (Liverpool, 1994), 38–42. The first person to demonstrate the untenability of the interpretation was A. H. Lloyd, *Numismatic Chronicle*, 15 (1935), 73–93, although some of his arguments—and almost all of his own interpretations of the coins—are equally untenable. For Apollo and Artemis in a chariot on the obverse of the tetradrachm (Kraay, plate 46 § 787) see Lacroix, op. cit. 30–3; Kraay, op. cit. 220; W. Lambrinudakis, *LIMC* ii/1. 267–8, esp. § 684a; L. Kahil, ibid. 715–17, esp. § 1214a. Also relevant to the interpretation of the two divinities riding on the chariot is the fact that in a later coin type from Selinus their place is taken by the figure of Victory riding in a chariot (Hill, op. cit. 134). For Heracles on the obverse of the didrachm (Kraay, plate 46 § 788) cf. L. Todisco, *LIMC* v/1. 60–1, esp. § 2316, and 66; below, with n. 85.

geographical considerations tend to suggest that the story of Empedocles diverting other streams into the river Selinus belongs more to the realm of legend than to the realm of actual fact.[82]

That would seem to be the end of the matter; but it is not. Quite apart from the problems associated with trying to locate Empedocles' diversion of the rivers, those who are reluctant to attribute any importance or significance to Diodorus' story are quick to point out that its final details—Empedocles manifesting himself as a god, then going off to throw himself into Etna—take us out of the ordinary realms of physical reality and 'should have inspired some degree of mistrust' in the rest of the story.[83] In fact, however, it is precisely these last details which provide the clue as to how we are meant to understand the story as a whole: not in primarily physical terms, but in terms of mythical symbolism.

While the reverse sides of the coins from Selinus show the town's two river-gods, the obverse of one of them portrays Heracles performing one of his famous labours.[84] The significance of this detail should be clear. Heracles was an extremely popular mythical figure throughout ancient Sicily and Italy, but in few places if anywhere was he more important than at Selinus—not just as a local hero but also as a divinity.[85] And the crucial point here is that the action which Empedocles is made to perform—rechannelling rivers and streams—is one of the most typical mythical labours accomplished by Heracles. Sometimes this mythical labour is performed for the specific purpose of purification and removing a plague; and the wording used by writers in describing Heracles' achievement, which at times is linked explicitly to his forthcoming apotheosis, is unmistakably reminiscent of the wording used of Empedocles in Diogenes Laertius. Variants of this feat were attributed to

---

[82] A. H. Lloyd, loc. cit., Lacroix, op. cit. 28.

[83] Lacroix, loc. cit.; cf. also Wright 12 ('These details ... throw doubt on Diodorus' authority').

[84] Kraay, op. cit., plate 46 § 788.

[85] For Heracles at Selinus cf. Calder 15, 20, 25, 29 (explicitly listed as a deity in front of Apollo); Lacroix, op. cit. 34–6; Kraay, op. cit. 220; J. de la Genière, *CRAI* (1977), 252–3; Jourdain-Annequin 37 with the refs. in n. 124, 206–7 n. 404, 298–9. On Heracles as a god see West (1966), 416–17; G. Donnay in T. Papadopoulos and S. A. Hadjistyllis (eds.), Πρακτικὰ τοῦ δευτέρου διεθνοῦς κυπριολογικοῦ συνεδρίου (Nicosia, 1985), 373–7; Jourdain-Annequin 119–27, 506–60; J. Boardman, *LIMC* v/1. 122.

Heracles across the whole width of the Greek world, from Syria in the East through to Italy and Sicily in the West.[86] The version of the story located in Syria is particularly significant because it provides a direct link with one of the oriental prototypes of Heracles: Nergal, who in his Akkadian form as Erragal very probably points to the real etymology of the Greek name Heracles, and who—on his Assyrian standard as well as elsewhere—is portrayed alone with two streams of flowing water.[87] As for the subsidiary details in the Empedocles story, they only serve to confirm the correspondence with Heracles: a hero who was well known not just as a popular saviour but also, more specifically, as a purifier from contagion, a freer from sickness, and a rescuer from plagues.[88] And finally, as we have seen, this Heraclean gesture is followed by the ultimate one: Empedocles dies in flames and ascends to the gods in heaven.[89]

In short, to concentrate on the physical details of the story told by Diodorus of Ephesus is to miss the point: far more important is the mythological information it conveys. In it and through it we are given a glimpse into a view of Empedocles identified with Heracles: of an Empedoclean legend calqued

[86] Herodorus, *FGrH* 31 F30; Diod. 4.13.3 (in anticipation of his immortalization ἐπαγαγὼν τὸν Ἀλφειὸν καλούμενον ποταμὸν ἐπὶ τὴν αὐλήν, καὶ διὰ τοῦ ῥεύματος ἐκκαθάρας αὐτήν . . .: cf. D.L., loc. cit., δύο τινὰς ποταμοὺς τῶν σύνεγγυς ἐπαγαγεῖν, etc.), 4.18.5-7, 4.21.5-22.2 (Campania), 4.23.1 (Sicily), 4.35.3-4; Schol. Theocr. 1.118b (Sicily) and 7.76/77h (68.16-69.2, 99.2-5 Wendel); Sen. *Hercules furens* 283-8, *Herc. Oet.* 1240; Lucan 6.343-77; Plin. *HN* 3.1.4; Paus. 8.19.4; Aristides 40.5-6; Oppian, *Cynegetica* 2.112-55 (Orontes in Syria); Philostr. *VA* 8.7 (. . . ἐπειδὴ σοφός τε καὶ ἀνδρεῖος ὢν ἐκάθηρέ ποτε λοιμοῦ τὴν Ἧλιν, τὰς ἀναθυμιάσεις ἀποκλύσας . . .). Cf. Gruppe 1009, Faure 272, J. Boardman, *LIMC* iv/1. 797; the relevance to the Empedocles story of Heracles' legendary cleansing of the Augean stables was noted in passing by Weinreich, ii. 181. Also worth comparing are Soph. *Tr.* 1012 and Eur. *Heracles* 225, 400-2 with G. W. Bond ad locc. (purifying); Pi. *Nem.* 3.21-6 and *Isthm.* 4.55-7 (ὃς Οὐλυμπόνδ᾽ ἔβα . . .). For Heracles in Syria see Gruppe 981-5; also H. Seyrig, *Syria*, 24 (1944-5), 62-80.

[87] Nergal and streams: E. Botta and E. Flandin, *Monument de Ninive* (Paris, 1849-50), i, plate 57, ii, plate 158; P. Amiet, *Glyptique susienne* (Mémoires de la Délégation Archéologique en Iran, 43; Paris, 1972), i. 232 § 1769 and ii, plate 162; S. Dalley and J. N. Postgate, *The Tablets from Fort Shalmaneser* (London, 1984), 40-1. For Nergal and Heracles see Appendix II.

[88] Cf. e.g. Eur. *Heracles* 225 with Bond ad loc., Diod. 4.13.3 (ἐκκαθάρας), Aristides 40.5-6, Philostr. *VA* 8.7 (ἐκάθηρέ ποτε λοιμοῦ); Gruppe 1013-15, Bayet 15-19, 410, Detienne 20 n. 2.

[89] Above, with nn. 9-11; cf. Diod. 4.13.3 (diversion of River Alpheius and cleansing of Augean stables prior to apotheosis: see Bieler, ii. 118-19), Pi. *Isthm.* 4.55-7. On the labour/apotheosis sequence in the case of Heracles see P. Holt, *L'Antiquité classique*, 61 (1992), 40-1.

on Heraclean mythology. Strictly speaking, this is neither surprising nor exceptional. Pythagoreans are presented as practising the 'imitation of Heracles' from the very beginnings of Pythagoreanism in the West right through to the Christian era. Milo of Croton—a man inextricably bound up with the history of Pythagoras and the first generation of his school—is portrayed as leading the people of Croton during their attack on Sybaris in 510 BC dressed in the costume of Heracles; while, over five hundred years later, Apollonius of Tyana modelled himself on the ideal image of Heracles the Averter—'the wise and courageous being' who diverted rivers, who purified cities from plagues—and thanks to his good works was subsequently accorded divine worship at Ephesus as an incarnation of Heracles.[90] And the imitation of Heracles is by no means ascribed only to Pythagoreans. For example we find it attributed, in various respects and varying degrees, to Pericles and Callias, to Alexander the Great, to the Roman Emperors Caligula, Nero, Domitian, and Commodus, to Maximian, and to the Cynic 'philosopher' Sostratus who lived to the end of his days on a diet of nothing but milk.[91]

These examples make it natural to suspect that behind Empedocles' Heraclean exploits as described by Diodorus of Ephesus lies nothing but either literary commonplace or personal vanity. And yet we have seen enough to know that that was not the case. The importance of Heracles in Pythagorean tradition, and in particular the fundamental significance of heroization in the gold plates, provide us with a firm historical basis for Diodorus' story. The details in it—like the other details in the legend about Empedocles' leap into Etna—are

[90] Milo of Croton: Diod. 12.9.2–6; above, n. 10. Apollonius of Tyana: Philostr. *VA* 8.7 (above, nn. 86, 88), Lactantius, *Divinae institutiones* 5.3.14; Dieterich (1911), 523, Detienne 20 n. 2. It is no doubt relevant here that Apollonius appears to have been particularly well informed about ancient Pythagorean tradition at Croton. Cf. W. Burkert, *Gymnasium*, 74 (1967), 459 n. 8; above, n. 10.

[91] Pericles and Callias: Weinreich, ii. 183. Alexander: Nock, i. 134–52, Edsman 240–1, Faure 209–76 and *passim*. Caligula, Nero, and Domitian: Edsman 240. Commodus: J. Lindsay, *Origins of Astrology* (London, 1971), 314–19. Maximian: ibid. 332. Sostratus: Reitzenstein (1906), 70–1 (for Heracles and milk cf. above, with n. 53), J. Schmidt, *RE* Suppl. viii. 782, and for earlier Cynics cf. Edsman 241, Smith 183 n. 57. In general, see also Weinreich, ii. 185–97; Bieler, ii. 113–20 (Dionysus and Heracles as exemplars of the god-man); and Vermaseren's comments, 56–7 (Heracles and Dionysus 'as benefactors of mankind became the perfect examples to imitate. *Sic itur ad astra*').

not just the product of a literary whim; on the contrary, they reflect a profound concern on the part of mystical circles in the West with initiatory values and with the mythical purpose and destination of life. Through his achievements, and above all by his legendary death on Etna, Empedocles was made to do something very precise: he was made to act out concerns that were not his alone by becoming a mouthpiece for the ideas of what was clearly an esoteric network, spanning southern Italy and Sicily and founded on cult practices and religious aspirations of the highest kind. The fact that these aspirations—as reflected above all in the gold plates and in the poetry of Empedocles himself—'go beyond everything else that is known from Greek mysteries of the Classical Age' and, in doing so, 'infringe upon the system of traditional Greek religion'[92] does not diminish the importance of this esoteric network, but makes it all the greater.

[92] Burkert (1985), 295 with n. 25.

18
Death on Etna

*

THERE is one final aspect of the Empedocles legend which deserves a few words: its location.

When touching on the story about his death in Etna, a recent editor of Empedocles brings to a close a list of groundless arguments for denying it any historical value with the statement that

finally, because of the geography of Mount Etna it would have been extremely difficult for anyone to cover the distance to the foot of the mountain, to make the climb of over ten thousand feet, and then to survive the intense heat long enough to approach the mouth of the crater.

The writer concludes with a reference to 'Strabo's detailed description of Etna, and his demonstration of the impossibility either for Empedocles to have leaped into the crater or for a sandal to have been thrown up by the fire'.[1]

The appeal to Strabo is no innovation. Bidez had already vindicated Strabo's scepticism over a century ago on the grounds that, by basing himself 'with reason' on the topography of Etna, the Greek geographer had delivered 'an absolutely convincing refutation' of the idea that there could be any historical basis to the legend about Empedocles' death.[2] And yet the truth is far from being so simple. To begin with, Strabo based his conclusion entirely on a report he had heard from certain people who had recently made the ascent of Mount Etna but been prevented from getting as far as the

---

[1] Wright 16 and n. 76; Strabo 6.2.8. For Timaeus' old argument, seriously repeated by Wright, that Empedocles could not have jumped into Etna because he 'had nothing to say' in his poetry about craters and volcanoes, see Kingsley (1990), 264; the claim is strictly speaking not even correct (above, Ch. 6). For the story as 'a typical invention of Heraclides', see Ch. 16.

[2] Bidez 40, 71.

Death on Etna

crater itself by the intense heat of the earth.[3] However, we know that the volcanic activity at Etna was never a constant phenomenon. The volcanic intensity ebbed and flowed, just as it does now—often making it possible to walk up to the crater's edge, sometimes not. There were scientifically minded people in antiquity who took a special interest in noting just how approachable the crater was at any given time; only the odd poet felt he had the literary licence to exaggerate the truth by portraying the mouth of the volcano as unreachable by humans.[4] And, even more pertinently, there is the case of Strabo himself. Those who continue to cite him as authority for rejecting the legend of Empedocles' leap into Etna appear not to have read on as far as the comments which he himself adds just a few sentences later. There, after discussing the other centres of volcanic activity around Sicily and summarizing the evidence he has gathered for the periodicity of volcanic phenomena, he concludes: 'If these data are to be believed, we probably shouldn't dismiss the mythical tales that are told about Empedocles after all.'[5]

Fortunately we have one other account of Etna which sets the seeming confusion of the ancient evidence in order. Appropriately, it occurs in the Latin poem that has come down to us under the name of *Aetna*. To illustrate a point he has just made about atmospheric conditions on top of the volcano, the author of the poem draws attention to the religious rituals practised at the very edge of the crater:

> Consider also the worshippers who propitiate the
> celestial deities with incense on the highest ridge,
> precisely at the spot which gives the most unrestricted
> view into the gaping mouth of Etna—provided, that is,
> that there is nothing to irritate the flames
> and that the depths remain in a state of stupefaction.

---

[3] οἱ δ' οὖν νεωστὶ ἀναβάντες διηγοῦντο ἡμῖν . . ., 6.2.8.

[4] Seneca, *Letters* 79, as opposed to Claud. *Rapt.* 1.160–1. Cf. E. H. Bunbury in W. Smith (ed.), *Dictionary of Greek and Roman Geography*, i (London, 1856), 62; C. Hülsen, *RE* i. 1112. For medieval ascents of Etna cf. J. K. Wright, *The Geographical Lore of the Time of the Crusades*² (New York, 1965), 220.

[5] εἰ δὲ ταῦτ' ἐστὶ πιστά, οὐκ ἀπιστητέον ἴσως οὐδὲ τοῖς περὶ Ἐμπεδοκλέους μυθολογηθεῖσιν, 6.2.10.

He then goes on to describe the rituals—performed by the priest and watched by the participants right at the crater's edge—as 'the smoke rises up to the heavens from the incensed altars'; but by way of contrast he is also careful to emphasize that during periods of volcanic activity the closest anyone can get to the crater is one of the nearby hills.[6]

This description of human as well as volcanic activity on Etna is significant for more than one reason. First, it presents a balanced statement regarding the accessibility or otherwise of the crater: a statement which automatically resolves all the apparent contradictions in our other sources.[7] But secondly, and even more importantly, here we are presented with evidence of ritual activity at the very mouth of the volcano. As has been pointed out, it is 'highly remarkable' that—contrary to what one might otherwise have tended to suppose—there was evidently 'a great deal going on' from a religious point of view right at the top of Etna, over three thousand metres high.[8] At the same time however, although this religious activity does seem remarkable considering the blanket of silence thrown over it in virtually all modern literature on the volcano,[9] one is bound to add that it is less than surprising when we bear in mind the very special nature of Mount Etna as a place simultaneously linked for ancient Greeks and Romans both with the underworld and with the heavens.[10]

As for the antiquity of the ritual practices performed at the edge of the volcano, there can be little doubt. We are not dependent for our knowledge of them on the *Aetna* poem alone: Pausanias, too, mentions rituals which involved throwing offerings right into the volcano's mouth and, as Frazer has pointed out, the antiquity of this type of ritual offering to a volcano is guaranteed.[11] Again, there is the use of incense and incense altars mentioned by the author of the *Aetna*. This, also, is hardly a Roman invention. On the contrary Cicero was

---

[6] *Aetna* 339–42, 351–7, 464–5.
[7] Cf. S. Sudhaus's comments, *Aetna* (Leipzig, 1898), 153.
[8] Ibid.
[9] So, for example, the Pauly–Wissowa article on Etna (*RE* i. 1111–12) says not a word about it as a focus of ritual activity.
[10] Above, Ch. 6 with nn. 10, 29; Ch. 16; Ch. 17 with nn. 6, 18.
[11] Paus. 3.23.9 with Sudhaus, loc. cit.; R. Ellis, *Aetna* (Oxford, 1901), 153; J. G. Frazer, *Pausanias' Description of Greece* (London, 1898), iii. 389.

certainly right when, in the first century BC, he spoke of the 'incredible' number of very ancient and very beautiful incense-burners that used to be found everywhere in Sicily but in his day were rapidly disappearing. This diffusion of the practice of incense-burning no doubt goes back, just like the mythology of Heracles' death and apotheosis through fire, to the early days of Phoenician influence and colonization of the island.[12]

We also have another source of information about incense being used for ritual purposes in the Etna region. The Latin poet Grattius has left a description of the use of incense by priests for the worship of Vulcan in the heart of the cavernous, volcanic regions of Sicily. A number of the details in his description make it clear that he is in fact referring to worship at the pre-Greek temple of Adranus, the god who became identified with Hephaestus: his temple was built out of lava on the south-west slope of Etna, and dates back to long before the time of Empedocles.[13] Taken together with the passage in the *Aetna*, the evidence would appear to suggest a coherent picture of ancient fire-worship based at a temple located close to—but also at a comparatively safe distance from—the main crater and carried up to the crater's mouth when circumstances and conditions permitted.

There is one more cult centre in the Etna region to consider. On the southern slope of Etna, a little to the east from the temple of Adranus in an anticlockwise direction, lay the small town of Hybla Geleatis. The town was dedicated to the cult of a pre-Greek goddess, a chthonic or underworld divinity; even at the peak of the Greek colonization of Sicily the inhabitants of the town, like the cult itself, remained non-Greek, 'barbarian'.[14] These people, and apparently one family in particular, were famous among the Greeks not just for their remarkable piety but, above all, for their skill in the interpretation of dreams and omens. For their skill as 'dream-prophets' they were renowned throughout the whole of Sicily and beyond. Here again we are dependent on the late authority of Cicero and Pausanias; and yet this time the person whom they cite as

---

[12] Cicero, *In Verrem* 2.4.21.46–24.54; Kingsley (1995*b*), § 2 with n. 47.
[13] Grattius, *Cynegeticon* 430–66. Cf. Kingsley (1995*b*), nn. 39, 45; Ch. 6.
[14] Paus. 5.23.6 = Philistus, *FGrH* 556 F57b; Freeman, i. 75, 159–62, 514–15, K. Ziegler, *RE* ix. 25, H. Hepding, ibid. 29, Zuntz 68–9.

the source of their information—Philistus—takes us back to the fifth century BC.[15]

There can be no doubt that, just as with the temple of Adranus, the local cult of the goddess at Hybla Geleatis was another symptom of the religious activity bound up with the volcanic phenomenon of Mount Etna as a whole. At the same time, Freeman was plainly right in connecting this local cult specifically with the active craters of water and mud on the outskirts of the town: striking manifestations of the mountain's intricate volcanic complex.[16] And here the family of dream-interpreters fits perfectly into place. In the ancient world a fixed centre for the interpretation of dreams invariably implied the existence of a dream oracle; and oracle centres in the volcanic regions of southern Italy and Sicily were inevitably associated with craters and openings viewed as points of access to the underworld. This was true in particular of oracle centres specializing in the interpretation of dreams.[17] That is perfectly understandable when we consider the chthonic, underworld nature of dream oracles in antiquity and their frequent location in caves, chasms, and openings into the underworld.[18]

Yet we have already come across all these details before in a different context: craters, especially water craters, dream oracle, point of descent to the underworld. We came across them in the context of the Sicilian mythology behind Plato's *Phaedo*, and of the background to Empedocles' own equation of Hades with fire; and we found them in a poem about Orpheus' descent to the underworld, apparently called the *Krater* and written by the fifth-century Pythagorean Zopyrus. In other words, from a modern point of view the strange details about cult and ritual in the immediate vicinity of Mount Etna may seem to introduce us into very unfamiliar—and certainly very neglected—territory; but from the point of view of Pythagoreanism in Empedocles' time, and of the Orphic poetry written by Pythagoreans in Sicily and southern Italy, it was extremely familiar. In short, even the location of Empedocles'

---

[15] Pausanias, loc. cit., and Cic. *Div.* 1.20.39 = Philistus, *FGrH* 556 F57; Kingsley (1995b), § 2.
[16] Freeman, i. 74–5.
[17] Above, Ch. 11 with n. 5.
[18] So e.g. Strabo 14.1.44 (Plutonium in north Caria: Ch. 15 n. 28); cf. also Deubner 6 n. 2, 31, and *passim*; Plut. *De gen.* 589f–592e and Paus. 9.39.4–14 with M. Hamilton, *Incubation* (London, 1906), 88 (oracle of Trophonius); Dodds 110–11; Zuntz 69.

Death on Etna 283

legendary death has something invaluable to tell us about the mythological and cult background both to his own interests and to the interests of Pythagoreans during his time. Far from being confronted with the physical impossibility of Empedocles jumping into the main crater at Etna we have, instead, stumbled on a veritable minefield of cult activity which provided the perfect setting for a ritual descent into the underworld. As we saw earlier, Zopyrus—poet, mechanic, Pythagorean—is able to throw a great deal of light on the historical phenomenon of Empedocles. Now, with his story of Orpheus descending into the fiery depths of the underworld via a dream oracle at a water crater, he has something more to show. Once again the historical Empedocles and the legendary Empedocles become entangled, inseparable from each other against a common background of Pythagoreanism in the West.

While on the subject of Pythagoreans and dream oracles it is worth establishing one final point. The similarities between the famous Pythagorean food taboos and the requirements of abstinence in Greek ritual and magic have often been noted; they are in fact so obvious that they were already commented on in antiquity. More specifically, attention has also been drawn to the similarities between those same taboos and Greek rituals and mysteries relating to underworld divinities such as Hecate and Demeter.[19] But what has not been given the full emphasis it deserves is the fact that the closest similarities of all are between the Pythagorean taboos and the requirements laid down for people using dream oracles.[20] This is not surprising. Contrary to what is often supposed, the immediate aspiration on the part of Pythagoreans for contact with the divine appears to have been directed in the first instance to the underworld, to

---

[19] Cf. e.g. Lobeck, ii. 897–904, Guthrie, i. 182–6, Burkert 176–91, and the statement in D.L. 8.33 = *FGrH* 273 F93: the Pythagoreans 'abstain from carcase-meat and flesh and mullet and black-tail and eggs and oviparous animals and beans and the other things forbidden by those responsible for performing the rituals in the temples' (οἱ τὰς τελετὰς ἐν τοῖς ἱεροῖς ἐπιτελοῦντες). In view of this evidence for a ritual background to Pythagoreanism itself it is, to say the least, dangerous to try to discover specifically Neopythagorean influence in late-attested food prohibitions at cult centres (so Nock, ii. 847–50, cf. Burkert 178 n. 91). On the general question of Pythagorean and Neopythagorean tradition see below, Ch. 20.

[20] The basic evidence is collected by Deubner, 15–16. Cf. also ibid. 30, and Burkert's more general remarks, 190.

the gods below rather than to the gods above;[21] and in antiquity the best way of actually making contact with divinities of the underworld was through the practice of 'incubation'—of awaiting a dream or vision while sleeping, as a rule, either on or even inside the earth.

In fact the surviving evidence leaves us in little doubt about how important dreams and dream-divination were considered to be by early Pythagoreans as means of providing them with knowledge, teaching, and guidance.[22] What is more, our sources specifically draw attention to the link between the Pythagorean food taboos and this concern with the divinatory power of dreams: with the ability of dreams to provide access to another world. There is, for example, the famous Pythagorean prohibition on eating beans—a prohibition vigorously shared by Empedocles. Cicero states as a basic reason for this prohibition by Pythagoreans the fact that beans cause flatulence, disturbing tranquillity of mind (*tranquillitas mentis*) and preventing them from having prophetic dreams; other writers say the same.[23] The modern tendency is to describe the

[21] Cf. esp. Burkert's comments, 185–6, also 112–13, 154–61; above, Ch. 9 with n. 8, Ch. 12 with nn. 50, 56–7, Ch. 17.

[22] Iam. *VP* 65 (from Aristoxenus: Rostagni, i. 138–40), 106–7, 114, 139 (cf. Burkert 185–6 with n. 147); D.L. 8.32 = Alexander Polyhistor, *FGrH* 273 F93 § 32 (ὀνείρους... καθαρμούς... μαντικήν); Ap. *Mir.* 46.3 (254–7 Giannini) = Clem. *Str.* 3.3.24.1 and D.L. 8.24 (cf. also Plin. *HN* 18.30.118); Nigidius Figulus, fr. 82 Swoboda (Liuzzi 28, 84); Cic. *Div.* 1.30.62, 2.58.119; Plut. *De gen.* 585e–f; Tert. *An.* 48.3; Philostr. *VA* 2.37; Olymp. *Prolegomena* 11.33–5 Busse; A. Bouché-Leclercq, *Histoire de la divination*, i (Paris, 1879), 283, 287; W. W. Jaeger, *Nemesios von Emesa* (Berlin, 1914), 54–5; Boyancé 110–11; Hopfner, *RE* viA. 2235–6; Detienne (1963), 43–6. Compare also Epimenides' famous 'teaching dream' (ὄνειρον διδάσκαλον, DK i. 32.15–17, cf. 28.1–2, 29.26, 31.21–23, 32.19–21) and, for Epimenides' links with Pythagoreanism, Burkert 150–2, West 49 and n. 44; Tert. *An.* 46.11, 47.3 (Epicharmus) = DK i. 207.14–17 with ibid. 194 n., 206.3–5, and Burkert 289 n. 58; above, Ch. 15 n. 28 (Velia); Pl. *Rep.* 571d–572b with Cicero, locc. citt. and Adam, ad loc., Boyancé 111, Bollack, iii. 460 and n. 5; Plut. *De Isid.* 384a. For the theory of dreams ascribed to Empedocles by Philoponus and Simplicius (DK i. 351.7–11) see W. J. Verdenius, *Parmenides* (Groningen, 1942), 20–1.

[23] Cic. *Div.* 1.30.62; cf. ibid. 2.58.119, Apollonius, Clement, and Pliny, locc. citt., Tert. *An.* 48.3, D.L. 8.24. The same explanation—that eating beans causes flatulence and interferes with the process of getting straightforward dreams (εὐθυονείριαι: not 'vivid dreams', LSJ) as opposed to disturbed and complicated ones (ὀνείρους τεταραγμένους: cf. Apollonius, Clement, and D.L. locc. citt.)—is attributed to 'the philosophers', οἱ φυσικοί, at *Geoponica* 2.35.3–4; to an anonymous 'they' at Plut. *Qu. conv.* 734e–f; and to Amphiaraus, the mythical founder of the famous dream oracle, at *Geoponica* 2.35.8 = DK i. 368.25. Cf. Deubner 15 and n. 1; Iam. *Pr.* 21 (150.18–20 des Places); and for the connection between disturbed dreams and flatulence, Arist. *De divinatione* 461ᵃ14–25. Regarding Empedocles and beans see Emp. B141 with DK

Death on Etna

explanation of this taboo on beans which links it with flatulence as a purely rationalizing and physiological interpretation—in sharp contrast to the more mythological explanations of the taboo which, with remarkable frequency, introduce symbolism relating to the soul and the world of the dead.[24] And yet, as Boyancé has pointed out, this dichotomy between rationalizing and mythological explanations is a false one. Cicero mentions flatulence specifically because of its detrimental effect on 'tranquillity of mind'—an important notion in Pythagoreanism, divination, and the mysteries—and he mentions it in the middle of a discussion about impediments to obtaining true and divinatory dreams: precisely the kind of dreams which were held by Pythagoreans to bring them into contact with the world of the dead. We find the same themes and ideas presented together by others who mention the taboo as well.[25] The explanation they give may be practical and pragmatic; but 'rational' in a narrow sense it certainly was not.

Against this background of ideas we are able to make sense of a rather disjointed section in Iamblichus' *Pythagorean Life* which presents a string of reasons, supposedly given by Pythagoras himself, for refusing certain foods. First, we are told, 'he rejected all foods that produce flatulence or cause disorder'; then, 'he prescribed abstention from all foods which interfere with divination' (ὅσα εἰς μαντικὴν ἐνεπόδιζεν); and finally, 'he refused anything that conflicted with ritual cleanliness (εὐάγεια) and rendered turbid the various purities of the

---

ad loc. (comparing *Geoponica*, loc. cit. and Aristophanes, *Frogs* 1032–3 = 'Orpheus', DK 1 A11: taboos, healing, oracles); Delatte 50–2; West 14–15.

[24] Cf. esp. D.L. 8.34 = Arist. fr. 195 (gates of Hades), Heraclid. fr. 41 (coffer or coffin: see Delatte 42 n. 1) with Burkert 183 n. 124, Varro *ap.* Plin. *HN* 18.30.118 (souls of the dead), Schol. T *Il.* 13.589, *Geoponica* 2.35.6 (πένθιμα γράμματα) with Varro *ap.* Plin. *HN* 18.30.119; Plut. *Quaestiones Romanae* 286d–e, Festus, s.v. *faba*, Delatte 37–41, Haussleiter 128–31. Against this background of connotations the story of the Pythagoreans who preferred death to walking in a bean-field assumes a special meaning (Iam. *VP* 191 = Neanthes, *FGrH* 84 F31b; cf. D.L. 8.40 = Hermippus, fr. 20 Wehrli).

[25] Boyancé 111 with n. 2; apart from Cic. *Div.* 1.30.62 cf. also Tert. *An.* 48.3 (... *apud oracula incubaturis*...) and the further refs. in n. 23 above. Burkert (184) surprisingly maintains the rational/mythological dichotomy and fails to note the divinatory link between the two types of explanation. For Cicero's 'tranquillity of mind' see Iam. *VP* 65 (ἡσύχους τε καὶ εὐονείρους) and 114, Pl. *Rep.* 572a. On the importance of tranquillity or ἡσυχία in Pythagoreanism see D.L. 8.7, 9.21, Luc. *Vitarum auctio* 3; for its connection with divination, M. Marcovich, *Estudios de filosofia griega*, i (Merida, 1965), 14; for its mystery-background, Burkert (1969), 28 and n. 62.

soul—especially the purity of the visions it sees during dreams.'[26] We are now in a position to see that these apparent alternatives are not alternatives at all: they are just three ways of stating the same point. Iamblichus' listing of non-alternatives as alternatives simply demonstrates his lack of understanding in this particular case of the materials he is transmitting, and at the same time gives us an insight into the long, complex process of exegesis and interpretation which stretches back into the centuries before him. Certainly this process of interpretation should not be over-simplified. It is clear, for example, that very early on in Pythagorean tradition—and quite probably with Pythagoras himself—the close link between beans and the world of the dead gave rise through a natural line of development to forms of interpreting the taboo which tied it in with the doctrine of reincarnation. However, the evidence of Iamblichus shows very well that the underlying affinity of the bean taboo with chthonic ritual, incubation, and divination continued to be acknowledged and appreciated.[27]

As we can see from Cicero, Iamblichus, and others, it was common knowledge in antiquity that Pythagoreans attached considerable significance to divination through dreams. But apart from the fragmentary references we possess to ongoing cult activity at dream oracles themselves, there is one textual source in particular which demonstrates that this idea of divination through dreams continued not just to be understood but also to be put into practice. That source is the Graeco-Egyptian magical papyri. There is no need to emphasize how saturated these papyri are with the concern for obtaining helpful dreams: a concern explicitly linked on one occasion to the theme of making an initiatory descent into the underworld within the framework of an invocation of Hecate and her magic

[26] *VP* 106–7.
[27] The best assessment of the bean taboo in later Pythagoreanism, and of its continuing relationship to dreams, prophecy, and divination, is H. Strathmann's, *Geschichte der frühchristlichen Askese*, i (Leipzig, 1914), 311–17. On the other hand Marcovich (op. cit. 8–10) radically over-simplifies—and falsifies—in postulating reincarnation theory as 'the real reason' for Pythagoras' ban on beans, on the grounds that Pythagoras himself had nothing to do either with cult practice or with magic: see below, Ch. 19. For a useful overview of the taboo and the various explanations provided for it see Haussleiter 127–40; for comments on the complexity of the tradition, Delatte 34; and note Long's remarks, 25–6. Regarding the theory of reincarnation and Pythagoras himself see Burkert 120–3; Philip 151–5; Schibli 12–13.

sandal.[28] This link could seem strange, or just accidental. But it makes perfect sense when we remember Zopyrus' description of Orpheus' descent into the underworld at the site of a dream oracle. It becomes even more understandable when we bear in mind that the practice of incubation provided the setting for those same Pythagorean rituals of death, descent to the underworld, and rebirth which were mentioned earlier for their relevance to the legend of Empedocles' death on Etna.[29] In this shared concern with dreams and dream oracles there is an obvious affinity between Pythagoreanism and the magical papyri. Naturally it is significant here that, as is well known, the writers and users of these papyri also continued to observe the same kind of ritual prohibitions observed by the Pythagoreans centuries earlier.[30] And yet to speak here of an exclusive indebtedness to Pythagoreanism on the part of the Egyptian magicians would be out of the question. When, for example, the Paris magical papyrus mentions the dream oracle of Amphiaraus, it is referring not to Pythagoreanism but to the cult background out of which the Pythagorean prohibitions themselves arose.[31] We must, also, seriously bear in mind the possibility of early Pythagoreans inheriting Egyptian magical traditions which are preserved quite independently in the Egyptian papyri: it is important to remember that the traffic between the Greek West and Egypt was not just one-way.[32] However, within this complex kaleidoscope of influences and interrelations there is also a sense in which we can, and must, speak in terms of a certain conscious continuity of tradition between Pythagoreanism and the magical papyri. When one of these papyri cites what it refers to as 'Pythagoras' request for a

[28] *PGM*² LXX.4–19; cf. above, Ch. 16 with n. 43 and, for the text, Ch. 16 n. 21. See in general Deubner 28–39 and *passim*; S. Eitrem in *MagH* 176–87.
[29] Ch. 17 with the refs. in n. 3; and cf. also above, n. 22, for refs. to incubation, magic, ritual death, and rebirth in the case of Epimenides. For some relevant comments on incubation, initiation, dream consciousness, descent to the underworld, death and rebirth see H. Corbin, *Avicenna and the Visionary Recital* (London, 1960), 220–2.
[30] So e.g. *PGM* IV.732–6, and for the following ἐν ἐκστάσει ἀποφοιβώμενος see Hdt. 4.13 φοιβόλαμπτος γενόμενος (of Aristeas) with Rohde 328 n. 109, Bolton 138, Burkert 147–50 and in *Gnomon*, 35 (1963), 238–9. Cf. Deubner 16, 28–30, Burkert 177 n. 86, 186 n. 146; also F. Böhm, *De symbolis Pythagoreis* (Berlin, 1905), 17.
[31] *PGM* IV.1446–7. The passage deserves to be read in its context; cf. also Deubner 31 n. 9.
[32] Cf. Ch. 16 n. 37 and, for Pythagorean indebtedness to Egyptian traditions, Kingsley (1994c), §§ I–II.

dream oracle', we may seem a long way from the southern Italy and Sicily of the sixth or fifth centuries BC. In a sense we are, both in time and space. We might also seem a long way away in terms of contents, because the procedure of the dream oracle includes allusions to astrology—were it not for the fact that astrological lore was evidently already used by Pythagoreans in the fifth century BC as a basis for their speculations about the gods.[33] The trouble is that here, as so often, the magical papyri hit the nail on the head and reveal aspects of early Pythagoreanism which modern preconceptions and interests have tended to stifle almost to the extent of making them disappear. As Deubner emphasized at the start of the century, the mention of Pythagoras and a dream oracle in the magical papyrus is bound to remind anyone who has assessed the evidence carefully 'of the numerous respects in which the Pythagorean way of life coincides with incubation rituals'.[34] What can be inferred and confirmed from the Pythagorean material itself is, once again, corroborated by the magical papyri.

[33] *PGM* VII.802–21; for the mention of the Pleiades in 828–32 cf. West (1971), 215–17. Regarding Pythagoreanism and astrology see F. Boll, *Aus der Offenbarung Johannis* (Leipzig, 1914), 40 n. 1 and *Kleine Schriften zur Sternkunde des Altertums* (Leipzig, 1950), 19–20, 382; Kingsley (1994*f*).

[34] Deubner 30. For the Ὀνειραιτητὸν Πυθαγόρου see *PGM* VII.795–6, Thesleff (1965), 243–4; for the coupling of Pythagoras' name here with the name of Democritus, Kingsley (1994*c*), § II with n. 32, and below, Ch. 20.

19
Sandals of Bronze and Thighs of Gold
*

IT is time to see just where we stand and start to draw out some of the main implications of the material considered so far. First, there is the question of Empedocles' leap into Etna. There is no point in denying that he could have died by jumping into the volcano: he could have. But by now it should be clear that whether or not he did so is more or less immaterial. What *is* important is that, interpreted rightly, the legend introduces us to a world of esoteric ideas and practices characteristic of certain mystical circles in southern Italy and Sicily. Paradoxically, Empedocles' unique talents and special charisma made him a perfect vehicle for conveying themes and possibilities which were by no means restricted in scope or relevance to one individual alone.

In short, the legend of Empedocles' death conveys a message that needs to be understood not just literally—as applying only to Empedocles—but also symbolically: as referring to an initiatory schema of ritual death, descent to the underworld, regeneration and transformation. Behind all the rationalizations and parodies, the details of the legend make up a message or 'code'; and in this respect the detail of the bronze sandal is particularly significant. From later sources, which themselves can be traced back to Sicily in the fifth century BC, we see that it was the magical 'symbol' *par excellence* of Hecate. Worn or held by the magician, it was the 'sign' of his ability to descend to the underworld at will.[1] There can be no doubt as to the meaning of these terms 'symbol' and 'sign': they refer to

---

[1] *PGM* IV.2292–3 (σύμβολον) with *GMPT* 79 n. 286, 2333–4 (σημεῖον); Porph. *ap.* Eus. *PE* 3.11.32 = fr. 359.69–70 Smith (σύμβολον); Betz 291 and n. 21, 293 with n. 33; above, Ch. 16 with n. 21.

'secret signs of the gods which give to those who know them power to control the gods'.[2] And the emphasis here is on the word 'secret'. Whether in the form of specific objects or spoken phrases, these *symbola* automatically establish a special bond not just with the gods but also with other people allied to the same mystical or magical tradition. They are effectively indistinguishable from the *symbola*, the 'passwords' or secret objects which—besides being central to the rites of regeneration and immortalization as described in the Bacchic gold plates, in the Hermetica, and in the magical papyri—played a major role among the earliest Pythagoreans as means of identification 'which are given the initiate and which provide him assurance that by his fellows, and especially by the gods, his new, special status will be recognized'.[3] Needless to say, a legend containing allusions to 'symbols' of this kind will be understood by other initiates; but for the outsider it will at best prove an enigma, at worst be treated as just some picturesque fantasy or garbled nonsense.

Equally important is the fact that there is no separating Empedocles himself from this initiatory context. As we have seen repeatedly, the background to the legend of his death is also the background both to the historical Empedocles and to his poetry. There is an obvious temptation here to adopt a conciliatory stance and try to describe Empedocles as someone who, admittedly, had affinities with this esoteric substratum of theory and practice but had risen out of and above it: who had 'refined' it, transmuted it, converted it into philosophically acceptable terms. The attempt may seem attractive, but it is ruled out by the evidence that survives. Empedocles' own words attach him to what, in modern terms, is the least acceptable level of this substratum. He was an actual, practising magician; and what is more, he himself talks in terms which demand that we understand his poetry in a magical context, as having a specifically magical purpose and aim.[4]

If Empedocles were the only early Greek philosopher

---

[2] Cf. M. Smith in R. S. Bagnall *et al.* (eds.), *Proceedings of the Sixteenth International Congress of Papyrology* (Chico, Calif., 1981), 648–50.
[3] Burkert 176, with the refs. in n. 78. Cf. Dieterich (1911), 95–6 and (1923), 64 n. 3; Boyancé 48–55; NF ii. 210 n. 20; Grese 85–8; Betz 292 with n. 26.
[4] Above, Ch. 15.

actively bound up in this scenario of initiation ritual, esotericism, and magic he would present a big enough challenge to accepted ideas about the fundamental nature and intentions of Presocratic philosophy. But he is not. Apart from what has already been said about the involvement of early Pythagoreans with cult practice and ritual—and, in particular, with initiatory practices focusing on a sequence of ritual death, descent to the underworld, and regeneration—there is also one other piece of evidence which needs to be mentioned. In much the same way that a remarkable legend about a bronze sandal attached itself to Empedocles, an equally striking legend attached itself to Pythagoras. This is the story describing how one day he happened to expose his thigh in public and people saw it was made of gold.[5] As was already noted in the mid-nineteenth century—and as has been pointed out more recently in greater detail—this story, too, contains a message waiting to be read. With almost the same definiteness as in the case of the bronze sandal, the symbolism of the golden thigh can be traced back to an initiatory scenario of death, descent to the underworld, and ritual dismemberment of the body followed by its reassembly and regeneration.[6] It is also significant that a background of metallurgical magic is clearly to be assumed here: one of much the same type as must be assumed as well in the case of the bronze sandal.[7]

The parallels with the Empedocles legend go still further. Just as the story of his leap into Etna represents a kind of code, understandable only by the few, so the legend about the golden thigh is one of several 'miracle stories' that attached themselves to Pythagoras; and in Burkert's words

there is always something enigmatic about the meaning of these miracles, which is apparently revealed to the insider but not explained to the uninitiated.[8]

[5] Ap. *Mir.* 6, Ael. *VH* 2.26, 4.17, and Iam. *VP* 140 = Arist. fr. 191, plus the further refs. in Burkert 142 n. 119, 159 n. 215.

[6] W. Mannhardt, *Germanische Mythen* (Berlin, 1858), 74; cf. Cook, ii/1. 221–31, Burkert 159–60 and (1969), 23–5. For comments on the sequence of ritual death, dismemberment, initiation, and regeneration in both shamanism and Greek myth cf. also Eliade 34–8, 53–8, 64, 429; West 144–6. Worth comparing as well are Plut. *De gen.* 590b with Culianu 44; Dante, *Paradiso* 1.13–21; Wind 171–6.

[7] See the material gathered and discussed by Lindsay 296–300; and for metallurgical magic and early Pythagoreanism, Burkert 376–7.   [8] Ibid. 144.

Finally, it is worth noting that one particular ancient text mentions Empedocles with his bronze sandal directly alongside Pythagoras with his golden thigh. In one of Lucian's *Dialogues of the Dead* the first two philosophers with whom the mock-hero Menippus enters into conversation during a descent into the underworld are Pythagoras, whom Menippus asks about his golden thigh, and Empedocles, whom he addresses as 'bronze-foot'.[9] The juxtaposition deserves considering from two angles. On the one hand, it needs to be assessed in terms of the deceptive nature of Lucian's mockery and satire. Underneath their brilliant surface they tend to conceal a great deal more information about ancient ritual and magic than meets the eye, and his familiarity with these matters is—in many respects—unparalleled in classical literature.[10] On the other hand, here in Lucian we have the ultimate stage in the reduction to parody of what originally were two cult legends: that is, legends having 'connections with religious or semi-religious communities' and that 'present the hero as the embodiment of the value system which he revealed to the community and by which they were sustained'.[11]

As with Empedocles and his involvement in magic, the attempt has been made to clean up the image of early Pythagoreanism and present Pythagoras himself as a philosopher who is more or less respectable in modern terms. The attempt has, in the main, been successful. Recent scholarship can for example still feel justified in concluding that, while 'Pythagoreanism later develops as a religious philosophy and as a specific way of life', there is no evidence from the fifth or fourth centuries to link early Pythagoreans with cult or ritual activities of any kind.[12]

---

[9] *DM* 6 (20).3-4.

[10] Cf. e.g. Dilthey 378 n. 3, 379 and n. 2, 382, 391 n. 1, 393 n. 3, 400 n. 1; Dieterich (1911), 22 n. 5, 41 n. 1, 105 n. 5; Kuster 12-13; Ganschinietz (1913), 13-16, 29, 42-3, 57, 61; Betz 289-90; L. Robert, *A travers l'Asie mineure* (Paris, 1980), 404; MT i. 168, 215.

[11] C. H. Talbert, *ANRW* ii.16.2 (1978), 1625-6, cf. 1634-5, 1641.

[12] Cole 238. Cf. e.g. A. Maddalena's attempt to 'rationalize' Pythagoras and early Pythagoreanism, *RFIC* 92 (1964), 115-17. The recent claim by Zhmud' that early Pythagorean interest in number belongs to 'natural science' as opposed to 'number mysticism' (291-2) requires that we misunderstand half of the evidence and ignore the rest; on the dangers involved in trying to impose sharp distinctions between the 'mystical' and the 'scientific' on the period down to the 4th c. BC see Kingsley (1990), 250 n. 34 and above, Chs. 7, 12-13.

Sandals of Bronze and Thighs of Gold 293

And yet the facts prove otherwise. The most obvious targets of this rationalizing trend have been the Pythagoras legends, usually dismissed as late and worthless fabrications. But first, the legend of Pythagoras' golden thigh in particular is just one aspect of the varied evidence for early Pythagorean involvement in ritual activity. It cannot be argued away without arguing away the rest of the evidence, which is impossible; and, on the contrary, the problems that writers in the fifth century BC—let alone the fourth—had in understanding the type of initiatory themes and motifs involved simply serve to emphasize the authenticity, and antiquity, both of the legends and of the ritual activities which they unquestionably imply.[13] Secondly, it is clear that—contrary to usual assumptions—the developmental trend in the history of stories and legends about Pythagoras was not one of increasing inventiveness but one of growing rationalization: of an increasing embarrassment in face of the bizarre or miraculous and a growing tendency to replace it with either the understandable or the merely sentimental. As for those elements of fantasy and blatant irrationality which occur in the biographical literature of late antiquity, they can often be shown to be no more than echoes of the very earliest Pythagorean traditions.[14] Thirdly it is also clear that, in the case of the golden thigh, the motif has the closest of affinities with Near Eastern—and specifically Anatolian—cult and ritual: enough has already been said about the transmission of cult practice from Anatolia and the Asiatic coastal regions to southern Italy and Sicily, and about its importance in forming the earliest strata of Pythagoreanism in the West.[15] What is more, in this particular case we have to consider the probability that the golden-thigh motif along with

---

[13] Cf. e.g. Burkert 156–61, with n. 215, on Hdt. 4.94–6 and Arist. fr. 191; and Faraone, *JHS* 113 (1993), 80 for parallel examples of misunderstanding.

[14] Cf. esp. Burkert 136–7, 141–6; also Rohde, *Kleine Schriften* (Tübingen, 1901), ii. 105–6, Burnet (1930), 87, 95 n. 3, O. Gigon, *Der Ursprung der griechischen Philosophie* (Basle, 1945), 131, Smith 189. So, to mention one relevant example, by the time the legend about Pythagoras' golden thigh came down to Porphyry (*VP* 28) and Iamblichus (*VP* 135, but cf. 140) it had been formally dismissed in the source they used as 'mere babble' (τεθρύληται); for some valuable comments on Iamblichus' own approach to the Pythagoras legends see J. Z. Smith, *Map is Not Territory* (Leiden, 1978), 197–204. On the later revival of early traditions see also Ch. 17 nn. 10, 90 (Apollonius of Tyana); and for the role played in their transmission by Bolus of Mendes, Ch. 20.

[15] Burkert (1969), 23–6; Ch. 15 with n. 28.

its ritual associations forms part of what, already in or around the end of the sixth century BC, Heraclitus of Ephesus scathingly referred to as the collection of tricks and swindles which Pythagoras put together and presumably took with him when he left the eastern Mediterranean for Italy.[16] It will be noted in this connection that Empedocles' bronze sandal, too, probably falls into the pattern of religious and mythological traditions transferred to the Greek West—along with the cult of Hecate—from Anatolia.[17] And finally, of course we have the invaluable test case of Empedocles himself: of someone who on his own admission was embroiled in the practice of magic and who, at the same time, plainly had the closest of ties with Pythagoreans in the West during the fifth century BC.

Potentially more successful than the attempt to strip Pythagoras and early Pythagoreans of involvement with magic, superstition, and ritual is the attempt to claim that, although such 'primitive' elements may have formed a part of their world, they somehow succeeded in transmuting and transcending them. At its simplest this line of approach is summarized in the statement that 'Pythagoras' success was clearly not that of a mere magician or occultist, appealing only to the feeble-minded and insecure.'[18] We also find it presented by Guthrie in a passage which deserves quoting in full.

---

[16] Heraclit. B81, 129; Burkert 130–1, 161. Ephesus' immediate proximity to Samos makes it almost certain that Heraclitus depended primarily on local tradition for his knowledge of Pythagoras' personality and activities. As for the terms in which he denounced him, it is worth noting how Alexander of Abonuteichus—who, centuries later, imitated Pythagoras by allowing people a calculated glimpse of his golden thigh—was also accused of trickery and sleight-of-hand. Cf. Luc. *Alex.* 40 and *passim*; Burkert 160 and (1969), 23; Phillips 2713, 2717.

[17] Note in this respect the long-standing tradition at the church and monastery of Archangel Michael in Mantamados, Lesbos, of providing the archangel with sandals of bronze, gold, silver, or iron: a tradition which is still remembered as being brought to Lesbos by pilgrims from Anatolia. I am grateful to Πρωτοπρεσβύτερος Eustratios Dessos for his kind help and information; see also his Το ιστορικό και τα θαύματα του Ταξιάρχη Μανταμάδου, i² (Mytilene, 1988), 46. Indicative of the antiquity of this tradition is the cult practice, attested in Babylonian sources of the 6th c. BC, of providing the god Adad and his wife Shala with sandals of silver: BM 77865 = J. N. Strassmaier, *Inschriften von Nabonidus, König von Babylon* (Leipzig, 1889), § 673; *Chicago Assyrian Dictionary*, M 38b and Š 291b. It is probably significant that the cult of Adad appears to have had its origins in the Semitic West. For another example of items of divine clothing imported into Sicily from Anatolia and beyond, see Dewailly 156 with n. 122.

[18] KRS 229.

Pythagoreanism contains a strong element of the magical, a primitive feature which sometimes seems hard to reconcile with the intellectual depth which is no less certainly attested. It is not on that account to be dismissed as a mere excrescence, detachable from the main system. All who work on the border-line of philosophy and religion among the Greeks are quickly made aware of a typical general characteristic of their thought: that is, a remarkable gift for retaining, as the basis for their speculations, a mass of early, traditional ideas which were often of a primitive crudity, while at the same time transforming their significance so as to build on them some of the most profound and influential reflexions on human life and destiny.[19]

These types of judgement—'mere magician', 'primitive crudity'—call for a few general comments. First, it is important to appreciate that those characteristics we associate most readily with modern science—observation, the collection of data, and the organization of knowledge—are not the prerogative of the scientist but, as noted by Marcel Mauss in the passage quoted earlier,[20] are part of his inheritance from the magician. Certainly major differences can be said to exist between ancient magic and modern science, especially in terms of attitude and approach; but when we look underneath what is often little more than a veneer of rhetoric and self-justification, we find that the similarities between the supposedly magical mentality and the modern, scientific mentality tend to be as significant as the dissimilarities.[21] Understanding of other cultures and civilizations—and of the history of our own—has moved on beyond the simplistic models still adopted without question by many classicists. As Peter Brown has recently pointed out with regard to the whole question of magic and so-called 'popular religion', there is now a growing need in scholarship to look for ways 'to answer a tradition of explanation that gives pride of place to physical insecurity and to unmodified fear of invisible forces in its rhetoric of explanation'.[22]

---

[19] Guthrie, i. 182. For a more balanced approach to several of the issues raised here cf. Lloyd (1983), 202–16.
[20] Ch. 15, with n. 43.
[21] Cf. e.g. Lloyd (1979), 48–9, 56 (on the Greek material); Krafft, pp. viii–ix and passim; K. Thomas, *Religion and the Decline of Magic*² (Harmondsworth, 1973), 799–800; Betz in *MagH* 247, with further refs.; Phillips 2677–773 and in *MagH* 260–1, 266–9.
[22] Brown 12–13.

As for what has been called the attitude of 'optimistic evolutionism'[23] implicit in Guthrie's somewhat unsubtle distinction between 'primitive crudity' and 'intellectual depth', we have already seen—and with regard to Pythagoreanism in particular—that it is precisely this type of evolutionary approach to the history of ideas in ancient Greece which is most likely to end up presenting a grossly distorted picture of the facts. At the same time, we have also seen how much can be assumed to have been *lost* in terms of profundity and fullness of approach as a result of the scholarly specialization which came into force at Athens during the fourth century BC.[24] As far as Empedocles is concerned, it is important to remember that—in Simone Pétrement's words—'there is nothing to prevent a magician from being a philosopher';[25] and it is worth repeating that there is no incompatibility at all between a magical outlook and those qualities which as a rule are valued most highly in Empedocles' poetry—his keen observation and attention to detail, his grand formulation of what he observes in terms of manageable principles that are capable of being grasped, and applied, in particular situations. As for the fact that he used poetry, this simply brings him even closer not only to the familiar figure in the ancient Near East and Mediterranean of the poet–magician, but also to what has been described as the 'cosmic poetry' found in the later magical papyri: poetry which presents

the belief that the world constitutes a divine whole, with its parts bound one to another by a kind of sympathy. All that is needed to succeed in any enterprise is to manipulate this influence with care.[26]

A substantial non-philosophical literature has sprung up around the figure of Empedocles as god-man and wonder-worker.[27] This has resulted in a fascinating situation. On the

[23] A. A. Barb in A. Momigliano (ed.), *The Conflict Between Paganism and Christianity in the Fourth Century* (Oxford, 1963), 100. [24] Chs. 9–10, 12.
[25] *Le Dualisme chez Platon, les Gnostiques et les Manichéens* (Paris, 1947), 319.
[26] Bernand 80–2. On the type of the poet–seer–magician in antiquity cf. e.g. Meuli, ii. 1029–30; F. M. Cornford, *Principium sapientiae* (Cambridge, 1952), 62–126; Morrison 55–7; Burkert (1983), 115–19. All these writers refer to Empedocles.
[27] Cf. esp. Weinreich, ii. 179–85; Nock, i. 250; Bieler, i. 1, 7, 17, 43–6, 67, 84, 99, 102–3, 122, 127, 130, 135–8, 149; Gernet 419–29; Smith 181–2 with n. 54; C. H. Talbert,

one hand we have a literature focusing on Empedocles as a 'divine man' and arch-magician; on the other we have a vast literature interpreting him, in essentially Aristotelian fashion, as a Presocratic philosopher. There was a time when this dichotomy would have been acceptable: a time when it was accepted that Empedocles wrote two poems—one religious and irrational, the other philosophical and rational—which had nothing in common with each other except the fact that they were both written by the same person. However, this is no longer the case: not because it is true, as is sometimes claimed, that Empedocles' two poems are really only one but because the most explicitly magical material occurs in the cosmological—and supposedly rational—poem.[28] And yet the next step has, strangely enough, never really been taken. That step involves seeing the need to understand the cosmological poem, as a whole, *in a magical context*. Theoretically, there has never been a better or more appropriate time for taking this step: recent literature over the past few years reveals an increasing appreciation of the need to view ancient philosophy in terms of a way of life and to understand philosophical documents against a wider contextual background.[29] What is more, we have already seen in a very specific way how Empedocles invites (or rather demands) that any reader of his cosmological poem approach it as an integral part of his magical activity, as a text that needs to be understood in a magical context.[30] In view of these specific and general factors there can no longer be any justification for continuing to set apart what belongs together and continuing to tear elements of doctrine out of their native

ANRW ii.16.2 (1978), 1629 with nn. 44, 46, 49, 1633–5, 1639, 1641, 1648. On the religious and magical faces of the 'divine man' see Smith (1973), 227–9. Talbert (op. cit. 1637 n. 82, 1639) is too dogmatic in his attempt to distinguish between the type of the 'divine man' and the type of man who, more dramatically, becomes an immortal; to claim that Emp. B112.4 could be understood, or was understood in antiquity, as describing the status of a 'divine man' rather than an immortal is absurd. Otherwise, his comments on the parallels between the story of Empedocles' death in Heraclides of Pontus and the story of Jesus' death in the Gospel of Mark are important, as are his comments on the parallel between Heraclides' story and the account of Heracles' ascent to heaven in Diod. 4.38 (1648; 1629 with n. 44, cf. 1639).

[28] Cf. M. Hadas and M. Smith, *Heroes and Gods* (London, 1965), 46–7; Lloyd (1979), 34–5. On the question of Empedocles' poems see Ch. 23.

[29] Cf. e.g. P. Hadot's programmatic statements in R. Goulet (ed.), *Dictionnaire des philosophes antiques*, i (Paris, 1989), 11–14.

[30] Ch. 15.

context. The alternative is not easy. It requires reassessing our entire approach to Empedocles from the ground up. It requires being prepared to concede that not all Presocratic philosophers were aspiring Aristotles, but could have had completely different aims and intentions. And, above all, it requires starting to become familiar with a literary genre which has only been explored by a very few and understood by even fewer: the genre of magical literature.

There is no denying that occasionally a connection has been noted between Empedocles the magician and some item of his teaching—as between his central doctrine of like being drawn to like and the principle, so basic to magical praxis the world over, of like influencing and attracting like.[31] But what has not been done is to see how fundamental these points of contact really are. To take the most obvious example: behind Empedocles' doctrine of like-to-like lies his major teaching of two cosmic principles, Love and Strife. This means that we have on the one hand Empedocles the magician, and on the other his basic doctrine of Love, or attraction and desire, and Strife, or hatred, contention, rivalry, ill will. A great deal has been written in the attempt to explain why Empedocles devised such an emotional, pungent dualism, but without the point being made that these forces of love or attraction and strife or repulsion are the fundamental governing principles of magical operations both in the ancient Greek world and elsewhere.[32]

We can also be a little more specific. The famous occult doctrine of sympathies and antipathies in nature which was to have such an enduring history from late antiquity through into the Islamic world and the Renaissance is associated in particular with the name of Bolus of Mendes, who lived in Egypt during the second century BC; with Bolus one of the

---

[31] Cf. Kranz 62, 91.

[32] Aspects of Empedocles' idea of strife can of course be traced back to Hesiod—cf. also Burkert (1992), 91–2—but this does not explain why he selected it or gave it such a prominent place alongside love. For the two basic categories of magical operation—procedures aimed at love, desire, and union and those aimed at hate, rivalry, and separation—see Bernand's overview, 285–334, plus e.g. S. Eitrem, *SO* 22 (1942), 73 and Gow's commentary on Theocritus' second *Idyll*; also Dubois 152–9 (Gela, 5th c. BC). For Empedocles' principles of Love and Strife cf. esp. B17.7–8, 20.4, 21.7–8, 26.5–6, 109.3. On these principles of his, the notion of sympathies and antipathies in Greek magic and alchemy, and the analogies to both in China, see Needham, ii. 39–40; v/4. 312 with n. e, 322–3; also below, n. 36.

primary practical applications of the doctrine was to the world of plants. The origins of this doctrine are usually claimed nowadays to be traceable back to Theophrastus' botanical works;[33] but the crucial point which has come to be overlooked is that Theophrastus' occasional and unsystematic allusions to what can broadly be described as the notion of sympathies, affinities, or mutual affections between plants are an intrinsic part of the lore he inherited from earlier generations of 'rootcutters', herbalists, and agricultural theorists.[34] Here we need to remember the special importance Empedocles attached to describing the working of the principles of love and repulsion in the world of plants; his self-declared expertise, as a sorcerer, in the field of plant remedies and magic; and the fact that, when he gives instructions to his disciple for mastering the magical power of his teaching, he specifically compares the process of nurturing his words with the process of tending and nurturing plants. Here the language of agriculture has blended imperceptibly with the language of the mysteries.[35]

[33] Cf. e.g. Fraser, i. 441–2, ii. 643, with further refs.
[34] On Theophrastus' predecessors cf. O. Regenbogen, *RE* Suppl. vii. 1455–72; Lloyd (1983), 120–35, 148; S. Amigues, *Théophraste: Recherches sur les plantes*, i (Paris, 1988), pp. xx–xxx, xxxviii–xl with R. W. Sharples, *CR* 39 (1989), 197. The single most important passage in this connection is *CP* 3.10.4 = Androtion, *FGrH* 324 F82, where Theophrastus explicitly names Androtion as authority for the idea of mutual affection or attraction between plants. For Androtion as a predecessor of Bolus cf. Wellmann 22, Kroll 232 and in *RE* Suppl. vi. 7, Jacoby in *FGrH* iiib (Suppl.), 108 with n. 12; for Androtion himself, E. H. F. Meyer, *Geschichte der Botanik*, i (Königsberg, 1854), 14–16.
[35] Emp. on plants: B33, 72, 77–81, 154c, plus fr. 152 Wright; A70, 72a, Plut. *Quaestiones naturales* 916d, Sext. *Math.* 8.286, Simpl. *Cael.* 586.8–10; Meyer, op. cit. i. 46–58, Wright 61, 63; and cf. also Theophr. *CP* 1.21.5–6 = DK 32 A5, where the Pythagorean Menestor apparently applies a biological principle of Empedocles specifically to plants. Menestor seems to have been indebted to Empedocles for other aspects of his plant theories as well: W. Capelle, *RhM* 104 (1961), 62 with n. 53. He was probably from the Tarentum region with—like Zopyrus (Chs. 11–12)—a good knowledge of Sicily (Capelle, op. cit. 57–8 with n. 40; Iam. *VP* 267 lists him as from Sybaris). Plant magic: B111 with Ch. 15 and nn. 14, 20–1. Words as seeds and plants: B110, Ch. 15 with nn. 47–8. For ἐποπτεύσῃς in B110.2 note the agricultural use of the verb in *Od.* 16.140 and Hes. *Op.* 767 (its only sense in Homer and Hesiod); for μελέτῃσιν in the same line, ibid. 377–82 (cf. Emp. B110.4 with B132) and 412, 315–16 (cf. Emp. B110.6), 443–5. For *epopteia*, agricultural imagery, and the symbol of grain see also Farnell, iii. 182–4 (Eleusis) with Ch. 15 n. 48. B110.4–5 must be translated, 'And from them [i.e. my words: Ch. 15 n. 47] you will obtain many other things as well, for they will grow each according to its inner disposition, in whatever way their natural tendency to growth dictates.' For the sense of *physis* here cf. Emp. B8.1–2 with Arist. *GC* 314ᵇ4–8 and Kirk 228–9, 228 n. 1, plus R. Guénon's comments, *The Great Triad* (Cambridge, 1991), 93–4 with n. 9. Comparison of teaching to agriculture and of words

Finally, to return to the question of Love and Strife, it seems never to have been noticed just how relevant a famous passage in Plotinus is to our understanding of Empedocles:

And how are magical operations (γοητείας) carried out? By sympathy, and thanks to the fact that there is a natural harmony between things that are alike and a natural opposition between things that are unlike.... For many things are 'drawn' to each other and enchanted without any third party deliberately working to bring the effect about. And the real magic in everything is the Love in it, along with the Strife. *This* is the primary magician and enchanter; it was when men observed *its* magic that they started using charms and spells on each other.

Plotinus then goes on to talk, in equally generalizing terms, about the natural power of love in the cosmos and the way that magicians apply it practically when formulating their incantations and charms. In the process, he compares this manipulation of the principle of sympathy in human affairs specifically to its application by farmers to plants.[36] The bearing of this passage on Empedocles—Empedocles the magician, Empedocles the teacher of the principles of Love and Strife—would in normal circumstances have been more or less self-evident. The failure to see its relevance is a dubious tribute to modern scholarship's success at splitting Empedocles in two: at dividing the 'rational' from the 'irrational', the philosopher from the magician.

As with Love and Strife, so with the elements. Occasionally the need has been felt to postulate a background of magic or mystery cult for Empedocles' doctrine of the elements— especially because of the similarities between the major role

---

to shoots or seeds was to become both a rhetorical and mystical commonplace: Antiphon, DK 87 B60, the Hippocratic *Law* 3, Plut. *De liberis educandis* 2b–c, *CH* 1.29, Matthew 13: 3–23, Mark 4: 2–20, Luke 8: 4–15, Iam. *VP* 161, Diels 178.

[36] Plot. 4.4.40. In the Loeb edition (iv. 260 n. 1) A. H. Armstrong dutifully refers for Plotinus' mention of Love and Strife to Empedocles (cf. also Rowson 238–9), but fails to say anything about the evidence for Empedocles as a magician. For Emp. as γόης cf. D.L. 8.59; Burkert (1962), 38 with n. 10, 48 with n. 60. On the alternative expressions used in antiquity for the famous doctrine of 'sympathies'—love or affection, φιλία; unification; mixture or κρᾶσις; strife; enmity or ἔχθος; separation—see J. Röhr, *Der okkulte Kraftbegriff im Altertum* (Leipzig, 1923), 36–7, 43–7, 51–9, 65, 75, plus n. 32 above: all these terms have their exact equivalent in Empedocles, and the direction of influence was certainly not just from him.

played by the four elements in the magic and mysteries of later antiquity and what Empedocles himself says about them with regard to purification and the soul.[37] But, as we have seen, this is more than just a hypothesis. Coming to grips with Empedocles' basic identification of elements with divinities has involved penetrating a world of mythology behind the veil of philosophy, and then penetrating a world of magic, cult, and ritual behind the world of mythology. In their 'interpretations' of Empedocles the mainstream traditions of Greek philosophy, both ancient and modern, have paid lip-service to the mythology and ignored the rest: their overriding interests lay elsewhere. On the other hand we have what appears to be a silly legend about Empedocles jumping into the fires of Mount Etna wearing a bronze sandal, symbol of the underworld: a legend which—not surprisingly, when we consider its origins in magic and ritual—provides a better guide to understanding Empedocles' equation of elements and divinities than the entire Greek philosophical tradition. There are some important lessons here about where we choose to learn from, and whom we choose to learn from about what.

As soon as we start to place Empedocles and his work in their mythological and magical context, some very striking results emerge: results that have a bearing on our understanding not only of Empedocles but also of much else besides. Here it will be worth mentioning just one example: the question of theurgy. Recent research has done much to show how important a place the practice as well as the theory of theurgy—that is, the 'performing of divine actions', chiefly with the aid of magical 'symbols' or *symbola*—occupied in Neoplatonic tradition.[38] Previous scholars, including Dodds, had tended to marginalize the phenomenon of theurgy, using all the familiar

[37] Cf. Eitrem, *SO* 5 (1927), 45–59, and Seaford, *HSCP* 90 (1986), 10–12, 17–23. For the elements in later mysticism and magic see e.g. *Korē Kosmou* 54–63 = NF iv. 18.11–20.26 (where the four elements are personified as divinities); the Paris magical papyrus (*PGM* iv) 487–537 with Festugière (1950), 307 n. 6; Delatte (1927), 420.2–15; Dieterich (1891), 56–62 and (1923), 54–61, 78–82; Reitzenstein (1927), 222–6 with the note at DK i. 358.3–5; Eitrem, *SO* 4 (1926), 39–59 and 5 (1927), 39–59; Griffiths 301–3; Seaford, op. cit. 1–23.
[38] Cf. esp. A. Sheppard, *CQ* 32 (1982), 212–24, with further refs. to the literature; Shaw (1985) and (1993). On the meaning of the term 'theurgy' see still Lewy's summary, 461–6.

terms of abuse such as 'rubbish' and 'spineless syncretism', because it represented an obvious deviation from the norm of 'Greek rationalist tradition'.[39] And yet here, as elsewhere, it is important to restrict oneself to the facts. On the one hand, the notion of a monolithic 'Greek rationalist tradition' is in certain respects just as mythical as it is real. Certainly one finds the nostalgia for such a tradition in some late writers, such as Plotinus during the course of his polemic against the Gnostics; but it is important to remember that this hankering after an 'ancient Hellenic tradition'[40] was based to a large extent on the often fundamental misinterpreting of the earlier literature in question.[41] On the other hand there is the fact that, inside Neoplatonism itself, acceptance of the role of theurgy was by and large more the norm than a deviation from the norm.[42] What is more, the elements of ritual and magic which were essential to the practice of theurgy were not viewed by many Neoplatonists in the way that modern scholars have preferred to view them: as a corruption of, or lower alternative to, rational philosophy. Instead they tended to be seen as the culmination of philosophy, as a way of attaining what thinking alone was incapable of ever achieving: the raising of men to the gods and the divinization of the soul.[43]

Viewed from this theurgical perspective, the similarities between the role attributed to ritual and magic by Neoplatonists and the role attributed to them by Empedocles and early Pythagoreans are striking. Neoplatonic theurgists were known as magicians and considered capable not just of extracting men's souls from their bodies but also of returning them, just like Empedocles; they were considered able to make rain and stop plagues, again just like Empedocles; they had visionary encounters with Hecate and in their rituals they used or

[39] Dodds 285-6, with my comments, Kingsley (1992), 343 n. 20.
[40] Plot. 2.9.6.
[41] Cf. e.g. Whittaker 63-95, and his further papers on the 're-reading' by Platonists of the *Timaeus* in his *Studies in Platonism and Patristic Thought* (London, 1984); J. Dillon, *The Golden Chain* (London, 1990), Ch. V.
[42] Cf. Shaw (1985), 2-3 ('... It is a curious fact that Neoplatonism today is identified with Plotinus and an intellectual mysticism which denied formal religious worship, for in the history of the tradition Plotinus stands nearly alone in this attitude...').
[43] Ibid. 4-28. At the same time the attempt made by Dodds and others to draw a sharp dividing line between theurgy and 'lower' forms of magic is on the whole unacceptable. Cf. ibid. 7-9; also Wasserstrom 160-6, Kingsley (1993*b*), 18-19.

actually wore the secret *symbola* of the gods, just like the Empedocles of legend; and they attached the greatest importance to the process of ritual immortalization, evidently preceded by a sequence of death and rebirth, just as we find in early Pythagorean circles and on the closely related gold plates.[44]

This list of similarities is instructive. In particular, it provides a powerful corrective to the—by now traditional—view of theurgy in late antiquity as no more than a symptom of the decay of classical civilization and of 'a visibly declining culture': as the 'seductive comfort' offered to 'the discouraged minds of fourth-century pagans', as the final 'refuge of a despairing intelligentsia'.[45] It also helps to show how misleading it is to describe theurgy as

this invasion of philosophy—a mental activity which the Greeks had always laboured to render rational—by this element of the extrarational based on revelation and ritual. . . . Plotinus alone appears to us as a heroic exception to this general crazy infatuation. . . .[46]

In fact when we view things in a broader perspective it becomes apparent that the Neoplatonists who were drawn into the world of theurgy were motivated, unconsciously as well as consciously, by a powerful urge for symmetry: by the wish to renew and maintain contact with the wellsprings of western philosophy, understood not just as a mental exercise but as a way of life. Certainly it is no coincidence that a close relationship evolved between the theory of theurgy and the collection of mystical poetry known as the *Chaldaean Oracles*: a title which was meant to indicate that the Julian who composed the oracles was a recipient of, and belonged in the hereditary line

---

[44] Theurgists as magicians: Dodds 283–99. Extracting soul from body and returning it: ibid. 285 with n. 21 (above, Ch. 15; cf. also Heraclid. frs. 76–86 with Bolton 153–6, Gottschalk 18–21). Rain-making: Dodds 285 with n. 22, 288 with n. 50; Saffrey 213–14. Stopping plagues: Dodds 285 with n. 24; Saffrey 211–12. Hecate: Dodds 288, 299. *Symbola*: ibid. 292–3, 295–6; Lewy 190–2, 436–41; and A. Sheppard, *CQ* 32 (1982), 220 with n. 26. Ritual immortalization, death, and rebirth: Dodds 291 with n. 66; Lewy 210, 415; Sheppard, op. cit. 222 and n. 36; Shaw (1985), 2 with n. 5, 10–16, 26.

[45] Dodds 287–8. Compare the title to another of Dodds's books, *Pagan and Christian in an Age of Anxiety* (as if every age does not have its anxieties); the theme has been well and truly worn into the ground.

[46] Saffrey 225.

of, Chaldaean magical tradition.[47] Here we need to understand that, already by the fourth century BC, the term 'Chaldaean' had come to denote the blending of Iranian and Mesopotamian traditions which occurred chiefly as a result of Cyrus' conquest of Babylon in 539 and the subsequent two-hundred-year occupation of the city by Zoroastrians; as it happens, genuine elements of both Iranian and Mesopotamian traditions can be detected in the *Chaldaean Oracles* underneath the surface of their Platonic terminology.[48] In other words the theurgists, far from just falling for the 'orientalizing craze' of the late Hellenistic period, were finding their inspiration in the same regions and types of lore that had provided much of the underpinning both for Empedocles' activities and for Pythagorean concerns over half a millennium earlier.

Apart from this common ground and circuitous relatedness, there is also a sense in which we must postulate a more direct line of affiliation: a sense in which the Neoplatonists who were theurgists were attempting to revitalize a centuries-old tradition of dry and—on its own—ineffectual rationalism by returning to the origins of western philosophy as embodied in early Pythagoreanism. It is no coincidence that that is exactly what Numenius, a figure of special significance in the subsequent evolution of Neoplatonism, claimed he was setting out to do in the second century AD: he insisted that the only way to rescue Platonism from the academic hair-splitting into which, according to him, it had degenerated was to trace it back to its origins in the cults, rituals, and doctrines of the Near East—aided by the guiding inspiration of early Pythagoreanism.[49] Similarly, it

---

[47] Cf. also expressions from Iamblichus such as 'the native doctrines of the Assyrians' (*De mysteriis* 1.2) which are collected, but poorly interpreted, by F. W. Cremer, *Die chaldäischen Orakel und Jamblich 'de mysteriis'* (Meisenheim am Glan, 1969), 9–11; the implications of the title 'Chaldaean' are better appreciated by Saffrey (220). On the meaning, and interchangeability, of the terms 'Chaldaean' and 'Assyrian' see Kingsley (1990), 253 with n. 47.

[48] On the syncretistic background to the term 'Chaldaean', and the blurring of the distinction between Iranian and Babylonian, cf. F. Rochberg-Halton, *JNES* 43 (1984), 115; Kingsley (1990), 253–6; and also Kuhrt 147–57. For oriental elements in the *Oracles* cf. e.g. Kingsley (1992), 339–46 with n. 20, and Lewy 399–433. On the motivations and dynamics behind the process of reformulating oriental traditions in Greek philosophical—and specifically Platonic—terminology during the first few centuries AD, see Kingsley (1993*b*).

[49] Numenius, frs. 1a, 24–8 des Places; O'Meara 10–13. For Numenius and theurgy cf. Leemans 69–77.

is well known how highly Iamblichus valued Pythagoreanism—not as a school of rational thought but as a total way of life.⁵⁰ These views and attitudes of Numenius and Iamblichus have become something of a laughing-stock for modern scholarship, as embodying the worst kind of romanticism and lack of historical realism on the part of individuals in late antiquity. But neither writer was quite the fool he has been made out to be; what have been dismissed as the irrational excesses and innovations of the Neoplatonists were in fact not their creation at all but, on the contrary, faithfully mirror—and perpetuate—the traditions of pre-Platonic Pythagoreans. Once again we find that, although at a doctrinal level the importance of Plato for the Neoplatonists was naturally fundamental, there were also respects in which he had no original contribution to make. Most Neoplatonists were themselves well aware of this fact, and said as much in emphasizing Plato's role as a mere link in the chain of transmission of earlier Pythagorean and Orphic tradition.⁵¹ It is simply a matter of taking them more seriously.

The idea that theurgy was some regrettable intrusion of the irrational into the intellectual life of late antiquity is just one manifestation of an underlying attitude which has paralysed attempts at understanding certain aspects of earlier western culture. Crucial to this attitude have been certain assumptions about the relationship between religion and magic. It used to be normal, and to a significant extent still is, to view magic as the degeneration and corruption of religion. From this perspective the magic known to us from late antiquity—and in particular from the magical papyri, with the tendency of their authors to 'pick up formulae of religious origin and utilise them for their own magical ends'—is little more than a rubbish-tip on which we can watch the infinitely purer conceptions of religion disintegrate and decompose.⁵²

⁵⁰ B. D. Larsen, *Jamblique de Chalcis* (Aarhus, 1972), 66–147; G. Fowden, *JHS* 102 (1982), 36–8; O'Meara 30–105; Shaw (1993); and on his 'imitation of Pythagoras', Shaw (1985), 27. For Iamblichus and Neopythagoreanism cf. Larsen, op. cit. 73–4, and for his interest in early Pythagorean cult, mystery, and use of *symbola*, ibid. 87–9.
⁵¹ Ch. 10, with nn. 61–2.
⁵² Dodds (1965), 73; Barb, op. cit. (above, n. 23), 100–25. Cf. e.g. Wilamowitz (1931–2), i. 10; Festugière (1981), 281–328; Nilsson, iii. 129–32; Bonner 181.

However, time has worked its changes in the last few years. There is a growing awareness that the dividing line between magic and religion is very thin, sometimes impossible to draw.[53] What at first sight may seem objective and distinctive categories often turn out, on closer inspection, to be no more than variables in the manipulative hands of social and political forces. As for the well-known distinction between magic as coercing the will of the gods and religion as simply appealing to it or supplicating, it is impossible to sustain.[54] The evidence shows that the 'magical mentality', if we can describe it as such, is perfectly capable of supplicating and praying rather than trying overtly to force the gods—which is hardly surprising when we bear in mind the bonds linking magic with rhetoric, and the powerful role which the Greeks ascribed to subtlety or persuasion as effective alternatives to outright force.[55]

These considerations naturally pose problems for any historical theory of magic as the degeneration and decay of religion. But such a theory also encounters problems of a purely chronological nature. Of course magic as generally understood is not only a late phenomenon.[56] Here it will be worth mentioning two examples which have a direct bearing on the material examined earlier. First there are the curses, inscribed on lead plates, which date from at least the early fifth century BC and were deposited either inside the Gaggera sanctuary at Selinus or just outside its walls.[57] What is significant about them is not only their age, but also the place

[53] Cf. e.g. A. F. Segal in R. van den Broek and M. J. Vermaseren (eds.), *Studies in Gnosticism and Hellenistic Religions Presented to Gilles Quispel* (Leiden, 1981), 349–75; Phillips 2711–32, 2772–3; Fowden 79–81; MT i. 1–2; Faraone, *JHS* 113 (1993), 77–8; and the various papers published in *MagH*.

[54] Smith (1973), 221–2 and in *Classical World* (1977), 396; H. S. Versnel in *MagH* 68–93; R. Kotansky, ibid. 122; F. Graf, ibid. 188–96.

[55] Cf. esp. Detienne (1973), and, for the links between magic and rhetoric, Ch. 16 n. 53.

[56] The impression of an increase in magical beliefs and practices after the 1st or 2nd c. AD is not only unquantifiable—cf. Lloyd (1979), 5—but also largely illusory. There are social as well as other reasons why earlier literary sources should have been selective and unrepresentative (Phillips, *passim*; below, with n. 91), and papyrus evidence would naturally be most prolific in the later Graeco-Egyptian period. The impression of such an increase must also be offset by the magical inscriptions from the classical period which are continually being discovered throughout the Greek-speaking world: for a convenient list of recent publications cf. Bernand 412–32.

[57] Jordan, *GRBS* 26 (1985), 175–7; *LSS* 125–9; Ch. 16 with nn. 34–5.

where they were found—together with the fact that they were dedicated to the same underworld divinities officially worshipped in the sanctuary. In other words, here in Sicily we have a perfect example of what would usually be described as typically magical practices being performed by people who no doubt also used the sanctuary as a centre of religious worship: clearly nothing strange was felt in asking the divinities worshipped to make one's curses effective.[58]

The second example also takes us back to Selinus. One of the most characteristic features of the magical papyri of late antiquity is said to be the way that they 'constantly operate with the debris of other people's religion';[59] and in his book on late magical amulets Bonner describes how some of the inscriptions on these stones

express a religious feeling, or at least repeat phrases and sentences of religious import—that is, liturgical fragments. Such bits of prayer and praise are worth little as indications of genuine religious feeling on the part of those who made or wore the amulets; at best, they serve to link the magicians and their customers, however loosely, to certain religious groups in which the liturgical material whence the amulet texts were derived was a vehicle of true religious feeling.[60]

One of the many problems here is that these standard depictions of 'theft' of religious material on the part of supposedly degenerate magicians in late antiquity describe perfectly what we know was already happening in the fifth and fourth centuries BC. The lead plate from Selinus which was mentioned earlier because of the verse fragments it preserves—fragments containing what appears to have been cult material relating to the worship of Demeter, Persephone, and Dionysus—had, itself, an obviously magical function: another section of its text, together with the parallel provided by a more or less contemporary lead plate from Crete, indicates that it was used as a protective charm (ἐπωιδή) for warding off 'a natural calamity or plague'. The probability of a reference here to the use of magical means for warding off some kind of a plague at Selinus

[58] For further evidence of the custom of depositing curse tablets in chthonic sanctuaries, with discussion, cf. Jordan, *Mitteilungen des Deutschen Archäologischen Instituts, Athenische Abteilung*, 95 (1980), 231 with n. 23, 236–8.
[59] Dodds (1965), 73.
[60] Bonner 181.

is, naturally, rather suggestive in view of the story examined earlier about Empedocles freeing Selinus from a plague.[61] Otherwise, in this fourth-century lead tablet we have a classical case of the process of magical appropriation and utilization of texts which supposedly characterized the 'degenerate' period of late antiquity. What is more, the suspicion is bound to arise that—as in the case of the curse tablets from the Gaggera—the people who did the appropriating were also to some extent involved in the cult worship from which they did the appropriating: in other words they were 'thieving' from themselves. As for the actual process of appropriating literary material for magical purposes, it is clearly important here not to forget the appropriation of lines from Homer as magical charms (ἐπωιδαί) on the part of early Pythagoreans: one other phenomenon already discussed for its relevance to Empedocles.[62] As for the question of 'genuine religious feeling' touched on by Bonner, that is an issue we will come back to shortly.

From the lead plates we need, finally, to turn back to the Bacchic and 'Orphic' gold plates considered earlier—and Zuntz's classic treatment of them in his *Persephone*.[63] No discussion of these gold plates can be complete without taking into account the gold plates of very similar appearance which date from Roman times, were used as amulets, and were inscribed with magical texts intended to protect the bearer against evil or misfortune. This use of gold plates as amulets is often referred to in the magical papyri.[64] Zuntz chose to emphasize 'the vast span of time separating the "Orphic" from the "magical" gold leaves', and portrayed these later amulets as

> product and symptom of the oriental superstition which engulfed the collapsing Roman Empire ... He who, for a while, has immersed himself in the quagmire of late-Roman superstition will ... find himself on familiar ground. On turning to the texts on the so-called

---

[61] R. Kotansky in *MagH* 111, 122, 127 nn. 27–8; above, Ch. 17.
[62] Ch. 16 with nn. 46, 52–3.
[63] For some general observations—which must however not be considered binding—on the significance of the metals used, cf. Zuntz 279; Martinez 2–6.
[64] Cf. *PGM* IV.1217–18, 1812–29, 1846–52, VII.579–82, X.26–8, XIII.888–933, 1001–25; Zuntz 279–86, West (1975), 231–2, R. Kotansky in *MagH* 114–16; and see further *Sepher ha-razim*, trans. M. A. Morgan (Chico, Calif., 1983), 54, 78–80.

'Orphic' Gold Leaves he must feel transferred into a different world. ... The practitioners of late-Roman witchcraft were so different a caste, and rooted in traditions so different, from those Greeks to whose mentality the 'Orphic' leaves witness.[65]

At first sight the evidence would seem to support him. For example, the gold plate found at Petelia in southern Italy can be dated to the fourth century BC; but it was found in a gold container together with a chain which date from around the third century AD. In other words it was either passed down from generation to generation, or at some point unearthed from the place where it had first been buried, and eventually used as a magical amulet over half a millennium after it had been made.[66] Again, there is the one piece of evidence that leaps Zuntz's 'vast span of time' and appears to play havoc with his chronological distinctions: a gold plate, dating from around the third century AD, which repeats formulae word for word from the much earlier Bacchic plates and plainly belongs to the same category of text. However, some of the wording on this late example—evidently described in the text as a 'gift' for the gods of the underworld—shows that it was used as a 'magic token' or amulet. Here, Zuntz noted, 'we are in the region of the Magic Papyri and Gnostic amulets'.[67]

Closer examination reveals a very different picture. Even though Zuntz saw this latest known example from the series of Bacchic and Orphic gold plates as testifying to some kind of continuity of literary and religious tradition, he also saw it as 'evidence of the same religion sadly debased', as revealing a 'fundamental change' of persuasion and religious attitude: 'the original, profound meaning' of the words had 'degenerated' to the point where the woman who used the plate, named on it as Caecilia Secundina, was simply able to 'rely upon its magic efficiency' to get her way with the gods. In short, 'what had, once, expressed a deep and metaphysical devotion has been turned into an amulet'.[68] What Zuntz fails to mention in his idealizing of the earlier gold plates (although he notes the point elsewhere) is the stupendous carelessness already exhibited in the inscribing of their texts: a laziness, a habit of repeating

---

[65] Zuntz 282–4.
[66] Ibid. 355–6; Ch. 16 n. 39.
[67] Ibid. 293 (dating), 333 (text), 335 (verdict).
[68] Ibid. 334–5.

words or phrases in the wrong place, and a tendency to produce outright nonsense which cannot be explained only in terms of semi-illiteracy on the part of the scribe but which also indicate that any real understanding of the text—and of its 'profound meaning'—was at best of purely secondary importance.[69] Particularly significant in this regard is one of the early gold plates—found folded up, very much like an amulet, inside another one in the Timpone Grande at Thurii. Zuntz wondered how 'a text so amazingly carelessly written' could 'have accompanied a prominent person into his grave,' conceded that here we have evidence of the gold plates 'gradually' being accepted as amulets and of their supposedly pure religious character starting to be eclipsed by magic, and added:

Perhaps this eclipse began when the words conveying a new vision of man and his destiny were first engraved on gold leaves—carriers, before and after, of magical potencies to combat demonic antagonists.[70]

However, his words 'gradually' and 'perhaps' contained a crucial element of deception: on Zuntz's own admission the two gold plates, one folded inside the other, which were found in the Timpone Grande were—at least at the time when he was writing—the most ancient examples known.[71]

In noting that even the early gold plates were used as amulets it is important, however, not to overlook the obvious. To write a text on a paper-thin gold plate, then fold it (unless its shape is of special significance) and place it on the breast of a dead person,[72] is to produce an amulet, a talisman, or a phylactery: the precise term is irrelevant. In this as in other respects the gold plate from Hipponium—first published in 1974 and seemingly the oldest discovered so far—has thrown an unexpected new light on the matter. The first line of its text makes no sense as it stands: yet another sign of the lack of understanding

---

[69] Ibid. 299, 345, 349, 353.
[70] Ibid. 353-4 (cf. 335 n. 2).
[71] Ibid. 297.
[72] So in the most fully documented of the finds so far, at Hipponium—G. Foti, *PP* 29 (1974), 97, 103—and in Thessaly (*TGL* 4). For details of the two gold foils, folded one inside the other, which were found by Cavallari in the Timpone Grande see Zuntz 290.

shown by those who inscribed the gold plates.[73] But what is clear is, first, that the opening line contained an attempt to define the function of the plate itself and, secondly, that its closest formal analogy is with the Caecilia Secundina text—written some six to seven hundred years later—which defines the plate on which it is inscribed as a 'magic token' or amulet. Zuntz realized the significance of this new evidence, and in his eventual account of the plate from Hipponium described it as a 'talisman'.[74] There is also another interesting detail about this plate. The first line of the text, with its reference to the time 'when someone is about to die', cannot originally have run straight on into the second with its description of the geography of the underworld.[75] Something has been missed out, again by someone evidently more concerned with the mystical potency of the words he or she was inscribing than with understanding what they say. Fortunately however, apart from its functional analogy to the Caecilia Secundina text, the first line of the Hipponium plate also corresponds to the very fragmentary text at the end of the plate from Petelia; and comparison of the relevant sections of the two texts from Hipponium and Petelia allows us to infer that they both contained special instructions for preparing a gold plate 'when someone is about to die'.[76] In short, use of the Petelia plate as an amulet when it was over half a millennium old appears to have been not only condoned, but also specified quite explicitly, on the original text itself.[77]

[73] The word which can be read as either ἔριον or ἠρίον was almost certainly intended by the writer to be ἠρίον, in spite of the fact that—regardless of G. Giangrande's claim to the contrary in A. Masarrachia (ed.), *Orfeo e l'orfismo* (Rome, 1993), 242–5—here it makes no sense. Cf. G. Pugliese Carratelli, *PP* 29 (1974), 117; Zuntz (1976), 134–5.

[74] Ibid. 132–5; cf. West (1975), 232, Pugliese Carratelli, *PP* 31 (1976), 459. On the magical status of the Hipponium plate (and, by implication, of the others to which it is obviously related: Ch. 17 with the refs. in nn. 29, 31) see West (1975), 231–2; Pugliese Carratelli, op. cit. 459–60; Janko 92, 97; R. Kotansky in *MagH* 130 n. 48, 131 n. 51.

[75] To have the first line run on into the second would be to say that, 'when someone is about to die' (ἐπεὶ ἄμ μέλλῃσι θανεῖσθαι), he or she will enter the house of Hades and find a spring of water on the right. This is plainly wrong: the person will enter the house of Hades *after* dying, not when only about to. Cf. Zuntz (1976), 136, M. Guarducci, *RFIC* 113 (1985), 391–2; Pl. *Phd.* 107d2–108a4, *Rep.* 614b–621b.

[76] West (1975), 231–2; Janko 91–2. It will be noted that West based his—generally plausible—reconstruction on passages in the magical papyri.

[77] The fact that, to judge from the gold container and chain, the plate was eventually used by a living person rather than just by someone about to die is of no significance: as

All these details should not allow us to forget that there is nothing at all surprising in the Greek gold plates being used as amulets from their very beginnings. For a number of reasons it must be considered extremely probable that at least part of the inspiration for them derives from Egypt, as a result of transmission via Phoenicians and Carthage. Formally, they bear direct comparison with the gold-plate amulets which have been recovered from tombs in Sardinia and Carthage: amulets which date from the seventh to fifth centuries BC and are indisputably Egyptian in inspiration.[78] Needless to say, in Egypt itself it is impossible to draw any meaningful distinction between the categories of magic and religion.[79]

Then, of course, there is the evidence of the texts on the Greek gold plates themselves. What can be considered the most typically magical quality of amulets—and this generalization is especially relevant to texts concerned with guidance and protection of the soul in the world of the dead—is the importance attached to the efficaciousness of the inscribed words: the confidence that what is described or stated will become the case.[80] On some of the gold plates this kind of confidence is particularly apparent; and here, as in other respects, the recently discovered plates from Thessaly throw some refreshing new light on the issue. Their declaration, 'Tell Persephone that Bacchios himself has freed you',[81] is one of the more blatant examples of pre-empting even the gods: of in effect manipulating them by assuming that they will do what they (in this case Dionysus) are said they will do.

Also worth noting is the line just before this line about Persephone and Dionysus: the opening statement on the plates from Thessaly, 'Now you have died and now you have come into being, thrice-blessed, on this very day.' The repeated

we can see from the magical papyri, it was the rule in the first instance to prepare gold-plate amulets for very specific occasions.

[78] Kingsley (1994c), § 1.
[79] Cf. e.g. Fowden 80 with n. 25.
[80] For the Egyptian material see P. Barguet, *Le Livre des morts des anciens Égyptiens* (Paris, 1967), 20; also Kingsley (1994c), § 1 with n. 18.
[81] *TGL* 10 a2, b2. For the ritual background cf. ibid. 12 ('No doubt the deceased lady on whose chest the two gold leaves had been placed aimed at convincing Persephone that she had undergone the proper Bacchic rites which secured the required and wished for' liberation).

Sandals of Bronze and Thighs of Gold   313

emphasis here on 'now, now, this very day' is formally identical to the refrains—crucial for the sake of efficaciousness—which occur repeatedly in the magical papyri and related literature.[82] To consider such a comparison irrelevant because on the gold plates it is a matter of immortalization whereas in the later literature it is usually a question of achieving far more ephemeral aims, is neither justified nor even correct. In terms strikingly similar to the opening formula on the Thessaly gold plates, the spell for ritual immortalization in the Paris magical papyrus repeatedly emphasizes—at the beginning of the spell—that the process of death and resurrection will occur 'on this very day' and—at the end of the text—that the process has, indeed, occurred 'on this very day'.[83]

As we can see in the case of theurgy, magic does not have to be trivial in its aims and aspirations. To consider it incompatible with what Bonner described as 'true religious feeling' is as arbitrary as attempting to claim that, because of its exalted subject-matter, the spell for ritual immortalization in the Paris papyrus must have a 'religious' rather than 'magical' origin.[84] As we have seen, the immortalization envisaged both in the Thessaly plates and in the gold plates from Italy is a *ritual* immortalization.[85] Modern prejudice against this kind of praxis has stood in the way of doing the evidence full justice: Dodds, for example, considered ritual divinization as described in the magical papyri no more than a deceptive sham, a dishonest utilization by the magician 'for his own ends' of mystical formulae of religious origin.[86] But in all probability he has inverted the true historical relation: there can be little room for doubt that the theme of mystical union with the divine arose, in the first instance, out of magical and ritual practice.

[82] Ch. 17 n. 27.
[83] *PGM* IV.516-18 (cf. 543-5), 685-7, and esp. 644-52.
[84] Cf. Fowden's observations, 82 with nn. 32 and 34, although his own tentative postulate of an 'evolution' in later magical texts towards more spiritual aims and goals is countered by the much earlier evidence we have been considering. The postulate is in fact part of the heritage from Festugière with which Fowden himself elsewhere takes issue: cf. Festugière (1981), 283 n., with Fowden 80 n. 22, 116 n. 1.
[85] Ch. 17. Cf. also Zuntz's observations on the *ritual* nature of the initiate's purity as emphasized at the start of the gold plates from Thurii (300–1.1, 302–3.1, 304–5.1, 307–8); and for the *ritual* background to the liberation alluded to in the Thessaly plates see *TGL* 12 (quoted above, n. 81).
[86] (1965), 72–3; and cf. Dodds 302 n. 35 on the 'infection of mysticism by magic'.

Finally, of course, we need to remember one other factor. That factor is Empedocles—who was plainly a magician, who considered his immortalization a fundamental prerequisite for his effectiveness as a magician,[87] and who in his description of his own immortality comes closer than any other person or text to the references to ritual immortalization preserved on the gold plates. This closeness, as we have seen, is no accident: the geographical, historical, and other factors involved have already been examined.

In portraying the Graeco-Egyptian magical papyri as belonging to 'a different world' from the early gold plates, 'rooted in traditions so different', Zuntz said both too little and too much. The Greek West in the fifth and fourth centuries BC was of course a different world from Greek-occupied Egypt some seven to nine hundred years later. But to the extent that he was elaborating a basic distinction between 'classical' religion and 'decadent' magic, Zuntz was mistaken. The magic was there to begin with. And, as we have seen repeatedly, it is not just that the world of the magical papyri helps to throw light on traditions and practices current in southern Italy and Sicily during the fifth century BC, or that in so doing it serves to clarify aspects of pre-Platonic magic, mysticism, and mythology which otherwise would have remained in the dark. What has also emerged is a fundamental *continuity* of ideas and traditions between fifth-century circles in the Greek West and the later practitioners of magic in the Graeco-Egyptian world. The close links between the magic practised by Empedocles, the legends that soon attached themselves to his name, and major strands of theory and practice preserved in the magical papyri are just a few examples of this continuity.

Zuntz himself was prepared to interpret the evidence of the gold plates as indicating the continuity of a 'specific circle, or group' persisting from the period of the earliest of the gold plates down to the time of the Roman Empire.[88] Yet in defining this continuity in terms of a descent from religion into magic he was falling prey to the same romantic vision of classical Greece which once provided a valuable impetus for study of the

---

[87] Cf. esp. B112; Ch. 15 with nn. 7, 11, 17, 23.
[88] Zuntz 286, 334, 335, 338 n. 6; Ch. 16 n. 39.

ancient world but has long outlived its usefulness. That is not to say that any kind of tradition extending over the better part of a millennium could possibly remain unchanged—although even here it is important to bear in mind that the process of transmission of magical lore and ritual is, as a rule, notoriously conservative.[89] With more material at our disposal we would no doubt be able to map out how and when the secondary influences, trends, and social factors came into play; and even on the basis of the evidence available it is possible to point to some general tendencies of a purely formal nature, such as the way that with time Greek magical texts seem to have become increasingly longer, less precise, more complex and verbose.[90] However, these are changes within the framework of magic, not evolutions into it or out of it.

This underlying continuity of magical lore between the Greeks of Empedocles' time and the Roman Empire naturally has significant implications for our understanding of antiquity. It shows little regard for established fields of scholarly specialization but cuts directly across them, and in doing so raises problems of a methodological as well as purely factual nature. What is required is not just a reorganization of details about the ancient world, but also a refocusing of the perspective from which we view and assess those details. Ironically, the main obstacle that has stood in the way of such a reassessment in the past has been not so much the lack of evidence as the nature of the evidence relied on. It is increasingly being realized that the standard classical texts which have been transmitted down to us via the familiar channels of manuscript tradition give a highly selective, and very partial, view of antiquity: a view that scholarship has been far too ready to accept as accurately representing the different aspects and strata of ancient society.[91] This is especially true with regard to magic: only with the comparatively recent discovery and publication of tablets, amulets, and papyri has a more balanced picture started to emerge. Yet because this newer evidence tends to conflict so

[89] Cf. e.g. Seaford 252–3; Wilson 1; S. West, *ZPE* 101 (1994), 11.
[90] Festugière (1981), 283 n.; K. Preisendanz in *Aus der Welt des Buches* (Zentralblatt für Bibliothekswesen, Beiheft 75; Leipzig, 1951), 232–3; P. Moraux, *Une défixion judiciaire au Musée d'Istanbul* (Brussels, 1960), 7–9, 44.
[91] Cf. Brown's remarks, 59; and for the relevance to magic and religion, Phillips 2688–709, 2716; Betz in *GMPT*, p. xli; Faraone 114–15.

openly with the more 'noble' and seemingly rational preoccupations exalted in much of the traditional literature, the process of giving it the attention it deserves has been a gradual and a difficult one.

Grasping just how slow and reluctant this process has been is not easy. A key figure in helping to make the study of the magical material respectable was Albrecht Dieterich; it is largely due to his influence that scholarship over the last hundred years has come to appreciate how important the magical papyri in particular are 'for a proper understanding of the religious history of the Empire'.[92] What is far less appreciated, however, is Dieterich's awareness that these same magical papyri have a great deal to tell us not just about the religious history of the Roman Empire but also about far older strata of myth, religion, magic, folklore, and cult in ancient Greece. The effect of this awareness is invariably dynamic because, in Barb's memorable words, it points to the conclusion that

Much that we are accustomed to see classified as late 'syncretism' is rather the ancient and original, deep-seated popular religion, coming to the surface when the whitewash of 'classical' writers and artists began to peel off.[93]

Just how relevant this type of insight is should by now have become apparent. Dieterich's approach to magic has been, and is being, fairly well applied to the period of late antiquity. But the possibility of applying it to an earlier period has more or less remained at the level of theory; and when it has been applied, the application has with only a few exceptions been partial and spasmodic.[94] Most of the work still remains to be done.

---

[92] Nock, i. 176. On Dieterich's contribution in this respect, and the problems he encountered, cf. Preisendanz in *PGM* i, pp. v–vi; Betz in *GMPT*, pp. xliii–xliv.

[93] A. A. Barb, *JWCI* 27 (1964), 4 n. 16.

[94] Cf. e.g. Betz's comments, *GMPT*, p. xlv, and J. Z. Smith's, *Map is Not Territory* (Leiden, 1978), 188; Gow's exemplary use of the magical papyri, ii. 35–63, and Rohde's, esp. 326–7 n. 107 (where his concluding remark would have been better translated, 'The abundance of such perversities in later ages will only be mentioned here to the extent that it helps to elucidate earlier reports'); and Faraone's more detailed study, *Classical Journal*, 89 (1993), 1–19.

20
Pythagoreans and Neopythagoreans
*

No discussion of the involvement in ritual, cult, and magic on the part of Empedocles and early Pythagoreans would be complete without saying a few words about the subsequent history of Pythagoreanism—and, in particular, about the phenomenon which has come to be known as 'Neopythagoreanism'. This word, like the terms 'Neoplatonism' or 'Middle Platonism', is a creation of modern scholarship; and as with all such constructs it is important to examine the assumptions and presuppositions which led to its creation.

The general consensus nowadays—largely embodied in an influential study by Burkert—is that the original, early Pythagoreanism died out in the 360s BC, while the Pythagorean 'revival' began at Rome three hundred years later with Cicero's friend Nigidius Figulus, 'the Potter'. And time was not the only factor separating the two movements: early Pythagoreanism and Neopythagoreanism were, to use Burkert's expression, basically 'heterogeneous'. As for the three-hundred-year period in between, Burkert was prepared to admit that it was a time when vast amounts of Pythagorean literature appear to have been produced, and yet he was adamant: during this intermediate period we have writers of pseudo-Pythagorean texts, but no Pythagoreans.[1]

To begin with the supposed difference in nature between early Pythagoreanism and Neopythagoreanism: the attempts made by modern authors at defining it are remarkably consistent. Yet as soon as we look more closely at what these authors say, the problems start to arise. For example Armand Delatte, perhaps the most prolific writer on Pythagoreanism in this century, felt nothing wrong in talking of

[1] Burkert (1961), 226–46, (1982), 22, 189 n. 105.

the period when certain Pythagorean sects were devoting themselves to magic and divination—i.e. starting from the first century BC in particular.[2]

The statement naturally implies its opposite for an earlier period. The evidence for involvement with divination, cult, and magic on the part of early Pythagoreans has once again been swept away, and what on an unbiased assessment of our sources is so striking a feature of original Pythagoreanism has been transferred instead to the Neopythagoreans.

The reason for this strange state of affairs is easy to isolate. Underpinning Delatte's reference to the rise of divination and magic in the first century BC lies a very specific view of intellectual history in the ancient world: a view well summarized in another writer's claim that the phenomenon of Neopythagoreanism appeared 'together with the blunting of the Greek critical habit ... and the growing thirst for mystical revelations'.[3] Or as Dodds was to write in his *Greeks and the Irrational*:

Many students of the subject have seen in the first century BC the decisive period of *Weltwende*, the period when the tide of rationalism, which for the past hundred years had flowed ever more sluggishly, has finally expended its force and begins to retreat. There is no doubt that all the philosophical schools save the Epicurean took a new direction at this time. ... Equally significant is the revival, after two centuries of apparent abeyance, of Pythagoreanism, not as a formal teaching school, but as a cult and as a way of life. It relied frankly on authority, not on logic: Pythagoras was presented as an inspired Sage ...[4]

The historical scheme evoked here is beguilingly impressive, and vastly misleading. What the reader can easily miss in Dodds's list of details characterizing Neopythagoreanism—a cult and a way of life as opposed to a mere teaching school, reliance on authority as opposed to logic, presentation of Pythagoras as an inspired sage—is that they apply perfectly to

---

[2] Delatte 41 (cf. 43).
[3] R. M. Wenley in J. Hastings (ed.), *Encyclopaedia of Religion and Ethics* (Edinburgh, 1908–26), ix. 320.
[4] Dodds 247.

Pythagoreanism in the fifth century BC.[5] We encounter the same problem when we turn to the portrayal of Neopythagoreanism by one other influential writer. Festugière has described how

> in this heightened disgust with theoretical knowledge and pure science, and perhaps also as a consequence of the laxity of contemporary morals, ... philosophy transformed itself into apocalypse, revelation. The sage dispensed oracles. No longer did people seek to understand: they just believed;

and he added the finishing touch to this historical melodrama about the last pre-Christian century by presenting the faculty of reason as 'lying in ruins'.[6] Nothing is said of early Pythagoreanism's profound affinities with oracle literature and revelation; of the reputation of Pythagoras' own home as a 'house of mysteries'; of the fundamental importance for the earliest Pythagoreans of ritual precepts or *akousmata*, 'which are not supposed to be understood but merely obeyed'; or of the fact that, when fifth-century Pythagoreans set about elaborating their tradition of revealed wisdom, they tended to look for guidance towards intuition, imagination, and mythology rather than to the elusive faculty called reason.[7] Again, nothing is said about the fact that Empedocles presented his teaching in the form of divine revelation; that he was a perfect embodiment of 'the sage dispensing oracles'; that he even presented his cosmological teaching itself as though it was an oracle; or that he expected his listener to accept and absorb his teaching rather than challenge or question it.[8] As with Delatte and

---

[5] Cult: Chs. 10–12, 17–19. Way of life: Pl. *Rep.* 600b, Rohde 374–6, Burkert 84, 96 n. 55, 190–2, 203–5, 240, 292–3. Reliance on the inspired authority of Pythagoras: ibid. 91, 135; Ch. 12 with n. 53; and cf. Delatte (1915), 95–6.

[6] Festugière (1981), 74–7, 77 n. 2.

[7] Early Pythagoreanism, revelation, and oracles: Pl. *Phlb.* 16c, Arist. fr. 191, Aristox. fr. 15 with Wehrli ad loc., Burkert 91, 141–3, 149. 'House of mysteries': Timaeus, *FGrH* 566 F131, Burkert 143 with n. 128, 155, 159. 5th-c. elaborations: above, Chs. 8–13. For the quotation about the *akousmata* see Burkert 178.

[8] Divine revelation: B23.11, 112–14, 124. Dispenser of oracles: B112.10–11 (Ch. 15 n. 7). Cosmological poem as oracle: B15.1, Arist. *Rh.* 1407ª31–7 = Emp. A25b, Ch. 4 with n. 26, Ch. 23 with n. 21. Acceptance of his teaching: Ch. 15 and n. 48. Festugière does mention in a footnote that according to Apollonius of Tyana Empedocles had become a god, but without adding that Empedocles had already said this of himself (op. cit. 82 n. 2; above, Ch. 15 n. 7); on Apollonius and early Pythagorean tradition see Ch. 17 nn. 10, 90, Ch. 19 with n. 14. The fallacies underlying Festugière's distinction

Dodds, Festugière has told only half of the story; facts have been rearranged, over-simplified, made subordinate to some preconceived historical scheme.

The same tendency towards over-simplification reappears to an alarming degree with regard to the issue of chronology. For Burkert, the production of texts attributed either to Pythagoras or to followers of his during the supposed three-hundred-year gap between Pythagoreanism and Neopythagoreanism implied a 'paradoxical' conclusion: Pythagorean texts, but no Pythagoreans.[9] However, in the absence of evidence to the contrary it must be assumed that writers of literature purporting to be Pythagorean will have had sympathies with Pythagoreanism and will very probably have considered themselves as standing in the line of Pythagorean tradition. As Nock has pointed out,

> Ancient pseudepigrapha in general appear to come from *milieux* linked to their supposed origin. ... If our Pythagorean pseudepigrapha were not written by Dorian-speaking dwellers in Italy, they come from men who would have claimed to stand in their succession.[10]

To dismiss the texts created by these authors as mere 'literary fictions', which meant nothing at all for them in terms of belief or style of life,[11] is not only gratuitous but is also—as we will soon see—to make somewhat over-simplistic assumptions about what it meant to be a Pythagorean in pre-Platonic times. In fact the very idea of a mere 'literary fiction' is by no means self-explanatory and, more than anything else, is reminiscent of modern reductionist tendencies to consider literary texts in an artificial void quite independent of their social context.[12] In

---

between faith or belief and knowledge or understanding are well exposed by Phillips, 2697–711.

[9] (1961), 234.

[10] Nock, ii. 523–4. Cf. also Morton Smith and H. Thesleff in *EH* xviii. 92, 95; and Fowden's comments, 95–6, 186–8.

[11] Burkert, loc. cit.

[12] Burkert's appeal to the case of Johann Valentin Andreae and the Rosicrucians (ibid. 234–5) is unfortunate, and if anything proves the opposite of the point he wished to make. See now the various papers published in *Das Erbe des Christian Rosenkreuz: Vorträge ... des Amsterdamer Symposiums 18.–20. November 1986* (Amsterdam, 1988).

such matters we must, of course, learn our lessons where we can. At the time when Burkert published his views on the Pythagorean pseudepigrapha it was fashionable to dismiss the Hermetica, too, as mere 'literary fictions'.[13] But recent developments, including the finds of texts near Nag Hammadi, have made it increasingly clear that even the most rarefied of the Hermetic texts are the products of specific circles of people belonging to a living tradition with its own cult practices and way of life.[14] Comparisons between Pythagorean and Hermetic literature are certainly very instructive. Both are so-called 'forgeries': writings methodically ascribed to influential spiritual authorities from antiquity.[15] There are many formal similarities, and overlaps in subject-matter, between the Pythagorean and the Hermetic pseudepigrapha.[16] And even the history of their treatment in modern scholarship is almost identical. In both cases the reduction of them to the status of mere 'literary fictions' was an understandable but unfortunate over-reaction to the tendency of certain scholars, earlier in the twentieth century, to present suspiciously tidy pictures of Hermetic 'congregations' and Neopythagorean 'conventicles'—both of them painted in rather blatantly Christian colours.[17]

There is one other point to bear in mind when considering the apparent lack of concrete evidence for continuity of religious or mystical practices, and this is the danger of formulating arguments *ex silentio*.[18] Here again, Burkert presented little more

[13] Cf. esp. Festugière (1950), 309 ('fictions littéraires'), with J.-P. Mahé's remarks, *Hermès en Haute-Égypte*, ii (Quebec, 1982), 24.

[14] Kingsley (1993*b*), 19 with n. 73.

[15] Ibid. 23; Fowden 186–8, 199–200.

[16] So e.g. with the 'plant books', listing magical and healing properties of plants and herbs: Thesleff (1965), 174–7 and (1961), 14 s.v. Kleemporos, 15 s.v. Lykon, 20–1 (Περὶ σκίλλης, *On the Effects of Plants*); Wellmann 34–6 and (1934), 1–5; F. Pfister, *RE* xix. 1448–9 (referring also to the magical papyri); Festugière (1950), 137–216; Burkert (1961), 232–3, 239. For contacts at a more 'philosophical' level cf. e.g. Burkert in *EH* xviii. 51–2; above, Ch. 11 with nn. 9, 11. On the interrelations between the 'technical' and 'philosophical' Hermetica see Kingsley (1993*b*), 18–19 with n. 71.

[17] Reitzenstein, *Poimandres* (Leipzig, 1904), 248 with Festugière's criticisms, (1950), 81–4, (1949), 32 (although it will be noted that Reitzenstein himself never used the term 'church'); Wellmann, *Die ΦΥΣΙΚΑ des Bolos Demokritos und der Magier Anaxilaos aus Larissa*, i (Berlin, 1928), 6, with Burkert's criticisms, (1961), 233–4. Note also the roughly simultaneous reaction that set in to the idea of a single Orphic 'church': here, as well, the reaction went too far (Ch. 10 with the refs. in nn. 21, 23).

[18] Cf. E. R. Goodenough's remarks, *By Light, Light* (New Haven, Conn., 1935), 17.

than a caricature of the situation when he concluded that the absence of archaeological evidence for Pythagorean 'conventicles' or 'monasteries' in, say, Hellenistic Egypt was enough to indicate the non-existence of a continuing Pythagorean tradition there. In fact, as just indicated, the very idea of Pythagorean monasteries is little more than a romantic fantasy dreamed up by Max Wellmann in the 1920s.[19] On the other hand, when we turn to the actual evidence it becomes clear that—at least as far back as the early fourth century BC— Pythagorean circles tended to be very small, sometimes no larger than a single household; and certainly they were not the sort of arrangements likely to advertise themselves in any archaeological remains.[20] Preference for this kind of social arrangement is easy to understand. With the dispersion of the Pythagoreans which evidently occurred in the mid-fifth century as a direct consequence of the dramatic attacks and oppression they suffered in southern Italy, it was inevitable that what communities had existed until then would become fragmented and, to some extent at least, go underground. As a result, any account of subsequent history which allows for the transmission of Pythagorean ideas from generation to generation on a modest, even one-to-one, basis must be taken seriously—all the more so because we know that this method of transmission dates back to the classical period of Pythagoreanism.[21] The view held by Cumont and others that, far from dying out, Pythagoreanism continued to lead a more or less 'underground' existence in southern Italy under the Romans makes perfect sense on this basis; and, what is more, it has been

---

[19] Burkert and Wellmann, locc. citt.
[20] See Burkert himself, (1982), 16–17, on the story of Damon and Phintias (Aristox. fr. 31); also ibid. 16 with n. 74, on the principle of financial assistance. Similarly, among the 'Neopythagoreans' Nigidius Figulus may have been at the centre of a Pythagorean group in Rome; but Anaxilaus of Larissa, to mention just one person, was evidently a quite solitary Pythagorean (ibid. 189 n. 105). See further below, with n. 43, on Neopythagorean individualism.
[21] For a vivid example of such an account see Iam. *VP* 252–3, esp. 253. The passage bears certain formal resemblances to the pseudepigraphic Lysis letter (Burkert 98 n. 5, Cassio 137); for the text of the letter, and comments on it, cf. A. Städele, *Die Briefe des Pythagoras und der Pythagoreer* (Meisenheim am Glan, 1980), 154–9, 203–51, Wilson 19–20. On the antiquity of such restricted methods of transmission see Burkert 179–80 and (1982), 19 with n. 96; above, Ch. 15 with nn. 12–13.

strikingly corroborated in the meantime by recent archaeological finds.[22]

Next we come to the question of the point at which the original Pythagoreanism supposedly died out: in the 360s, according to the famous—and very influential—claim by Aristoxenus that he personally knew the 'last' of the Pythagoreans.[23] However, to accept everything that Aristoxenus says as representing historical truth would, to say the least, be naïve. In general, it does not require much familiarity with the evidence to appreciate his special fondness for rewriting history. With him history-writing was a creative enterprise; repeatedly one finds him turning the facts on their head and altering chronology to suit his own purposes, especially when dealing with the history of Pythagoreanism.[24] In this particular case his claim that he personally knew the 'last' of the Pythagoreans is blatantly polemical, and demonstrably false. On the one hand, this assertion formed an integral part of his strategy for making his own—highly unorthodox and idiosyncratic— view of Pythagorean tradition sound plausible by depriving those who would be liable to contradict him of their credentials and so cutting away the ground from under their feet. On the other hand, we happen to know that Pythagoreans and Pythagoreanism did go on living—in Athens as, no doubt, elsewhere—regardless of his claims and assertions. The truth of the matter was seen, and stated very clearly in another context, by Burkert himself.[25]

Also worth noting while on the subject of Aristoxenus is the way that he helps to throw light on another issue. His testimony shows that by his time a clear schism had developed between those—the so-called *akousmatikoi*—who adhered to Pythagorean tradition literally and faithfully and those—the *mathematikoi*—who in a sense were more sophisticated because

---

[22] Fabbri and Trotta, *passim*. On the affiliations between the Velia finds and Pythagoreanism cf. G. Pugliese Carratelli, *PP* 18 (1963), 285–6, Burkert (1969), 22, 28; for the chronology, Fabbri and Trotta 72–3, 76, and esp. V. Nutton, *PP* 25 (1970), 212. For earlier views on the continuity of Pythagorean tradition in Italy cf. e.g. Nock, ii. 621; Cumont 151. Zuntz arrived at much the same conclusion on the basis of the south Italian gold plates (338 n. 6).

[23] D.L. 8.46 = Aristox. fr. 19; Iam. *VP* 251 = fr. 18; Diod. 15.76.4; Burkert 200, and (1961), 226 with n. 1. [24] Kingsley (1990), 261–2, cf. 252–3.

[25] Burkert 107, 198–205 (cf. Méautis 9–10). See also Burkert (1989), 210–11.

they placed a greater emphasis on inner understanding, even to the extent of dispensing with the literal performance of certain ritual practices. This second trend plainly goes back to before the time of Plato. It can already be seen at work in men such as Philolaus and Archytas: men who concentrated on allegorizing Pythagorean myth and lore and, in so doing, inevitably helped to shift attention away from the literal. In other words, already during the late fifth century BC, to be a true Pythagorean might in some circles have been seen more as a matter of inner understanding than of 'living the Pythagorean life' of ritual actions or abstinence.[26] Such a state of affairs has an obvious bearing on our assessment of the writers of the later Pythagorean pseudepigrapha: there is little point in applying to them criteria which are possibly inapplicable to some of the greatest Pythagoreans in earlier times.

While Aristoxenus' report about ancient Pythagoreanism dying out in the fourth century is worth little because of the subjective and personal factors involved, so is Cicero's report[27] about Pythagoreanism being revived three hundred years later. And it is worth little for much the same reason. Just as Aristoxenus claimed he personally knew the 'last' of the Pythagoreans, so the man whom Cicero names as responsible for 'reviving' Pythagoreanism happens to have been a close friend of his, Nigidius Figulus: a remarkable coincidence, and one which becomes all the more suspect when we bear in mind Cicero's notorious vanity and love of putting himself at the centre of attention. That vanity and love were so strong that, even on his own admission, they made him throw to the wind any concern or respect for mere historical accuracy in his reports and descriptions of events.[28] For him truth was no object or objection, and deliberate exaggeration no obstacle,

---

[26] For *akousmatikoi* versus *mathematikoi* see Burkert's helpful summary, 166–208. On literal application of the detailed instructions, or *akousmata*, as equivalent to living 'the Pythagorean life' cf. ibid. 180–3, 190–1, 204–5; for liberation from performance of the *akousmata* through allegorizing interpretation of them, 173–6. On the development of Pythagorean allegory before Plato's time see above, Chs. 9–10, esp. Ch. 10 n. 48.

[27] *Timaeus* 1.1 (Liuzzi 20 § VI); Burkert (1961), 226.

[28] Cf. *Fam.* 5.12.3 ('So I ask you quite openly, again and again, to embellish and eulogize my actions even to a more excessive degree than you might perhaps feel inclined to, and to turn a blind eye to the laws and stipulations of history'); J. P. V. D. Balsdon in T. A. Dorey (ed.), *Cicero* (London, 1965), 202–3. For Cicero's friendship with Nigidius see *Fam.* 4.13 (Liuzzi 18–20 § V).

when it came to winning for himself a place in the annals of history and 'enjoying his little bit of glory'.[29] What is more, when we turn to the specific passage in question and read Cicero's words about Nigidius a little more closely we see that even he, with all his pomp and rhetoric, felt obliged to qualify his assertions in a way that provides a clearer pointer to reality. He does not present Nigidius' 'revival' of Pythagoreanism as an accepted fact, but simply as his own generous opinion; and his choice of wording also suggests that the famous 'extinction' of ancient Pythagoreanism which he refers to was not as absolute as has usually been assumed.[30] Probably the first person to be surprised that anyone would take this kind of assertion literally, and ignore his qualifications, would have been Cicero himself. After all, his words about Nigidius occur at the opening of an encomium, where exaggeration of the facts combined with adherence to established forms of rhetorical expression was considered obligatory.[31] This is not to say there is nothing to be learned from them. On the contrary, they allow us to infer that Nigidius played a role in bringing Pythagoreanism into the limelight at Rome and making it a focus of popular attention among the high-flying social circles which Cicero used to move in; but that is all.

Well after Aristoxenus, and in the century before Cicero and Nigidius, we encounter Bolus of Mendes from the Nile delta in Egypt. Very little is known about him, but that little is particularly significant. On the one hand the evidence suggests that he played an important role as a transmitter of Pythagorean traditions, and specifically of miracle stories about wonder-workers of antiquity—with Pythagoras himself (and details such as Pythagoras' golden thigh) taking centre-stage.

[29] *Fam.* 5.12.3, 7, 9.
[30] *Timaeus*, loc. cit. ('sic *iudico* post illos nobiles Pythagoreos quorum disciplina exstincta est *quodam modo* . . . hunc exstitisse qui illam renovaret'). *exstincta est quodam modo* does not mean 'was somehow extinguished' (Dillon 117) but 'was to a certain extent extinguished'.
[31] See on this point Della Casa 38–45, who draws the correct conclusion that in the oratorical encomium at the start of his *Timaeus* Cicero made sweeping statements which are 'not to be viewed in the light of historical truth'; and ibid. 37 n. 2, where Della Casa notes that Cicero's encomium of Nigidius at the beginning of his own version of the *Timaeus* is a formal and artful imitation of Plato's encomium of Timaeus at *Tim.* 20a. Cf. also Liuzzi 78–9.

What is more, Bolus also produced works very closely related to the pseudo-Pythagorean literature considered earlier. On the other hand he appears to have been a crucial figure in shaping the subsequent development of Graeco-Egyptian alchemy, and it is no doubt largely as a result of his influence that we find Pythagoreanism and alchemy starting to overlap. He also shows the closest of affinities with the world of the magical papyri, particularly on the matter of dream oracles and divination. His resulting position as a type of mediator between earlier Pythagorean traditions and the magic of the later papyri is obviously of major significance in the light of the evidence already noted for similarities between early Pythagoreanism and Graeco-Egyptian magic—and, what is more, for the actual transmission of Pythagorean lore down into Egypt.[32]

Problems have arisen in categorizing Bolus. In one of its entries the *Suda* lexicon calls him a 'Pythagorean'; but elsewhere he is referred to as a 'Democritean'.[33] The generally accepted way of explaining the apparent contradiction has been to play down the *Suda*'s reference to him as a Pythagorean. This is done by claiming that in late antiquity the term 'Pythagorean' was just a synonym for 'esotericist' or 'occultist', so that the label 'Democritean' is the one that really matters.[34] For a number of reasons this is not correct. First, it is not true that by late antiquity 'Pythagorean' had come to mean nothing but an esotericist or magician: the term always indicates someone who at the very least felt a conscious affinity with Pythagoras and with Pythagorean tradition.[35] Second, the

---

[32] For full refs. to the points in this paragraph see Kingsley (1994c), § II.

[33] Pythagorean: *Suda*, s.v. Βῶλος Μενδήσιος. Democritean: Schol. Nicander, *Theriaca* 764, Stephanus of Byzantium, s.v. Ἄψυνθος, and cf. the *Suda* s.v. Βῶλος Δημόκριτος (probably better retained as an anomaly than emended to Δημοκρίτειος). There can be no doubting that the two *Suda* entries refer to one and the same person: Weidlich 17, J. H. Waszink in *Reallexikon für Antike und Christentum*, ii (1954), 502, Kingsley (1994c), nn. 27, 42.

[34] Kroll 230–1; Burkert (1961), 233–4; Halleux (1981), 63–4.

[35] In support of the supposedly looser meaning of the word Burkert (op. cit. 234) appealed to Pliny's denomination of the fictitious Egyptian king Nechepso and his high priest Petosiris as 'Pythagoreans'. But this is simply a mistake: the text in question, where Pliny lists the authors he used for the second book of his *Natural History*, must be read *Sosigene, Petosiri, Nechepso, Pythagoricis, Posidonio* ('Sosigenes, Petosiris, Nechepso, Pythagoreans, Posidonius') and not, as in several editions, *Sosigene, Petosiri Nechepso Pythagoricis* ... ('Sosigenes, Petosiris and Nechepso the Pythagoreans'). For the Pythagoreans in question cf. Plin. *HN* 2.6.37, 2.20.84, and compare also the straight-

Suda's description of Bolus as a Pythagorean is not to be dissociated from his probable role as a transmitter of Pythagorean traditions, or from the affinities between his own writings and pseudo-Pythagorean literature. Third, his title of 'Democritean' clearly has to do with the fact that in some of his works—but only some of them—he set out to gain extra authority for his ideas by putting them into the mouth of Democritus.[36]

Finally, however, and most important of all, there is the basic point whose immediate significance for the case of Bolus has scarcely been appreciated. This is the fact that from very early on, and even during Democritus' own lifetime, there was a tradition linking Democritus with Pythagoreanism and even giving him Pythagorean teachers. In all probability the tradition has a basis in historical fact.[37] What is more, as soon as we put aside the somewhat anachronistic and plainly inadequate distinction between a 'rational' Democritus and 'mystical' Pythagoreans, it becomes apparent that the points on which the historical Democritus was most closely aligned with the Pythagoreanism of his time were their shared interests in the relation between physical constitution, lifestyle, and ethics, and—in particular—the tendency of both Democritus and Pythagoreans to consider philosophy in terms of medicine and therapeutics.[38] This mesh of correspondences helps to

---

forward mention of Petosiris and Nechepso—without any Pythagoreans—in the listing of authors used for bk. 7. For the case, also cited by Burkert, of Aeschylus as a 'Pythagorean' (Cic. *Tusc.* 2.23 = Aesch. test. 159) see W. Rösler, *Reflexe vorsokratischen Denkens bei Aischylos* (Meisenheim am Glan, 1970), 25–37; M. Griffith in R. D. Dawe *et al.* (eds.), *Dionysiaca* (Cambridge, 1978), 109–11, 129–31. As for Kroll's claim (231) that in Roman times the word 'Pythagorean' had simply come to mean 'occultist' or 'magician', it is neatly contradicted by Schol. Bobiensia on Cic. *In Vatinium* 6.1 (Liuzzi 9, 20 § vii; Della Casa 34–5, 44).

[36] Kingsley (1994c), § ii (*On Sympathies and Antipathies; Cheirokmēta*).
[37] Cf. Glaucus of Rhegium, Apollodorus of Cyzicus, and Thrasyllus *ap.* D.L. 9.38, Duris, *FGrH* 76 F23, Thrasyllus *ap.* D.L. 9.46 (Democritus wrote a work on Pythagoras); E. Zeller, *Kleine Schriften*, i (Berlin, 1910), 470–1, Burkert 110 n. 2, 147 and n. 146, 215 and nn. 25–6, 228–9, 259 and n. 101, 289–90, 292, H. Steckel, *RE* Suppl. xii. 199, 208–9. The relevance to Bolus of this early connection between Democritus and Pythagoreanism is noted by Wilson, 22.
[38] Cf. esp. F. Wehrli, *MH* 8 (1951), 55–61; also Boyancé 176–82. Burkert (292) has some significant comments on the 'embarrassment' caused by the 'inescapably certain ... relationship between Democritus and certain Pythagoreans': embarrassing because of the way that it cuts across the most familiar stereotypes about Presocratic philosophy.

explain why, in later sources, we find not just Pythagoras and Democritus but also Pythagoras, Democritus, and Empedocles listed together as experts in medicine—and especially in the healing power of plants.[39] In short, it is a fundamental mistake to treat Bolus' titles of 'Democritean' and 'Pythagorean' as in any way incompatible. On the contrary, in view of the traditions Bolus inherited, the first would actually tend to point to the second.

Far from being a problem to be removed or disposed of, this apparent ambiguity in Bolus' status has some major implications for our understanding of Pythagoreanism in antiquity. We find much the same problem elsewhere: for example in the case of Numenius of Apamea, described by some ancient writers as a Platonist and by others as a Pythagorean. In his case both descriptions are appropriate: Numenius saw himself as a continuer of Pythagorean tradition and also as a defender of Plato's real teaching against what he considered the over-intellectualism of the later Platonic Academy.[40] And here too, as with Democritus and Pythagoreanism in the case of Bolus, the merging of Pythagoreanism and Platonism had its historical justification. On the one hand, as we have seen, Numenius was fundamentally correct in his perception of Plato's—and the early Academy's—indebtedness to earlier Pythagoreanism; while, on the other hand, it is still possible to detect some of the ways in which the early Platonists deliberately attempted to pass off their own teachings as Pythagorean.[41] It is important not to misunderstand this second point. To portray the Platonizing reinterpretation of Pythagoreanism as an aberrant departure from the 'true', 'pure' pre-Platonic Pythagoreanism is to overlook the essential fact that—before Plato's time as well—Pythagoreanism was perpetually changing, reformulating itself, consciously adapting to incorporate new developments, new trends, and new ideas.[42] The Platonizing of

---

[39] For Pythagoras and Democritus cf. e.g. Plin. *HN* 24.99.156, 24.102.160 (the Democritus of the *Cheirokmēta*—see above, with n. 36—was 'the most diligent student of the Magi after Pythagoras'); for Pythagoras, Empedocles, and Democritus, Celsus, *De medicina*, proem 7, Plin. 30.2.9.

[40] Numenius, frs. 1, 7, 24–8 des Places; cf. Leemans 14–16, Puech, i. 45–7.

[41] Above, Chs. 7–13; H. J. Krämer, *Der Ursprung der Geistmetaphysik* (Amsterdam, 1964), 44–56.   [42] Kingsley (1990), 261 with n. 99.

Pythagoreanism in and through the early Academy was a natural extension of this process, not a radical departure. These points bring us face to face with an even more fundamental issue: an issue which is almost invariably overlooked when viewing the Pythagorean movement as a self-contained phenomenon. The so-called Neopythagoreans of the first century BC onwards were on the whole noticeably lacking in any flair for organized orthodoxy; they appear as a rule to have been strikingly individualistic.[43] But is it true, as is usually stated or implied, that in this respect Neopythagoreanism was essentially different from Pythagoreanism down to the fourth century BC? In fact, once again, the evidence points to an underlying similarity and continuity between early Pythagoreanism and Neopythagoreanism, rather than to any basic difference. Particularly in the later fifth and fourth centuries BC Pythagoreanism was, as we have seen, characterized by a vigorous strain of individualism and non-orthodoxy. Contradiction of Pythagoreans by Pythagoreans was unexceptional and unexceptionable; the emphasis was clearly placed on the importance of creative, original contributions. Even fundamentals could be questioned, and reinterpreted. Understood from this point of view, the famous split between *akousmatikoi* and *mathematikoi* is not just explicable but also predictable.[44]

The full implications of such a situation have rarely been appreciated, let alone properly grasped. The overriding desire for simple classifications has virtually obliterated the fluidity and complexity of what really happened. Far from being unique to modern scholarship, this desire for tidiness is also to a large extent a reflection of the facile schematizing adopted by ancient writers and biographers. When we pick our way through these over-simplifications we repeatedly encounter the scenario of Pythagoreans teaching highly creative individuals who can in a sense be described as perpetuating aspects of Pythagorean tradition through their own writings, but who themselves can only questionably be referred to as 'Pythagoreans'. This is the case with Parmenides, with Empedocles,

---

[43] Cf. Festugière's comments (1981), 75 n. 2; H. Dörrie, *Der Kleine Pauly*, iv. 86.
[44] Chs. 8 with n. 15, 12 with n. 42, 13 with n. 37; Zhmud' 273–4, 279–81, and esp. 288–9.

and with Plato.[45] Writers in antiquity took up the challenge posed by such hard-to-classify phenomena—and with a vengeance. They had Empedocles and Plato, in particular, dramatically barred from Pythagorean circles for 'stealing doctrines' (λογοκλοπία) and revealing them in their writings.[46] The crude fictitiousness of these stories is, however, more or less transparent. Partly they are the product of an overtly hostile attitude on the part of Greek biographers to philosophy and philosophers in general. And partly they are based on the inability to understand how someone could legitimately publish writings which might in any way be related to the proverbially secret Pythagorean teachings: Empedocles' tendency to conceal his real meaning behind riddles and enigmas, or Plato's diplomatic ability to remain silent on certain key subjects, were too subtle to attract the attention and consideration they deserved.[47] Certainly, in the case of Plato, we can be sure that the truth about his relationship with the Pythagoreans who inspired and taught him was a great deal more complex than the stories of plagiarism and banishment would allow.[48]

[45] For Parmenides see Sotion *ap.* D.L. 9.21 (DK 28 A1), with Burkert (1969), 28; for Empedocles cf. Timaeus, Neanthes, and Hermippus *ap.* D.L. 8.54–6 (DK 31 A1), with Ch. 10 n. 1; for Plato, above, Chs. 8–10, 12–14.

[46] Timaeus, *FGrH* 566 F14 = D.L. 8.54, Neanthes, *FGrH* 84 F26 = D.L. 8.55; Riginos 67; and cf. Burkert 223–7, 454–5; Ch. 9 with n. 38.

[47] See Riginos's general comments, 67 with nn. 24, 27. For Empedocles' riddles see Chs. 4, 23–4; and for Plato's silence cf. e.g. *Tim.* 55c with L. Robin, *Greek Thought* (London, 1928), 227, de Santillana 108–10. The *Timaeus*' silence about the dodecahedron is certainly to be understood against the background of Plato's repeated and serious emphasis on the idea that solid geometry can only be taught in a one-to-one situation, under the personal guidance of a teacher who is able to communicate the needed information at the right time (πρὸς καιρόν) in accordance with the capacity of the pupil's soul. Cf. esp. *Laws* 968c–e, where Plato draws a careful distinction in this context between knowledge that is an unqualified secret, not to be divulged at all (ἀπόρρητα), and knowledge that can be divulged but only to those who have been suitably prepared (ἀπρόρρητα); also *Rep.* 528b–c, H. J. Krämer, *Arete bei Platon und Aristoteles* (Heidelberg, 1959), 401. Even this very idea of a graded system of education was no doubt inspired at least in principle by Plato's encounter with Pythagoreans: cf. J. S. Morrison, *CQ* 52 (1958), 211–12, and also Burkert 84–92. For the tradition of secrecy surrounding the properties of the dodecahedron in pre-Platonic Pythagorean circles, see ibid. 450–1 with n. 16, 457 and n. 54, 460–1; and for the undoubtedly Pythagorean background to the mention of the dodecahedron in the *Timaeus*, Ch. 8 with n. 17 above.

[48] Lloyd 159–74. On the reports about the banishment of Hippasus for betraying Pythagorean secrets cf. Burkert 206 with n. 73, 454–62.

This looseness, this flexibility, this blurring of Pythagoreanism at the edges, is ultimately a reflection of—and a reflection on—Pythagoras himself. It is all too often forgotten that Pythagoras was a collector: a collector of ideas, a collector of traditions. That is how he is presented to us in the earliest sources,[49] and that is how we must approach him—as well as his successors. Only by doing so can we begin to appreciate the remarkable breadth of influences and diversity of components which went to make up Pythagoreanism during the first hundred and fifty years or so of its existence. To mention just a few examples, we find mathematical and astrological material adopted from Babylonia; practices and ideas incorporated from Greek incubation ritual; mystery traditions taken over from Anatolia and, almost certainly, from Crete and Egypt—not to mention from southern Italy and Sicily.[50] Even Pythagoras' apparently innovative transposal of Ionian medical traditions across to Croton in southern Italy, or his practical concern with the links between healing and diet, turns out on closer examination to be far less innovative than one might have wished to suppose.[51] This is not in any way to diminish Pythagoras' immense importance, or to deny his obvious originality in other respects; but it does mean that, from the very outset, Pythagoreanism had in its founder a major paradigm for interacting with, borrowing from, and contributing to other religious, cosmological, or medical traditions. A hard core of esoteric theory and practice was essential if the identity of Pythagoreanism was to be preserved; but otherwise

[49] Heraclit. B40, 129; Ion of Chios, DK 36 B4; E. Zeller and R. Mondolfo, *La filosofia dei Greci*, i/2 (Florence, 1938), 316–18.

[50] Babylonia: A. Aaboe, *Episodes From the Early History of Mathematics* (New Haven, Conn., 1964), 39, Burkert 429, 433, 441–2, 454, Kingsley (1990), 256 n. 66, (1994*f*). It is also worth noting that, as Stephanie Dalley has pointed out to me, the famous Pythagorean table of opposites—odd/even, right/left, male/female, light/dark, etc. (Arist. *Met.* 986ª22–6; Burkert 51–2)—bears striking similarities to Babylonian rules for interpreting auspices in divination. Cf. I. Starr, *The Rituals of the Diviner* (Malibu, 1983), U. Jeyes, *Old Babylonian Extispicy* (Istanbul, 1989), 51–186 and, for early Pythagoreanism and divination, Ch. 18 above. Regarding transmission of Mesopotamian divinatory techniques westwards see also Burkert (1983); Starr, op. cit. 2. Incubation lore: Ch. 18. Anatolia: Ch. 15 with n. 28, Ch. 19 with nn. 15–16. Crete: Burkert 151–2; Ch. 17 n. 17. Egypt: Kingsley (1994*c*). Italian and Sicilian traditions: Burkert 112–13 with n. 21 *ad fin*; Chs. 7–13.

[51] Burkert 292–4. Cf. also Ch. 15 and n. 28 for the transfer of healing, incubatory, and oracular techniques from Asia Minor to southern Italy.

the frontiers with other movements, trends, traditions, and individuals were in one way or another always open. The term 'eclecticism' could, for the most influential of early as well as later Pythagoreans, hardly have had the negative connotations it has come to assume in scholarly circles.[52]

In much the same way that Near Eastern traditions supplied Pythagoreanism with some of its practices, ideas, and formulations, Pythagoreanism in its own turn came to supply them. We have already seen this happening in the case of magic (Chs. 16–19). On the one hand, early Pythagoreans plainly participated in the broad world of ancient Greek magic and ritual; but on the other hand, they also came to contribute in later centuries to the even broader world of Graeco-Egyptian magic. This pattern of reciprocity helps to explain how a magical spell explicitly attributed to the Pythagorean Apollonius of Tyana finds a place in the Graeco-Egyptian magical papyri; and why another Pythagorean, Anaxilaus of Larissa, is also known to have exerted an influence on the magical praxis of late antiquity.[53] Examples of this kind have even led some scholars to suggest that one or several Pythagoreans were responsible for shaping and codifying the corpus of Graeco-Egyptian magical papyri.[54] However, such a hypothesis reflects an exaggerated estimate both of the literary unity of the magical papyri and of the extent of any recognizably Pythagorean influence.[55] The truth is, rather, that Pythagoreans to some extent influenced the flow of magical tradition because they themselves were swimming in the stream—not because they were directing it. In this respect Lucian presents a highly significant, and almost symbolic, scenario with his literary portrayal of a man called Arignotus. Arignotus was a Pythagorean, a magician, and a man who had learned his occult skills from a famous Egyptian priest-magician called Pachrates who hardly spoke Greek, was immersed in the temple traditions of his native country, and is

[52] Cf. J. M. Dillon's comments in Dillon and A. A. Long (eds.), *The Question of 'Eclecticism'* (Berkeley, 1988), 125.

[53] Apollonius: *PGM* xia.1. Anaxilaus: Nock, i. 188 with Kingsley (1994c) and n. 41.

[54] Nock, loc. cit.; H. Parry, *Thelxis* (New York, 1992), p. xi n. 4. Parry's references elsewhere (ibid. 35, 42 n. 45) to 'the publication of the Greek Magical Papyri' during 'the early centuries AD' are not only over-simplistic but radically misleading: the papyri were of course never 'published' as such prior to the 19th century.

[55] Cf. Ch. 18 with the refs. in nn. 19, 31–2; and the note above.

referred to independently in the magical papyri.[56] In the web of interactions between Pythagoreanism and Graeco-Egyptian magic the Pythagoreans influenced and were influenced; they taught, and learned. Pythagoreanism wrapped itself around other traditions, and other traditions wrapped themselves around it.

This looseness and flexibility of Pythagoreanism's relationship to other, apparently distinct traditions is especially manifest in the case of its relationship with Hermetism and Hermetic texts. Several examples of this connection have already been mentioned; and on the face of it, the connection could hardly be less surprising. On the one hand, enough has already been said about the passage of Italian, Sicilian, and more or less specifically Pythagorean traditions down into Egypt—and particularly into the Graeco-Egyptian magical papyri. On the other hand, in recent years it has become increasingly clear how closely the writings ascribed to Hermes Trismegistus—themselves very much a product of Greek culture in Egypt—are related to these same magical papyri.[57] On the surface, Pythagoreanism and Hermetism are two quite distinct and different movements: the one formally acknowledging its origin in Pythagoras, the other in Hermes. But, as noted earlier in this chapter, the overlap between them in interests and concerns was at times immense. To an extent we are certainly justified in talking of common sources of ideas and inspiration, no doubt oral as well as literary, which were shared by Pythagoreans and Hermetists alike. And yet, just as in the case of the Graeco-Egyptian magical papyri, we must also allow more specifically for the transfer of Pythagorean traditions into Hermetism. Such a transfer may, superficially, be problematic because it leads to difficulties in categorization; however, it is important to bear in mind that the ability of even the most esoteric lore to jump from one tradition to another is well attested in antiquity—even in cases where the exoteric attitude of the one tradition to the other was one of explicit

---

[56] Luc. *Philopseudes* 29–36. For Arignotus cf. Dilthey 382; for Pachrates/Pancrates, *PGM* IV.2446, Kingsley (1993*b*), 13 with nn. 51–2. For Arignotus' remarks in *Philopseudes* 29–30 cf. Arist. fr. 193, Iam. *VP* 139, 148, J. Z. Smith, *Map is Not Territory* (Leiden, 1978), 201.

[57] Kingsley (1993*b*), *passim*, with the further refs. in n. 12.

hostility.[58] We are brought back in this respect to the fundamental point which has been well demonstrated by Burkert: that pagan mystery traditions never shared the strict exclusivity of Christianity but instead tolerated each other, encouraged a certain sense of individualism in their initiates rather than imposing on them a group identity, and as a rule openly permitted diversities of allegiance even among the priests and organizers.[59]

The ancient religious world was a fluid as well as an open one. Ironically, one of the implications of this openness is that in considering the survival of Pythagoreanism we must not just restrict ourselves to the so-called 'Neopythagorean' movement but must also take Hermetism into account as well. The same point applies, almost predictably, in the case of Empedocles. Early in this century Eduard Norden drew attention to striking similarities—not only in ideas but also in the way these ideas are formulated—between Empedocles and the Hermetica; Norden himself was clearly aware that the examples he gave were just the tip of an iceberg.[60] And yet, in spite of their importance, Norden's observations seem never to have been followed up; or if one were to be cynical one might be tempted to say that they have never been followed up precisely because they are so important. In fact the similarities in question are—as we will see shortly—pointers not just to a shared store of literary banalities but to a far deeper, and far more significant, continuity of ideas, traditions, and approaches to life. The fact that this continuity cuts across the recognized denominational and national boundaries, and as a result is so 'confusing to scholars who look for neat systematization',[61] is all the more reason to take it seriously rather than put the evidence aside as if it did not exist.

[58] Cf. e.g. Kingsley (1992), 343–6.
[59] Burkert (1987), 43–53.
[60] Norden 130–3. Cf. Ch. 11 with n. 11.
[61] Burkert (1987), 49.

21
'Not to Teach but to Heal'
*

THE figure of Bolus of Mendes is also helpful in the way that he raises another perhaps even more important issue. As already noted, Bolus' writings were concerned to a large extent with the occult doctrine of sympathies and antipathies in nature, and in particular with the practical application of this doctrine to the world of plants for purposes of healing, magic, and ritual.[1] The intimate relation between this literary output of Bolus and the 'Neopythagorean' plant-books of the first century BC onwards, and in certain circumstances Bolus' influence on this later Pythagorean literature, are undeniable.[2]

Modern scholarship has been unanimous in its characterization of this type of literature on the healing and magical properties of plants—and monolithically consistent in assessing how it originated and evolved. Giuseppe Messina well summarized the general verdict some sixty years ago:

> The blending of medicine with magic, of scientific observation with superstition, points automatically to a particular intellectual trend in antiquity which is well known to us: that is, to the Neopythagorean school ... This new school broke with the older trend of natural science and research—which derived from the Peripatetic school of Aristotle and was based on the exact observation of natural forces and of their corresponding effects—and, instead, established as its goal the discovery of the occult magical powers of plants, animals, and stones and the explanation of the effects of these powers via the

---

[1] Ch. 19 with nn. 33–4. Various details of Wellmann's at times over-imaginative reconstruction of Bolus' ideas and literary output are certainly questionable; but Bolus' basic involvement in these areas, and the influence he exerted on later writers, are beyond doubt. Cf. e.g. DK ii. 212.2–3; 6–11, with 13–15 and Kingsley (1994c), § II, for the attribution of the *Cheirokmēta*; 212.13–213.15; Beck 561. The most balanced discussion of the reasons for ascribing to Bolus a work *On Agriculture* remains Oder's (76–7); cf. also Weidlich 14–15.

[2] Kingsley, loc. cit.; and cf. above, Ch. 20 with n. 16.

principle of sympathy and antipathy. With this innovation it was perfectly inevitable that medicine would pass over into magic; the man who knew the magical powers of nature was, simultaneously, a healer. And so as to procure some prestige for this new line of approach, the Neopythagoreans felt the need to pass off their forgeries as the work of famous men from remote antiquity . . .[3]

The same general strategy of portraying the literature produced by Bolus and late Pythagoreans as a reaction to, and betrayal of, Aristotle and his Peripatetic school has become as unquestioned as it is common. Peter Fraser, for example, has no qualms about stating that

in the physical sciences the theoretical speculations of Theophrastus and Straton were replaced by a partly scientific, partly occult approach to the physical world, based especially on the notion of sympathy . . . This development is particularly associated with one name, that of Bolus . . . who stands with one foot in the Peripatetic atmosphere of the third century, and the other on the edge of the occult world of later antiquity, and who on that account occupies a position of the utmost importance in the history of the decline of Alexandrian and indeed of Greek science;

or, more simply, about summarizing Bolus' interests as 'the perversion of the Peripatetic' literature on the particular subjects which concerned him.[4] Even more recently, this same theory of historical corruption and degeneration has been used to support additional hypotheses about the nature and origins of the interests exemplified by Bolus and current in the late Hellenistic period:

A fundamental characteristic of this Hellenistic wisdom is that it was intensely *practical*: it aimed at control of the world, not at disinterested understanding. That indeed distinguishes it from the great rival tradition of Aristotle, in which *theoria*, the knowledge and contemplation of things for their mere beauty and order, is the goal of science. Practical arts lie at the origins of Hellenistic wisdom, and it was the interaction of the Greeks with the cultures and skills of the lands which Alexander had won that brought them into being. . . . [This new] science was always intensely practical and exploitive—Nature's sympathies and antipathies were there to be *used*—and that is why its

---

[3] G. Messina, *Der Ursprung der Magier und die zarathustrische Religion* (Rome, 1930), 22–3.
[4] Fraser, i. 439–40, 442.

manifestations ... seem more magical than scientific even in a debased sense.[5]

These judgements, these developmental schemes, may seem innocent enough. But underneath their convincing appearance they contain some essential misunderstandings about the course taken by ancient Greek science and philosophy. To begin with, it is a major fallacy to talk of a magical approach to the understanding of nature as just a perversion or corruption of the pure, rational science associated with Aristotle and his school. The fact of the matter is that the 'blending of medicine with magic' which Messina discounted as a decadent and chronologically secondary phenomenon dates back, in the Greek world as of course elsewhere, to a period far earlier than Aristotle. Here we touch the bedrock of folklore—and, with it, enduring traditions of healing techniques, specialized knowledge, and the methodical understanding of nature which it is impossible to isolate from, or place in unqualified opposition to, the domains of Greek philosophy and science.[6]

We can also be more specific. There can be no doubt that the type of plant literature which Bolus of Mendes himself produced—or which he influenced, or just had close affinities with—is generically related to the type of 'root-cutting' manuals that were dependent on a combination of observational, magical, and medical traditions dating back to centuries before Plato or Aristotle. The fourth-century physician Diocles was almost certainly one of the more important links in this sequence of traditions; and he is all the more important for our purposes because he is known to have been profoundly indebted in several respects to Empedocles, whose work dated from the century before him.[7] The full significance of this broad genealogy starts to become clear when we remember the point

[5] Beck 496, 559–60. Elsewhere (500–1) Beck notes in passing the role played by Pythagoras and Empedocles as 'precedents' for these later Hellenistic trends, but overlooks the significance of this disconnected observation to his general evolutionary scheme.

[6] Cf. Lloyd (1979) and, with more specific reference to plant and herbal lore, Lloyd (1983). On the difficulties inherent in distinguishing between the domains of magic and science see Ch. 19 with the refs. in n. 21; and, with specific reference to botany, Edmund Leach's comments in Raven 169–70.

[7] F. Pfister, *RE* xix. 1448 (on the relation of the plant-books produced under the names of Pythagoras and Democritus to the earlier *rhizotomika*); Ch. 15 n. 14.

already noted: that the famous magical doctrine of sympathies and antipathies which is so closely associated with Bolus has an obvious (even if almost totally neglected) antecedent in Empedocles' doctrine of Love and Strife.[8] And, as we have also seen, it is a major mistake simply to discover 'the seeds of the notion of sympathy' in the work of Aristotle's successor, Theophrastus, and then present Bolus as perverting this doctrine to his own dubious ends.[9] The truth is simultaneously more complex and more simple. Certainly Bolus appears to have used the botanical and zoological works of both Aristotle and Theophrastus; yet when we look more closely at the way in which he used them, we see that his interests, as well as his methods of selection, were very specific. Basically, he seems to have read both Theophrastus and Aristotle for the kind of detailed information contained in their writings about the views held—on precisely such subjects as the theory of sympathies—by other, less rational or positivistic, authorities from before their own time. A very significant pattern emerges: where Aristotle or Theophrastus cites a predecessor's view only to pass judgement on it—whether implicitly or overtly—as superstitious, illogical, or unfounded, Bolus takes careful note of the earlier writer's view and methodically ignores Theophrastus' and Aristotle's objections.[10] What we have here is not 'perversion' or 'debasement' but a conscious selectivity coupled with a sense of continuity, of upholding a prior tradition, and an obvious disregard for the viewpoint or concerns of the Peripatetics.

We are led to exactly the same conclusion by that other aspect of the medico-magical literature associated with Bolus and the later Pythagoreans: the aspect of it which 'was intensely *practical* ... Nature's sympathies and antipathies were there to be *used*'.[11] This emphatic concern with practicality, with control as opposed to disinterested understanding, is not just a distortion of some supposedly pure Hellenic science

[8] Above, Ch. 19; Needham, v/4. 312 n. e.
[9] So e.g. Fraser, i. 442.
[10] See Oder's comments, 75; also Wellmann 22; and for Bolus' knowledge of Theophrastus, Kingsley (1994c), § 11 with n. 52. Worth comparison here is Pliny's way of reproducing traditions that he found in Theophrastus while glossing over or ignoring Theophrastus' own sceptical comments: cf. Lloyd (1983), 145–6.
[11] Beck 496, 556, 559–60.

'Not to Teach but to Heal'   339

due to the contaminating influence of foreign cultures in the last few centuries BC. On the contrary, it is already exemplified in the fifth-century figure of Empedocles—and, what is more, it is reflected in what we are able to understand about both Empedocles and early Pythagoreanism down even to the smallest of details: emphasis not only on practical application but also on detailed knowledge and observation as means to achieving understanding and mastery of cosmic principles, involvement in magic and ritual, and the attachment of primary importance to healing in all its facets.[12]

What, then, of the alternative: the *non*-practical approach to the study and understanding of nature? By way of answer, it is enough to refer to what was said earlier about Plato, Aristotle, and their successors: about their decisive replacement of earlier—and notably Pythagorean—approaches to the philosophical life as an integrated combination of understanding and practice, wisdom and the application of it, with the new role-model of the disinterested student and theoretical scholar.[13] Later Pythagoreans may have departed very consciously from this new norm, but that is a point of only secondary significance. The most important point, and the necessary starting-point for any discussion of the subject, is the fact that in doing so these later Pythagoreans were simply remaining true to the initial impetus of Pythagoreanism: a fact which gains even further in its implications when we bear in mind that the notion of 'philosophy', and indeed the word itself, very probably originated in early Pythagorean circles.[14] Historically, of course, the significance of this accord between early and later Pythagoreanism is further underlined by the evidence already considered of Pythagorean and related traditions passing directly from southern Italy and Sicily down into Hellenistic Egypt. Of course Athenian intellectual culture, including Platonism and Aristotelianism, exerted a major influence as well in Alexandria and on Graeco-Egyptian society; and yet that influence is unlikely to have arrived by the same channels. Modern scholarship's

[12] Above, Chs. 15, 19.
[13] Ch. 12 with nn. 36, 38.
[14] Cf. e.g. Cameron 33–4, and Gottschalk's discussion, 29–31. For Heraclitus' use of the term in fr. 35—where the tone is transparently ironic: '"philosophical" men are going to have to do a heck of a lot of inquiring'—cf. his frs. 40, 108, 129 with Gottschalk 30 and n. 59; and for the meaning of *historia* in Heraclitus, Kingsley (1994c), 1–2.

conscious as well as unconscious Athenocentrism explains, but does not justify, the deeply ingrained tendency of classicists to view and judge ancient history from the exclusive perspective of Athenian interests or values.

This Athenocentrism is, undoubtedly, responsible to a large extent for the failure to observe and appreciate the various aspects of continuity between early and later Pythagoreanism. But we must also be more specific. One of the major barriers to uncovering these strands of continuity has been the remarkable tendency to overrate the importance of Aristotle—virtually to the extent of mythologizing him. Earlier we saw some of the misunderstandings about the history and evolution of Pythagoreanism which have resulted both from the tendency to exaggerate the philosophical significance of Plato, and from the corresponding failure to look back in time to what really happened in the period before Plato when assessing developments that supposedly occurred only well after his lifetime.[15] Now, in the case of Bolus and Neopythagoreanism, we are faced with the equally misleading tendency to exaggerate the historical significance of Aristotle for the ancient Greeks. There is almost a palpable element of heresy involved in making such a statement; Giorgio de Santillana remains very much a lone voice in his criticism of what he has called 'the massed power of Platonic and Aristotelian scholarship, which tends unconsciously to see the classical landscape through the eyes of its favorite authors'.[16] But it does not need too much reflection to show the implausibility—and indeed naïvety—of attaching so much importance to one single individual, or two or three, by dividing the history of ancient Greek philosophy into a 'pre-Socratic' and a 'post-Aristotelian' phase and then assuming that all philosophers in the second period were merely reacting to the achievements of Socrates, Plato, and Aristotle while slipping further and further into the irrationalities of late antiquity. This is not to deny that Aristotle as well as Plato contributed enormously to the scope of later Greek philosophy, especially through the new language and terminology which they handed down to later generations of philosophical writers throughout the Greek-speaking Near East and Mediterranean.

[15] Chs. 9–11.
[16] de Santillana 190.

What is almost invariably overlooked, however, is that the influence of this often very characteristic vocabulary may have been far more superficial than one might be led to suppose. What always needs to be considered is how later individuals *used* this language or lingua franca, and adapted it to their own means and ends. Mere verbal similarities can be deceptive; superficial signs of influence or indebtedness may conceal vastly different interests and concerns.[17] And when—as we so plainly must in the case of the traditions we have been considering—we start to put the words into context and approach ancient philosophical literature as just one facet of a way of life, the picture changes radically. Terminologies may have altered and new affiliations been formed, but that is not to say they altered the underlying issues of vocation, attitude, and approach to life. Certainly Plato and Aristotle were neither the beginning nor the be-all and end-all of ancient philosophy; and all roads did not necessarily pass through Athens.

One other obstacle to understanding the prehistory of the plant literature produced by Bolus and later Pythagoreans has been the strange failure to piece some equally basic items of evidence together into a coherent whole. Essential to the texts in question is the idea of the philosopher not just as a magician but also as a healer. Festugière was simply voicing what for a long while had been, and still is, the common opinion on the subject when he claimed that this idea of the philosopher as healer was taken over by the 'Neopythagoreans' from the Cynics.[18] And yet, here again,[19] Festugière has left out half of the story. To begin with, it is well known that there was from the very outset an intimate relationship between Pythagoreans and certain key figures among the earliest Cynics, with the Cynic philosophy or way of life itself being profoundly influenced by Pythagorean ideas and traditions. There is even a sense in which it is possible to talk of certain basic strands of Pythagoreanism being 'absorbed' into Cynicism and so reappearing in a form 'which suited the demands of the age': yet another example of Pythagoreanism's tendency to break

---

[17] Kingsley (1993*b*), esp. 18–20.
[18] (1981), 74 with n. 1.
[19] See above, Ch. 20 with nn. 6–8.

frontiers and spill over into apparently distinct traditions.[20] And, to be more specific, there can be no doubting the fundamental importance of this idea of the philosopher as healer in early Pythagorean circles. One is reminded, to begin with, of the figure of Empedocles: philosopher, healer, and evidently expert in the knowledge and use of magical and medicinal herbs.[21] But to the Empedoclean evidence we also have to add the unambiguous evidence for the interrelationship between the role of philosopher and the role of healer in early Pythagoreanism itself. This Pythagorean evidence takes us back to well before the time of Plato, and to some extent at least must be assumed to go back to Pythagoras himself. What is more, it is clear that the type of healing and medicine involved had a great deal to do with the world of incantation, magic, and ritual.[22]

In this context one particular passage from Aelian is especially worth mentioning. The passage as a whole almost certainly derives from material contained in Aristotle's monograph on the Pythagoreans, and can safely be assumed to reflect some of the earliest strata of Pythagorean tradition. In it we find Pythagoras drawing attention to, among other things, the special 'sacred' quality of the leaves of the mallow plant: a clear reference to the type of concern with folk tradition and ritual that, as we have seen, lies at the roots of Pythagoreanism. And the section comes to a close with an account of how

as he went around from town to town the word got about that Pythagoras was coming—not to teach but to heal.[23]

The picture these words evoke brings us back once again to Empedocles, who himself describes how he used to wander around from town to town followed by huge crowds of people 'enquiring about the road to gain, some of them consulting me

[20] Burkert 202–5. Cf. e.g. Reitzenstein (1906), 45 n. 2; Detienne 32–7; also Ch. 17 with n. 91.   [21] Ch. 15.
[22] For full discussions of the primary evidence, including the famous Aristox. fr. 26 and the Pythagorean praise of *iatrikē* or the healing art in Iam. *VP* 82 (taken from Aristotle: Burkert 166–7 with n. 5), cf. E. Howald, *Hermes*, 54 (1919), 203; Rostagni, i. 135–61; Boyancé 93–153 and *passim*; F. Wehrli, *MH* 8 (1951), 56–61; Burkert 212–13, 292–4.
[23] Ael. *VH* 4.17 (70.21, 71.1–3 Dilts) = Arist. fr. 174 Gigon. For the antiquity of the mallow tradition cf. Burkert 185; and for the antiquity of the material in *VH* 4.17, ibid. 141–7, 167 n. 5, 168–71, 181 with n. 114.

for oracles and others asking to hear the well-healing utterance for all kinds of sickness and disease'. One also notes that according to the Platonic *Seventh Letter* even Plato, as wise man and philosopher, found himself the centre of attention for people who approached him as a human oracle assuming he could answer their questions about physical and spiritual health as well as financial gain.[24] And so we return to the point about the antiquity, certainly in Pythagorean circles, of the emphasis on the intense *practicality* of the role of the philosopher. The aim in these circles was plainly to bring one's own life into order, and the lives of others as well; and it is in the light of this aim that we need to understand Empedocles' various roles as philosopher, magician, and healer.

As we have seen, it was precisely this spectrum of roles that the so-called 'Neopythagoreans' continued to honour and maintain. There are aspects here of early Pythagoreanism which modern scholarship has shown little interest in understanding, but which an individual such as Bolus perfectly appreciated. In other words, in the passage of Pythagorean traditions down from Sicily and southern Italy into the Graeco-Egyptian world of later antiquity we have a line of intellectual and spiritual transmission which may not be 'philosophical' in any narrow sense, but which is of capital importance for placing Empedocles and early Pythagoreanism in their real historical setting. Earlier we saw the pressing need to reassess the work of both Empedocles and pre-Platonic Pythagoreans by viewing them in a primarily magical and practical context. Now we see where we should have looked for a better understanding of them: to their self-declared successors, the much-maligned Neopythagoreans.

The issue of Pythagoreanism's relationship to the Hermetica has already been touched on earlier. This is not the place to go into the evidence in detail, but it is an appropriate place to mention one particular point of contact in passing.

[24] Emp. B112 with Ch. 15 n. 7; *Letters* 7.330c–331d. The very human touch in the Platonic passage, plus the striking Empedoclean precedent, is a strong argument for considering the material in the *Seventh Letter* authentic rather than (as claimed by Lloyd, 160) inauthentic. In Emp. B112.7 the obvious restoration is τοῖσιν δ' ὧν ἂν ἵκωμαι ἐς ἄστεα ... Cf. e.g. *Il*. 1.272 (τῶν οἳ ...) and 9.592, *Od*. 9.128; and for the MSS' ἅμ see, apart from Zuntz's explanations (191), the very next line of B112.

The *Korē Kosmou* is one of the more exotic of the surviving Hermetic texts. It has particularly close affinities with alchemical literature, and the influence of native Egyptian religious tradition is pervasive as well as profound.[25] This exotic content of the *Korē Kosmou* made it understandable, and to an extent even inevitable, that one very prominent feature of the text—its coupling of philosophy (*philosophia*) with magic (*mageia*)—would be interpreted as an integral aspect of such a typically Graeco-Egyptian, orientalizing milieu. So, for example, Cumont explained the text's use of the word 'philosophy' as meaning

a religious 'wisdom' conceived of as a secret tradition, an esoteric doctrine only revealed to an initiated élite. At the end of antiquity the term *philosophus* was used as the common designation for a doctor of occult sciences; and, for the initial origin of the deviation or perversion that altered the sense of this word, we need to look to the notion of 'philosophy' entertained by the Hellenistic clergy—

a notion, as Cumont went on to explain, that 'coupled philosophers with magicians, initiates, and seers'.[26] Yet we have already encountered this idea of Hellenistic 'perversion' elsewhere, in a case where the supposed abuse turned out to be no perversion or distortion at all but a continuity of early Pythagorean tradition stretching back to the fifth and even sixth centuries BC. As for the coupling of philosopher and magician, we have seen that it is one of the most important keys to understanding Empedocles in particular and early Pythagoreanism in general. The connotations of the word 'philosopher' elaborated by Cumont were by no means just Hellenistic distortions or innovations: they were essential aspects of it from the very outset, even if temporarily eclipsed in certain more rationalizing and narrowly intellectual circles.

This is not all. In another place in the *Korē Kosmou* a list is given of the highest and noblest incarnations: 'the most just among you, anticipating the transformation into divinity' (τὴν

---

[25] For the links with alchemy cf. Festugière (1967), 230–48; for the Egyptian background, H. Jackson, *Chronique d'Égypte*, 61 (1986), 116–35, T. McA. Scott, *Egyptian Elements in Hermetic Literature* (Ann Arbor, Mich., 1992), 38–125.

[26] F. Cumont, *L'Égypte des astrologues* (Brussels, 1937), 122 with n. 4 *ad fin*. For the same general verdict cf. e.g. Fowden 117–18.

εἰς τὸ θεῖον μεταβολὴν ἐκδεχόμεναι),[27] will enter into human bodies and become 'just kings', 'true philosophers', 'authentic prophets', and 'genuine root-cutters'.[28] Certainly the role of root-cutters referred to here—that is, the role of herbalists and, by implication, magical healers—was a recognized feature of the Graeco-Egyptian world: a role well known to us for example from the magical papyri, as commentators on the passage have pointed out.[29] And yet there is more to the reference than that. The idea of the noblest incarnations points to Pythagoreans, and back to Empedocles in particular. In fact there are surviving fragments of his poetry which describe precisely how 'at the end' (εἰς τέλος)—just before being entirely divinized—souls in their final incarnation become 'rulers', 'prophets', and 'healers' (ἰητροί).[30] The correspondence is unmistakable; but even more remarkable is the light that the *Korē Kosmou*'s mention of 'root-cutters' throws on Empedocles' own brief reference to 'healers'. As we have seen, one of the main ways in which Empedocles' idea of 'healing' needs to be understood is, precisely, in the context of the Greek traditions of root-cutting and the use of magical and medicinal herbs. The *Korē Kosmou* supplies explicitly what, in the case of Empedocles, was otherwise to be inferred.

That, too, is not all. After listing the various kinds of highest human incarnation, the *Korē Kosmou* goes on to give examples of the highest kinds of animal incarnation for various different species. Festugière has noted the main signs of indebtedness here to the Graeco-Egyptian literature associated with Bolus of Mendes.[31] But it was left to Zuntz to point out the full significance of the passage by noting how closely it corresponds—once again—to Empedocles' teaching, and by concluding that in this respect the *Korē Kosmou* can actually allow us to reconstruct some of the features of Empedocles' doctrine of reincarnation.[32] Zuntz was also able to draw another important conclusion.

---

[27] 'Capable of receiving' (NF iv. 13) for ἐκδεχόμεναι is too weak; for the idea cf. esp. *CH* 4 and 13 with Ch. 11 and n. 11.
[28] *Korē Kosmou* 41–2 (NF iv. 13.14–18). For the reference to *rhizotomoi* or root-cutters here cf. also *I Enoch* 7 with NF iii, pp. ccxv–ccxviii; Needham, v/4. 342 and n. a.
[29] NF iv. 38 n. 144, with further refs.; cf. Ch. 15 with nn. 15–16.
[30] Emp. B146–7.
[31] *Korē Kosmou* 42 (NF iv. 13.20–14.12), with NF iii, pp. cxcviii–cxcix, ccvii.
[32] Zuntz 232–4; cf. also K. Alt, *Hermes*, 115 (1987), 402–4.

Festugière had supposed that, as a typical piece of mythologizing about the soul produced in late antiquity, the philosophical and mythological content of the *Korē Kosmou* passage could be derived from the famous myths in Plato's *Phaedo* and *Republic*; it was in much the same spirit as this that Burkert underlined the close correspondences between details in the *Korē Kosmou* and features found in late Pythagorean literature, post-Platonic Pythagoreanism being for Burkert little more than Platonism thinly disguised.[33] However, Zuntz has shown that the Hermetic text differs on the most fundamental points from the Platonic myths, and that the details it preserves must have been derived from early Pythagorean tradition quite independently of Plato.[34] Early Pythagorean, and specifically Empedoclean, ideas have been preserved—and, naturally enough, further elaborated—in one of the most typically Egyptian of the Hermetica.

Sometimes it is difficult to draw basic implications from texts when that would involve re-evaluating equally basic assumptions. But there are two points that need making here. First, we see once again how wrong it is to suppose that Plato was the necessary route for all earlier traditions passing down to later antiquity. To define later Pythagoreanism as 'Platonism with the Socratic and dialectic element amputated', to present Plato and Plato's myths as 'the principal source for all later Pythagoreans',[35] is to ignore the fluidity and complexity of Pythagorean tradition by fragmenting it in a way impossible to justify; to overestimate the extent of Athenian influence on the Graeco-Egyptian world; and radically to underestimate the role played both by oral tradition and by texts now lost to us in the preservation and transmission of what, for many, were not just theoretical ideas but pointers and guides to a way of life.

Second, we saw earlier on where *not* to look for a real understanding of Empedocles' poetry, and of the mythological and mystery traditions that underlie it and make it comprehensible: to Aristotle, Theophrastus, and the following generations of scholars or 'doxographers' who merely repeated and elaborated *ad infinitum* what those early Aristotelians had said.

---

[33] NF iii, pp. ccii–ccv; Burkert in *EH* xviii. 51–2.
[34] Zuntz 233–4 with 233 n. 4.
[35] Burkert 96.

Now we are in a position to see where we can look, and need to look, for a basic appreciation of the context to Empedocles' poetry and to the man as a whole. This of course is not to say that we can afford simply to ignore the information about Empedocles preserved in Aristotle, Theophrastus, and the writings of their successors. What they have to tell us is always useful, and often invaluable. But, just like Bolus of Mendes did, we need to learn to read and use them with discrimination.

22
Nestis

*

EMPEDOCLES the magician, the healer, the expert in esoteric lore, and the initiate throwing himself into Mount Etna: these are all so many connected aspects of fact and legend that we have encountered as a direct result of following up Empedocles' own equation of divinities with elements—and, above all, his equation of Hades with fire. But there is one last detail which needs looking into, and that is the identity of the goddess equated by Empedocles with the element of water. Interestingly, in this case the limitations of our information about the goddess will prove to be almost as significant as the extent of it.

Plainly he saw her as important. Whereas in fragment 6 he gives just one line to the divinities of *aithēr*, earth, and fire all together, he gives this fourth and final deity another whole line to herself.

> Hear first the four roots of all things:
> Dazzling Zeus, life-bearing Hera, Aidoneus, and
> Nestis who moistens the spring of mortals with her tears.

Who is this Nestis? Greek literature provides little in the way of a direct answer. In the century after Empedocles the comedian Alexis, apparently from Thurii, seems to have described Nestis as a Sicilian goddess. Alexis may well have had independent knowledge of the existence of such a goddess; however, it is impossible to be sure that he was not simply referring to Empedocles' poem—and making fun of him in the process.[1]

Yet this is by no means to suggest that the name is just an Empedoclean invention. Empedocles did not create his mythological entities out of the blue. On the contrary, he was

---

[1] Photius, s.v. Νῆστις = Alexis, fr. 323 Kassel–Austin; and cf. Eust. *Il.* 19.208 (iv. 314.15–16 van der Valk). Note Alexis' mockery of the Pythagoreans (Méautis 10–11; Burkert 198–201).

profoundly indebted to previous religious and literary traditions; and it was his ability to use these pre-existent traditions by moulding them subtly to his own ends and aims that made his poetry meaningful.[2] There is another point, as well, which has nothing to do personally with Empedocles at all. Anyone familiar with Sicilian religious history is only too aware of how little we know about the island's gods and, above all, its goddesses. About the greater part of religious tradition and practice on the island we are totally ignorant; if anything, it is more the rule than the exception to find Sicilian deities that were known to the Greeks referred to very fleetingly only once—or at the most twice—in the surviving body of classical literature.[3] But there is one other factor here which, for our purposes, must be considered more or less decisive. In an important, although forgotten, note Paul Kretschmer has drawn attention to evidence indicating that the same -*ti*- suffix and -*tis* ending as in the name Nestis were a regular feature of the Indo-European, pre-Greek names of goddesses in Sicily and southern Italy; Kretschmer himself used this evidence, with clear justification, to confirm the credentials of Empedocles' Nestis as a genuine Sicilian goddess. In other words Empedocles' originality lay not in inventing the name, but in transposing and applying it by equating it with one of his elements—just as we saw him do earlier with Zeus, Hera, and Hades.[4]

There is also one further point to bear in mind in the case of Nestis. This is the well-known phenomenon in ancient Sicily of famous Greek gods and goddesses coming to be referred to by local, and unusual, names as a result of Greek colonists assimilating their own religion to the indigenous, non-Greek traditions they found on the island.[5] As to what is most likely to have been the more familiar name of the goddess in question, writers over the last two hundred years have repeatedly come

---

[2] Cf. Ch. 4; Ch. 19 n. 32; Bollack, i. 277 with n. 5.
[3] Cf. e.g. Photius (θεὸς Σικελιακή) and Hesychius (θεός. Σικελοί) s.v. Λάγεσις, with P. Kretschmer's comments, *Glotta*, 12 (1923), 278.
[4] Ibid. 278 with n. 1; above, Chs. 4, 6.
[5] Cf. e.g. Freeman, i. 183–6; Chs. 6, 18 (Adranus/Hephaestus); and, in the case of Nestis, Andò 39–41. For the same kind of phenomenon elsewhere cf. Burkert (1992), 78–9. On the general issue of Greek colonization and cultural assimilation in Sicily see A. J. Domínguez Monedero, *La colonización griega en Sicilia* (Oxford, 1989), with A. Johnston's comments, *CR* 42 (1992), 357–8.

up with the same answer. In Empedocles' fragment 6 the first divinities named, Zeus and Hera, are of course husband and wife. From a formal point of view this is an obvious incentive for suspecting that Hades and Nestis are another divine couple, with the name Nestis—no doubt used deliberately out of deference to Sicilian tradition and so as to preserve a sense of religious exclusivity—a local title for Persephone.[6] As we will soon see, this is almost certainly correct.

All there is to say about the etymology of the name Nestis can be said very briefly. As a Greek word it simply means 'fasting' or 'not eating'. However, this did not stop ingenious writers in late antiquity from coming up with more ingenious etymologies. Simplicius gives a derivation of the name from the verb *naein*, 'to flow', which of course is appropriate for the element of water—and etymologically quite impossible. That has not, unfortunately, prevented it from gaining a foothold in modern scholarship, with even the great Wilamowitz attempting to justify it.[7] Yet it is important to keep to the essentials. 'Fasting' is all that the word can mean; that is how any straightforward Greek would have understood the word and that is how, in the absence of any indications to the contrary, we are bound to understand it.[8]

None the less, to draw this simple conclusion is certainly not to say there is no more complex reason for a goddess ending up with such a strange name. There is in fact one other factor very often involved in situations of this kind, even though in the case of the name Nestis its relevance seems not to have been noted. That is the phenomenon of double etymology: a phenomenon which occurred when Greek colonists or explorers took a foreign name and—usually by altering its pronunciation very slightly, often just modifying the sound or sequence of the vowels—transformed it into an already-existing Greek word that sounded similar. To an unsuspecting person, this Greek

---

[6] C. G. Heyne in D. Tiedemann, *System der stoischen Philosophie* (Leipzig, 1776), i, pp. viii–ix; Sturz 213, 550; B. ten Brink, *Philologus*, 6 (1851), 733–4; Bollack, iii. 175; Andò 34–51.

[7] Simpl. *De anima* 68.13–14 Hayduck, Wilamowitz (1909), 348–9, (1931–2), i. 20; and cf. e.g. Sturz 212–13 for further impossible etymologies, such as from *neein*, 'to swim' or 'float'.

[8] So, sensibly (although with an unnecessarily forced explanation), Hipp. *Ref.* 7.29.5–6; cf. also the *Suda*, s.v. Νῆστις.

word would seem to provide the obvious etymology for what was in fact a foreign word to begin with. The phenomenon of double etymology was common throughout antiquity wherever Greeks encountered foreign languages and cultures; it was particularly common in ancient Sicily. And one of the more significant aspects of the phenomenon as far as we are concerned is that the secondary, Greek form of the name often had a special appropriateness, however flippant or playful, to the person or object named. 'Folk etymology', as the phenomenon of double etymology has misleadingly come to be known, has a surprisingly informative value and logic of its own.[9]

From a formal point of view, as we have seen, Empedocles' listing of divinities in fragment 6 would tend to suggest that Nestis may well be the wife and partner of Hades by analogy with the partnership of Hera and Zeus. Bearing this in mind, it is hardly a coincidence that fasting and a festival of fasting—called, precisely, the Nesteia—were a major part of the ritual and mythology associated with the mysteries of Persephone. We know this from the Homeric *Hymn to Demeter* as well as from later writers, and the same connection of Persephone with fasting appears also to be referred to on the 'Orphic' plates found at Thurii: a point of particular significance considering the intimate links between these gold plates and Empedocles. In fact fasting lay at the heart of the Persephone mysteries both at Eleusis and in the Greek West. Her abduction and forced marriage to Hades in the underworld cause fasting on earth; not only does Demeter fast during her absence, but she herself fasts until the last moment while in the company of her husband.[10] This convergence, yet again, on the wife and

[9] So, for example, Apollo Asgelatas (from the Akkadian *Azugallatu*, 'great lady doctor') became Apollo Aiglatas, 'the radiant one'; the Egyptian magician Pachrates (the divine 'child') became Pancrates, 'the master of everything'. For examples and general discussion of the phenomenon see Kingsley (1993*b*), 11–15; for its occurrence in Sicily, Freeman, i. 53, 462–72, 559–64, Kingsley (1993*b*), 12 with n. 46, (1995*b*).

[10] *Dem.* 47–50, 200–10, 371–416; M. P. Nilsson, *Griechische Feste von religiöser Bedeutung* (Leipzig, 1906), 313–25; L. Deubner, *Attische Feste*² (Berlin, 1966), 55–6; Richardson 22–3, 165–7, 213, 217–26, 345–6; and cf. Zuntz 352 with n. 2. For the gold plates see Olivieri 23, 25, Zuntz 347.5, 6, and loc. cit.; the text of this plate is particularly uncertain and, as Martin West has pointed out to me, Zuntz's νήστιας (347.6) could be a vocative Νῆστι in line with the other gods in the vocative immediately following. In general, cf. also Clem. *Pr.* 2.21.2 (ἐνήστευσα . . .); Philicus, *Hymn to Demeter* 37 (*SH* 323 § 680); Callimachus, fr. 21.10 Pfeiffer (νήστιες); and for the evidence relating to Sicily,

partner of Hades is clearly significant: regardless of its ultimate origin or history, the name Nestis is a singularly appropriate epithet for Persephone and—as a number of scholars have recently noted—was very probably one of her alternative titles.[11]

However there is still one other factor that needs considering. This has to do with Empedocles' description of Nestis 'moistening the spring of mortals with her tears' (δακρύοις τέγγει κρούνωμα βρότειον). Springs, streams, and wells played a fundamental and ancient role in the mysteries of Persephone.[12] In Philicus' *Hymn to Demeter* we even have a reference—specifically, it will be noted, in the context of the fasting associated with Persephone's disappearance—to Demeter 'creating a spring of running water with her tears' (σοῖς προσανήσεις δακρύοισι πηγήν). It is possible that with these words Philicus had Empedocles' line about Nestis in mind, but not necessary or even likely; instead, we very probably just have two alternative allusions to the mysteries of Persephone and Demeter.[13] Are we to suppose from the parallel in Philicus that Empedocles' Nestis must be Demeter, and not Persephone? The answer is definitely no. Details in both the legend and the ritual associated with the Persephone mysteries were notoriously subject to variation from cult centre to cult centre, not to mention from poet to poet. We also need to remember— surprising as it may seem—that the roles and images of Persephone and Demeter were by no means always firmly differentiated, but often became blurred and sometimes indistinguishable. This point is especially relevant to Sicily, and in particular to the cult centres at Syracuse and Acragas, where the supremacy and overshadowing prominence of the role ascribed to Persephone were probably greater than anywhere else in the Greek world.[14]

Yet underpinning these general considerations is the more

the Platonic *Letters* 7.349c–d and Diod. 5.4.4–7 with Nilsson, op. cit. 315 n. 7, Richardson 165, and Ch. 8 with n. 22.

[11] Bollack, iii. 175–6; Gallavotti 173–4; Andò 37–8.
[12] Richardson 18–19, 179–82, 250, 326–8.
[13] Philicus, *Hymn to Demeter* 37–40; Gallavotti 174.
[14] On the blurring and even merging of Demeter and Persephone cf. e.g. Farnell, iii. 121–2, Harrison 272–4, Zuntz 75–8, Richardson 14; Burkert (1979), 131 (Acragas); for Persephone's prominence in Sicily, Zuntz 70–4, Andò 41.

specific fact already mentioned: that in origin it was Persephone who had the most direct link with springs and streams. In mythology she was leader of the nymphs who were the daughters of Ocean and lived in springs and streams. At Eleusis her worship was very closely associated with the local spring and cave: the cult that grew up around her also 'grew up around the spring and grotto'.[15] At Syracuse, as we have seen, according to local tradition she actually disappeared with Hades into the spring called Kyane. Ritual links between this spring and Persephone herself were, here again, very close: when animals were dropped into the spring as sacrifices, they were dropped in as sacrifices to Persephone.[16] Later writers certainly offer a wealth of stories personifying the spring as a nymph called Kyane who was transformed into water, according to most versions while trying to prevent Persephone from being abducted by Hades. Here, too, we find the idea that the spring was created from her tears.[17] But even in these stories Kyane belongs firmly to Persephone's entourage; and the sheer variety of them together with the overlap of functions ascribed to Persephone and Kyane—as for example in the theme of Kyane herself being raped, like her mistress—tends to suggest that behind the poetic elaborations the nymph may have been little more than a doublet of Persephone herself.

At the famous cult-centre near Andania in Messenia we again find an explicit link between Persephone and the spring of water on which the mysteries focused. Significantly, she is referred to in the inscriptional evidence not by name but as *Hagne* or 'Holy': one of Persephone's more familiar cult titles. The connection here is so close that the spring has in recent times been referred to, mistakenly, as 'the spring Hagne' and actually identified with Persephone.[18] But in fact our ancient sources are only clear about the point that the spring rose up directly alongside the statue of Persephone–Hagne and was

---

[15] Richardson 18–19; and ibid. 181 for refs. to the same connection elsewhere.
[16] Diod. 4.23.4, 5.4.1–2, Zuntz 71–2; above, Ch. 9 with n. 7.
[17] For Kyane's tears cf. Ov. *Met.* 5.427; and for refs. to the Kyane stories, with comments, K. Ziegler and C. Lackeit, *RE* xi. 2234–5. On the specific connection between tears and underworld rivers see Stewart 103.
[18] Richardson 18, 181, following O. Kern, *RE* xvi. 1268.

traditionally referred to as 'the spring of Hagne'.[19] Further details about the precise relationship between this spring and Persephone are, understandably, not forthcoming. The mystery-inscription which has been preserved is, as one would expect, categorical about the secrecy of the rites and the exclusion of non-initiates; and Pausanias too, for whom the mysteries at Andania were 'second in sanctity only to the Eleusinian mysteries', is equally explicit in stating that the details and significance of the rites performed there were subjects on which he was forbidden to speak (ἀπόρρητα ἔστω μοι).[20]

This is not the first time we encounter the theme of secrecy and self-conscious reserve: as noted earlier, there was no doubt a deliberate reticence and element of the enigmatic in Empedocles' use of the name Nestis. What is more, in the case of Empedocles such secrecy can hardly be considered unexpected: we find much the same type of religious reticence in the way Parmenides presents his own underworld goddess—and in fact, as Burkert has pointed out, this allusiveness on the part of Parmenides has a close parallel of its own in the allusive reference to Persephone as 'Hagne' in the mysteries at Andania.[21] The evidence brings us almost full circle, with the details sliding into place to produce a remarkably coherent picture. The coupling of Nestis with Hades, the meaning of her name, her connection with tears, springs, and streams, and even Empedocles' allusive style all point—due allowance made for the factor of secrecy—to the very probable conclusion that Nestis was a cult title for Persephone.

There could be few things less surprising than to find Persephone being given such an important place as one of the four elements in Empedocles' teaching. We have already encountered her at virtually every turn while tracing the back-

---

[19] Paus. 4.33.4; *SIG*[3] ii, § 736.84 = H. Sauppe, *Ausgewählte Schriften* (Berlin, 1896), 279.84, with Sauppe's note ad loc. and commentary (280, 299–307). For Persephone as ἁγνή cf. Paus., loc. cit.; Zuntz 317; Ch. 16 n. 35.
[20] Sauppe, op. cit. 274.34–275.45 = *SIG*[3] ii, § 736.34–45; Paus. 4.33.5.
[21] For Empedocles' reticence with regard to the name Nestis cf. Andò 51; above, with n. 6; and Ch. 4 with n. 26. For Parmenides see Burkert (1969), 13–14 with n. 31 and, on the location of his goddess in the underworld, Ch. 5 with n. 15, Ch. 17 n. 6.

ground to his teaching in Sicilian mythology, in Pythagorean and Orphic tradition, and, last but not least, in Sicilian magic: she, and the mysteries associated with her, occupy a central role in all of them. And here the link with Pythagoreanism holds in another respect as well. Making the element of water—including, by implication, the water of the seas[22]—the tears of a divinity is a striking idea. Its closest analogy in early Greek literature is with the Pythagorean saying that 'the sea is the tears of Kronos': an analogy which is made even closer by the fact that both Kronos and Persephone were subterranean gods *par excellence*.[23] Otherwise, an in some respects even more meaningful parallel to Empedocles' idea of all water as the tears of Persephone is provided by the Gnostic image of 'all watery substance' (πᾶσα ἔνυγρος οὐσία) as the tears shed by Sophia:[24] a parallel which assumes additional significance when we bear in mind the extensive resemblances between Empedocles and Gnostic ideas. The Gnostics had their fall of the soul into an alien and frightening world, their lived ideal of the magician-liberator, their cosmic mixture, separation and 'roots', and their mythological traditions of oriental origin.[25]

Quite apart from the specific role evidently attributed by Empedocles to Persephone as one of his elements or 'roots', it is important, as well, not to overlook the broader consequences of this attribution from a structural point of view. Equating the four elements with two divine couples—Zeus and Hera, Hades and Persephone—may seem insignificant from a modern philosophical point of view; however, it provides a surprisingly vivid example of a phenomenon studied in detail by Carl Jung and referred to by him as the 'marriage *quaternio*'. Jung traced

[22] Emp. B22.2, 27.2, 38.3, 115.9–10; Bodrero 78, 80, Wright 23.
[23] For the 'tears of Kronos' cf. Porph. *VP* 41 = Arist. fr. 196, Plut. *De Isid.* 364a, Clem. *Str.* 5.8.50.1, Stob. i. 77.9–16 = Hermes Trismegistus, NF iv. 99.4–11; see also *OF* fr. 33 (= DK i. 19.2–3 = Clem. *Str.* 5.8.49.3), and Traglia 68 with n. 98. For Kronos as subterranean deity cf. e.g. Ch. 6 n. 1.
[24] Irenaeus, *Adversus haereses* 1.4.2 (ii. 66–7 Rousseau-Doutreleau); Tert. *Adversus Valentinianos* 15.2–5.
[25] For Empedocles and Gnosticism see the—still preliminary—observations by H. Leisegang, *Die Gnosis* (Leipzig, 1924), 84–7, R. Crahay in U. Bianchi (ed.), *Le origini dello Gnosticismo* (Leiden, 1967), 324–32, 337, and Mansfeld, xiv. 261–314. For the oriental background to the Pythagorean, Orphic, and Gnostic ideas of water as the tears of a divinity cf. esp. Dieterich (1891), 20–31; also West (1971), 217–18, Olerud, *passim*.

the often fundamental importance of this double-marriage structure in folklore and mythology, but above all in Gnosticism and alchemy; and in the case of Gnostic and alchemical texts he drew particular attention to correspondences between this *quaternio* phenomenon and the traditional four-element theory.[26] Needless to say, what is most significant about the recurrence of this mythological structure in Empedocles is not that it provides us with just one more example of the phenomenon, but that here we would seem to have the double-marriage structure embedded in the first known formulation of the traditional four-element theory: at the very origin of an idea which was to prove extraordinarily fertile and enduring in the subsequent history of western culture.

With these analogies and broader implications noted, it is also important not to wander too far from our starting-point. Scholars who recently have drawn the conclusion that Empedocles' Nestis must be a veiled reference to Persephone have missed the full significance of this conclusion because they wrongly equated Empedocles' Hades either with earth or with air. In fact, as we have seen, the husband of Persephone—the element of water—was the element of fire. Just how appropriate this ascription of water and fire to an underworld couple is, becomes clear as soon as we turn back to the point from which we began: the physical phenomena of Sicily. To quote again Strabo's succinct statement, already cited for its relevance to Empedocles' teaching about the elements, 'the entire island is hollow beneath the earth, filled with rivers and fire'. Similarly, in the Sicilian-based myth at the end of his *Phaedo* Plato repeatedly emphasizes that the underworld consists of vast rivers of water and fire: the correspondences between these subterranean rivers of water in the *Phaedo* myth and the Sicilian mythology of Persephone were noted earlier, as well.[27] In short, in these streams of water and fire we have all the basic materials

---

[26] Jung (1968a), 209–11, 226–65, and *The Practice of Psychotherapy*² (London, 1966), 219–35. It is worth noting that Jung also pointed to the almost continual presence of the incest motif in the marriage *quaternio* structure, between one partner in one of the couples and the partner of the opposite sex in the other. Zeus' incest with his daughter Persephone was of course an event that had vast repercussions according to the Orphic theology which, through his Pythagorean contacts, Empedocles was very probably familiar with (above, Chs. 10–12, and cf. West 108 with n. 75).

[27] Strabo 6.2.9, Pl. *Phd.* 111d5–e2, 112e4–114b5 (above, Chs. 6–7); Ch. 9.

for a magnificent allegory: the immense masses of fire and water below the earth's surface explained as the marriage of the two great divinities of the underworld. And in this context it is worth remembering two points. First, in his world-scheme Empedocles portrayed the subterranean water as lying above the subterranean fire: a detail he used to help explain the phenomena of evaporation and condensation at sea or at the site of hot springs. Second, another cyclical phenomenon in Sicily coincides rather well from a seasonal point of view with the traditional times of Persephone's stay on earth, her abduction, and her awaited return. That is the streams, oscillating dramatically between massive torrents and dried-up channels, which are very probably referred to in the account of oscillating subterranean waters that occupies the centre-stage in Plato's *Phaedo* myth—with its constant allusions to the mysteries of Persephone.[28]

There is one last point that demands our attention. However easy it is to justify Empedocles' equation of elements with gods from a geological or geographical point of view, we need also to give due emphasis to another aspect of the matter which, if anything, is even more crucial. This, as we have seen repeatedly, is the continuous interaction between Sicilian geography and mythology, and in particular between the island's geography and its mysteries. The sacred marriage of Zeus and Hera was a familiar theme throughout the ancient Greek world; yet in the case of a coupling of Hades and Persephone some more specific factors come into play. It was Sicily, as Pindar says, that was given to Persephone by Zeus as her wedding gift. The island was in fact renowned for its festivals devoted to celebrating her marriage to Hades; and it appears to have been Acragas, Empedocles' home town, that held pride of place among the cities of Sicily in celebrating this mythical marriage.[29] In other words, the coupling of water and fire in Sicily

---

[28] For evaporation, condensation, and the location of Empedocles' water and fire see Chs. 3, 5; for the streams and the *Phaedo* myth, Ch. 7 with n. 17, Ch. 11. On the seasonal correspondences of the Persephone mysteries cf. Richardson 12–20; the Sicilian streams will doubtless have been at their fullest in early spring.

[29] Cf. esp. the festival-list in Pollux 1.37 (i. 11.5–6 Bethe); Pi. *Nem.* 1.13–14 plus Schol. *Nem.* 1.16–17 (iii. 13.8–25 Drachmann), Schol. *Ol.* 2.15d (i. 62.21–3) (Acragas), Schol. Theocr. 15.14 (307.8–9 Wendel), Diod. 5.2.3, Plut. *Timoleon* 8.4; Zuntz 71, Jourdain-Annequin 278–9.

was not just geologically self-evident: interpreted as the coupling of Persephone and Hades it was mythologically self-evident as well.

That—with the extent, and nature, of the evidence at our disposal—is as far as we can go. Even so, it is far enough for us to be made aware once again of the degree to which it reflected the mythology and mysteries of his native island.

23
'Conceal My Words in Your Breast'
*

EVEN though it is possible to guess fairly safely at who she was, the identity of Empedocles' Nestis remains—significantly enough—shrouded in uncertainty. And yet, Nestis apart, we have already seen the need to situate Empedocles' teaching about his elements in the context of Sicilian mythology, mystery, and magic: that is, to understand his teaching in an initiatory context. We have also seen that it is not just a question of generously making allowances for this mythical and magical background by portraying it as some kind of primitive mire which Empedocles tried, but ultimately failed, to rise out of in his aspiration to become a rational philosopher. On the contrary, the very framework as well as the purpose of Empedocles' poetry was clearly initiatory and magical. We are faced here with a choice of either making Empedocles a pawn in the simplistic game of interpreting early ancient philosophy as a one-directional evolution from so-called 'irrationality' to 'rationality', from *mythos* to *logos*; or of attempting to understand him in his own context and on his own terms. This second option is for the historian and scholar; the first is for the fiction-writer or speculative philosopher.

In recent literature on Empedocles it has become routine to raise the issue of his equation of gods and elements in fragment 6, hastily propose a set of correspondences between the four gods and the four elements which on closer analysis turns out to be wrong, and then to dismiss this 'naturalistic theology' of his as of 'only secondary significance', as a question which 'fortunately is of little importance for Empedocles' thought'.[1] But this dismissal of Empedocles' primary equation of elements with gods as insignificant, incidental, and little more

[1] Olerud 83; Guthrie, ii. 146.

than arbitrary is simply a way of cutting away the ground from under his feet: of failing to take seriously what he clearly took seriously, of taking from his poetry what suits one's own preconceptions and ignoring the rest. The end result, as we have seen, is a spurious Empedocles, a pseudo-Empedocles. The fact that this pseudo-Empedocles has the seal of approval of Aristotle and Theophrastus is neither here nor there. It has the seal of their approval because it is to a very large extent the product of Aristotle and Theophrastus: a faithful reflection of their particular interests, interpretations, and concerns. Modern writers on Empedocles invariably take Aristotle's and Theophrastus' treatments of him as the starting-point for their own accounts of his philosophy on the grounds that these are the most reliable and relevant sources of information available. But they are not. Aristotle and Theophrastus both have invaluable information to offer, especially when they quote from Empedocles verbatim; yet our most important and reliable sources of secondary information are of quite another kind. These other sources are the literary, historical, geographical, mythological, and religious texts which may not even mention Empedocles' name but which allow us to view his own words in a far more meaningful perspective; and the picture that emerges from the use of these sources is a very different one indeed from the Aristotelian and Theophrastean version.

To say that Empedocles' poetry was firmly rooted in Sicilian and south Italian mystery tradition is one thing; but we must also take the matter a step further. His poem to Pausanias was initiatory: initiatory not only in the sense that he himself was an initiate, but also in the sense that he was an initiator.[2] And here another factor comes into play which has a very specific relevance to this initiatory context. This is the way that, right at the start of his poem to Pausanias, Empedocles introduces his four elements in the form of a riddle. The riddling quality in his initial equation of elements with divinities is self-evident: it was noted in antiquity, has been noted again in more recent times, and the dynamics of the enigma can still be analysed in detail.[3] But what appears not to have been noted at all is the relation between this riddling introduction to the poem and the poem's

---

[2] B110–11; above, Ch. 15.
[3] Ch. 4, with n. 26.

background of mystery and initiation. In the ancient world riddles, mysteries, and initiation went hand in hand. To be more specific, the posing and solving of riddles represented a particular kind of initiation in cases where esoteric techniques were passed on from master to pupil, which is exactly what Empedocles states he is doing in relation to his pupil Pausanias.[4] There is evidence from Crete for this use of initiatory riddles in the transmission of divinatory and magical lore: a significant provenance, considering that we have already seen other examples of correspondence between Empedoclean and specifically Cretan divinatory or magical lore—just as we have also seen why these correspondences are more than accidental.[5] But, Crete apart, the connection between riddles, mysteries, and initiation was a widespread and well-known phenomenon in antiquity.[6] It is also worth noting that the evidence at our disposal points to its special importance in the Dionysiac mysteries, and in the ritual traditions associated with the south Italian gold plates: two further areas of the most immediate relevance to our understanding of Empedocles.[7]

Where the riddling aspect of Empedocles' fragment 6 has been noted—invariably just in passing—by modern scholars, the tendency has naturally been to smooth the matter over and play down its significance. So, for example, van Groningen has well observed how Empedocles launches into the main body of his poem not with any argument or any philosophical reasoning but with a dogmatic assertion of the existence of four elements, cloaked in the form of a riddle; and yet, he goes on to add, Empedocles will not keep us in suspense for long.

The poet, as interpreter of the gods, can allow himself a certain degree of obscurity; but the philosopher will soon supply the clarifications required.[8]

However, this optimistic assessment, this reassuring image of Empedocles the 'philosopher' coming to the aid of Empedocles the 'poet', has no foundation in fact; and, what is more, it is

---

[4] B111.2; Ch. 15 with nn. 12–13.
[5] R. F. Willetts, *Klio*, 37 (1959), 26; Ch. 15 with nn. 20, 27–8.
[6] Pl. *Phd.* 69c; Demetrius, *On Style* 100–1; Seaford 254–5.
[7] Ibid., with Chs. 17 and 19 above.
[8] van Groningen 204–5.

certainly wrong. It is wrong to begin with because, if Empedocles really had gone on to spell out which of his divinities corresponded to which of his elements, some writer or commentator would have been sure to notice the explanation and put an end to the perpetual guessing that continued from the fourth century BC down to the very end of antiquity. Admittedly not many ancient writers on Empedocles, or for that matter on any other of the Presocratics, knew or had studied his work as exhaustively as one might now wish to suppose;[9] but on such an important point someone somewhere is bound to have noted the relevant passage. Even this argument, though, is only of secondary importance. The main reason why van Groningen's assessment is wrong lies in the portion of Empedocles' poem that survives rather than in what has been lost. We have already examined how Empedocles presented his equation of elements and gods in fragment 6: he did so implicitly, 'pregnantly' so to speak, communicating in the riddle itself the solution to the riddle.[10] In typical oracular mode he is neither explicit nor simply silent; instead he adopts a middle course, providing the seeds of the solution for the discriminating hearer. And from the other surviving fragments we can see that this is precisely how Empedocles chose to communicate. His emphasis was on the 'growth of understanding': an image which was far from just some isolated metaphor because it refers to his own basic perception of his words as seeds that need assimilating, absorbing, tending, and assiduous nurturing on the part of the person hearing and receiving them.[11] In brief, what Empedocles himself says is only the start of the matter. What then needs to happen is for the listener to take in the words, to make them his own and in fact make them a part of himself, to allow them to grow and—in the process—actually transform him. The initial riddle of the elements is itself a seed containing the potential of its future growth, an enigma capable of being resolved through a process of pondering; the hearer provides the solution, not Empedocles. As for

---

[9] Regarding the degree and nature of Theophrastus' acquaintance with the works of the Presocratics about whom he wrote so much cf. e.g. J. B. McDiarmid in D. J. Furley and R. E. Allen (eds.), *Studies in Presocratic Philosophy*, i (London, 1970), 178–238, esp. 237; Kingsley (1994*b*).

[10] Ch. 4.

[11] B17.14; B110; Ch. 15 with nn. 42, 47–8; Ch. 19 with n. 35.

'Conceal My Words in Your Breast' 363

the poem as a whole, it is an initiatory text in that its declared purpose is not just to provide facts and information but to induce a process of inner transformation. Empedocles himself emphasized this initiatory and mystery dimension through his conscious adoption of mystery terminology. As was mentioned earlier, the very imagery of seeds and natural growth that he uses belongs to the language of ancient mystery tradition.[12] We can now add that his deliberate use of enigma represents one further aspect of this initiatory or mystery background.

It is worth noting that the teaching style or method involved here has a direct bearing on an issue much discussed by Empedoclean scholars in the past few years. That is the issue of whether we have the remains of one major poem by Empedocles, or two.[13] This is not the place to examine the issue in detail; but the salient facts are as follows. First, when quoting from Empedocles ancient authors attribute several of the surviving fragments explicitly to works with two different titles. One of them is *Physika* or *Peri physeōs*—best translated into English as 'How things came to be what they are': clearly a cosmological poem—and the other is *Katharmoi* or 'Purifications'. It has been argued[14] that, in line with the practice in antiquity of occasionally referring to one and the same literary work—or various parts of the same work—by different names, these two titles could simply be alternative names for the same poem. This is a theoretical possibility, but no more. Second, no ancient author attributes any of the fragments to 'the *Peri physeōs*, also known as the *Katharmoi*', or vice versa; and no ancient writer attributes any fragment to a poem with one title while another writer attributes it to a poem with the other. This makes the theoretical possibility even less likely. Third, in one passage Diogenes Laertius mentions the two titles side by side as referring to two quite separate poems. It could be argued

---

[12] Ch. 15 with n. 48. Note in this context Iamblichus' double comparison of the enigmatic form of teaching used by Pythagoreans not only with the utterances of the Pythian oracle but also with the fertilizing power of nature, which respectively 'bring into manifestation endless and incomprehensible amounts of ideas and things through just a few choice words *or tiny seeds*' (*VP* 161).

[13] For the other works ascribed to Empedocles see Wright 17–20; Kingsley (1995*d*). Regarding the survival of the *Katharmoi* see Mansfeld, *Mnemosyne*, 47 (1994), 79–82.

[14] Osborne 25–6 and in *CQ* 37 (1987), 27; Inwood 14.

that Diogenes has here been misled by the various sources of information on which he relied into mistaking what were two alternative titles for one poem as separate titles for separate poems; but his reference to the two separate poems together in the context in which he mentions them almost certainly derives from an earlier source and is not disposed of so easily.[15] And fourth, we are told—again by Diogenes—that the *Peri physeōs* and the *Katharmoi* were addressed to two different audiences: the first to an individual called, by Empedocles himself, Pausanias and the second to a plurality of listeners identified by Empedocles as fellow-citizens in his home town of Acragas.[16] This difference in addressees is plainly significant, and cannot simply be disregarded. Here, too, we are obliged on balance to accept what Diogenes says in the absence of any genuine reasons for doubting him. And what is more, Diogenes' implicit differentiation between the two poems in terms of addressees, far from being contradicted, is independently corroborated by one further significant fact: of the various fragments cited by other ancient authors as coming from Empedocles' *Peri physeōs*, most contain addresses to a person in the singular—but not one of them an address to an audience in

---

[15] D.L. 8.77; D. Sedley, *GRBS* 30 (1989), 271 n. 8; and, on Diogenes' indebtedness to Lobon of Argos, Zuntz 237–8. The supposed problems in reconciling Diogenes' data here about the lengths of the poems with the evidence of the *Suda* in no way undermine the authority of the Diogenes passage. It is not the text of Diogenes but the MS evidence of the *Suda* which is either uncertain or contradicted by our other sources; and, if any mistake or textual corruption has occurred, it is very probably in the much later *Suda*. Cf. e.g. Wright 20–1; O'Brien (1981), 4–13; Mansfeld (1992), 228 n. 67. The attempt by Osborne (25, 29) to dispose of Diogenes' evidence is mere special pleading; at the end of the day she is forced either to assume a radical misunderstanding on his part or, with equal lack of justification, to alter his text.

[16] D.L. 8.54 (αὐτὸς ἐναρχόμενος τῶν Καθαρμῶν φησιν· ὦ φίλοι ...) and 8.60 (Παυσανίας ... ὧι δὴ καὶ τὰ Περὶ φύσεως προσπεφώνηκεν ...). Inwood admits that according to the evidence at our disposal 'the purifications' differs from 'the physics ... in its addressee', but goes on to claim that this does not help us 'to establish anything about the poems' (13 and n. 29); and yet admitting the difference has already allowed us to establish something. For Pausanias see, apart from D.L. 8.60 = Emp. B1, Plut. *Qu. conv.* 728e (reading τῶι Παυσανίαι), Ch. 15 with n. 5, Ch. 16 n. 8. While lines such as B2.3–8 (spoken to the Muse) are mere passing appeals, and irrelevant, the formal address to the people of Acragas in B112 is entirely different from the poetic 'asides' discussed by Osborne, *CQ* 37 (1987), 32. The generous phrasing of B112.1–2 indicates that Empedocles' intended audience was people with whom he was acquainted as fellow-citizens of Acragas—and perhaps even, as members of the aristocracy, related to—rather than a select group of intimate friends or followers: cf. Zuntz 188–9 (on DK's B112.3), Wright 265 and also, for the public nature of the poem, n. 26 below.

the plural.[17] The evidence is clear, and its implications straightforward; all the objections which have been raised to accepting that Empedocles wrote two separate poems are completely groundless.[18] There are other considerations which also need taking into account: for example that the teaching delivered to Pausanias was plainly intended to be esoteric,[19] and that the notion of Empedocles beginning a poem with a public address to the people of Acragas (as we know from Diogenes that he did) but then launching into an esoteric doctrine for a strictly limited audience is altogether implausible. Once again, this is a case where we must allow the evidence to speak for itself.

Here we come to the core of the matter. It has become clear that Empedocles' poetry needs to be viewed against a background of mystery and initiation. But in an initiatory context the particular audience addressed is naturally all-important: what is said will vary according to the status of the hearer. Several scholars have pointed out that—in addition to the singular form of address—some of the lines intended for

---

[17] So, in addition to Diogenes' quotation of B1 (Παυσανίη σὺ δὲ κλῦθι), we have Tzetzes' quotation of B6 (ἄκουε), ps.-Plutarch's of B8 (τοι ἐρέω), and Simplicius' of B17 (κλῦθι, τοι, σὺ δέρκευ, ἦσο, σὺ δ' ἄκουε) and 62 (κλύ'); plus, now, the further singular forms of address on the extension to B17 as reconstructed from papyrus by Alain Martin. Cf. also B21 (δέρκευ) and 23 (σ', ἴσθι, ἀκούσας) with Simpl. Ph. 157.27, 159.12 and 27–8, and Wright 77–8.

[18] For the strange conclusions which have been drawn from Plutarch's statement that B115 came 'at the start of Empedocles' philosophy', see Kingsley (1995d). The argument that the failure of some ancient writers to specify from which poem of Empedocles they are quoting a particular fragment shows Empedocles only wrote one poem is null and void: clearly they viewed all of Empedocles' poetry as reflecting his teaching. In fact not only the practice of quoting his poetry without naming the work, but also the practice of quoting from one poem to elucidate an idea in the other, would be to follow well-established tradition in antiquity: cf. e.g. Labarbe 39–42 and passim; Mansfeld (1992), 228–9, 283–4, and in Mnemosyne, 47 (1994), 79 with n. 9; Procl. Th. Pl. 1.2 with Saffrey and Westerink ad loc. and above, Ch. 4 n. 9. Regarding the papyrus discovery, by Alain Martin, from Akhmīm, it only serves at the most to confirm that (as is to be expected) there was an overlap of themes and interests between the two poems; but what is doubtless also significant is the specific use to which the papyrus material appears to have been put, as 'support' or 'underlay' for a copper crown—indicating a funerary purpose. For metal crowns, initiatory and funerary practices, and Empedocles' own poetry, see Wilson 7–10; for the use and religious significance of funerary crowns in Egypt (as symbols of 'justification' for the soul in the other world), K. Parlasca, *Mumienporträts und verwandte Denkmäler* (Wiesbaden, 1966), 59–60, 143–5; and for death-crowns made from copper or bronze in particular, R. Pagenstecher, *Archäologischer Anzeiger*, 33 (1918), 117 with refs.

[19] See n. 20; Ch. 15 with n. 13. Osborne's caricature of the cosmological poem as an 'objective' and 'secular' work (31) misrepresents the surviving evidence entirely.

Pausanias specifically include the instruction to keep what he hears to himself, to 'conceal my words in your dumb breast'; and they have correctly concluded that the original work in question was, as just mentioned, an esoteric one in contrast to the more exoteric address to the citizens of Acragas.[20] Yet in one important respect this conclusion appears to present us with a paradox. The topic of reincarnation and the issue of the life of the soul after death are dealt with openly in the public address to the people of Acragas, whereas the evidence indicates that in his address to Pausanias Empedocles only referred to these subjects indirectly and in passing, allusively and enigmatically.[21] In other words, as Kahn put the matter in an influential paper, it is the author of the *esoteric* address to Pausanias who merely 'hints at his belief in immortality' whereas in his exoteric address he is far more explicit about the 'mysterious doctrine of punishment, purification and rebirth'. As Kahn went on to argue,

There is an obvious contrast between the religious attitude of the two poems, since one suppresses any direct reference to the doctrine which the other openly proclaims. And since the more revealing of the two is also the more public, it is difficult to explain this difference in terms of the audience to which they are addressed.[22]

[20] B5; Kahn (1971), 6, Bollack, iii. 39–40. The text of the fragment is corrupt, but the emphasis on maintaining privacy and silence about the teaching is clear from a combination of two words which can be recovered with certainty: the verb *stegein* and the adjective *ellops* (demanded by the context of Plutarch's citation, *Qu. conv.* 728e). The attested convergence of both these words on the sense of 'silence' (for στέγειν cf. e.g. Tiresias in Soph. *Oedipus Tyrannus* 341, LSJ s.v. στέγω B III.2; for ἔλλοψ, Bollack, iii. 40, Gemelli Marciano 62) must, quite apart from Plutarch's introductory comments on *echemythia* or Pythagorean 'reserve', be considered decisive in determining the primary meaning of both words and of the fragment as a whole. Diels's reconstruction of the text—στεγάσαι φρενὸς ἔλλοπος εἴσω, 'conceal [my words] in your mute breast' (or, less probably, 'mutely in your breast')—is preferable to Bollack's (ii. 12–13, iii. 39–40) from the point of view both of sense and of closeness to the MSS. For the action implied in the words cf. B110.1–2; and for the meaning of φρήν here, Wright 269, Onians 25–30. An obvious parallel to the saying from later literature, and one that fully justifies Plutarch's comment on its 'Pythagorean' recommendation of silence, is Nicomachus' description of knowledge as something 'carefully guarded' by Pythagoreans 'silently in their breast', ἄρρητος ἐν τοῖς στήθεσι διαφυλαχθεῖσα (Porph. *VP* 57 = Iam. *VP* 252; Burkert 461).

[21] Cf. B8, 11, 15; and for reincarnation and the *Katharmoi*, Kingsley (1995*d*). The deviously enigmatic phrasing in B15—'A man who was wise in such matters would not divine (μαντεύσαιτο) in his heart . . . '—is particularly significant: on oracles, divination, and riddles cf. Arist. *Rh.* 1407ᵃ31–7 with Ch. 4 and n. 26; Iam. *VP* 161 with n. 12 above.   [22] Kahn (1971), 7–8.

Kahn's own solution to the problem was to follow the path already trodden by a number of scholars in the late nineteenth and early twentieth century, and explain this difference in developmental terms as due to a change in Empedocles' understanding and outlook: at the time when he wrote his formally more esoteric address to Pausanias he still felt compelled for certain reasons to restrict what he said to a 'limited' and 'partial' revelation of the truth, and it was only years later that he plucked up the courage to speak out about such esoteric topics—in an exoteric poem.[23] However, this forced idea of a 'limited revelation' in an esoteric poem must be rejected for a number of reasons. To begin with, such an idea of a merely partial or limited revelation is based entirely on the interpretation of a couple of fragments that appear to have formed part of Empedocles' introduction, or proem, to the teaching he addressed to Pausanias. This is not the place to analyse what remains of this proem; but here it will be enough to say that the idea of Empedocles promising Pausanias no more than a restricted revelation of truth is the end result not just of mistranslating Empedocles' Greek but, what is more, of accepting a Greek text which at two crucial points is at variance with the perfectly sound reading of the manuscripts.[24]

Second, this idea of the poem addressed to Pausanias as a limited revelation in comparison to the more popular poem becomes even more implausible when one notes how exactly the contours of Empedocles' two separate addresses—the one to Pausanias, the other to the citizens of Acragas—correspond to what appears to have been a basic physiognomy of the mysteries and stages of initiation in the ancient Greek world: first *katharmos* or ritual purification, second *paradosis* or transmission of the esoteric doctrine on a one-to-one basis, and thirdly *epopteia* or 'overseeing', which is when the teaching has been transmitted and the stage of learning comes to an end.[25] The title of Empedocles' address to the citizens of Acragas was

---

[23] Ibid. 7–8, 20–2, 5–6.
[24] B2.9, where πεύσεαι· οὐ πλεῖόν γε must be retained (with the pause after πεύσεαι, as in B111.2); B3.8, where θάρσει must be understood as the verbal imperative and not made (with G. Hermann and subsequent editors) into a dative θάρσεϊ.
[25] Ch. 15 n. 48, with refs.

Katharmoi, or 'Purifications'.²⁶ The thrust of his address to them was to urge them to attain what he considered a state of ritual purity by getting them to stop animal sacrifice, bloodshed, and the eating of certain prohibited foods; even his use of mythology, and of the grand theme of reincarnation, was essentially a secondary device designed to bolster and justify his instructions for ritual purification by frightening his audience into doing what he asked of them.²⁷ In stark contrast, when we turn to Empedocles' address to Pausanias we see that a level of purification is assumed and forms the basis for the teaching to be transmitted. Pausanias has already attained to a stage of purity, *katharsis*, in thought and action; he is now to become recipient of the *paradosis* or initiatory transmission; and Empedocles pointedly alludes to Pausanias' potential success in mastering and assimilating this initiatory teaching by using the term *epopteia*.²⁸ In other words, all the tangible signs simply serve to reinforce the conclusion that what Pausanias will hear will be a more advanced form of teaching than the one given to the people of Acragas.

In the third place there is the simple but significant fact that the supposedly mysterious and secret doctrine of transmigration was, in itself, not mysterious or esoteric at all. Poets were already making fun of it either during or very shortly after Pythagoras' own lifetime; it is mentioned openly in the fifth century, and by Aristotle in the fourth; and later in the fourth century Dicaearchus apparently acknowledged that the Pythagoreans 'maintained a secrecy which was quite exceptional, but that the teachings of theirs *which were best known to everybody*' (μάλιστα γνώριμα παρὰ πᾶσιν) were the doctrines of the immortality of the soul and reincarnation.²⁹

There is one final consideration, however, which is the most important of all. Wherever one encounters the phenomenon of

---

²⁶ D.L. 8.54, 62–3; above, n. 16. It is interesting to note that according to ancient writers it was the *Purifications*, and not the poem to Pausanias, which was read out in public at Olympia: Dicaearchus, fr. 87 Wehrli, D.L. 8.63 (Favorinus), cf. 8.66 (Timaeus). For Sicily's close connections with Olympia see *LSS* 27, 123 with n. 2; for Empedocles', Wright 6.
²⁷ Cf. Kingsley (1995*d*).
²⁸ B110.2, 111.2; Ch. 15 with nn. 13, 48.
²⁹ Xenophanes, DK 21 B7 (KRS 219–20), Hdt. 2.123, Arist. *De anima* 407ᵇ20–6, 414ᵃ22–5 (cf. Emp. B125–6), Dicaearchus *ap.* Porph. *VP* 19; Burkert 120–3.

esoteric traditions one invariably comes up against the same basic emphasis on the point that it is not ideas or doctrines in themselves that matter, but the ability to discover the reality of these ideas and teachings inside oneself and then make them one's own. At a theoretical level, this often involves emphasizing that supposedly esoteric ideas about the universe and man are an open secret because true esoteric teaching aims not at filling the disciple or pupil with mere fascinating theories but with providing opportunities for making these ideas and theories real in his own experience. In other words its aim is to provide him with less and less by, more and more, giving him nothing but the basic nuts and bolts. And at a more practical level, when one looks at modern first-hand reports of encounters with esoteric traditions and adherence to an esoteric discipline, one finds this provoking the frequent complaint that one is actually receiving *less* in the way of any formal doctrine or teaching than outsiders are.[30] This means that, from whichever way one approaches the matter, there is nothing at all surprising in Empedocles using the doctrine of reincarnation within the framework of an exoteric poem as a kind of window-dressing, not only to frighten his audience but also to titillate them; whereas, paradoxically, in the more esoteric document he might appear from the formal point of view to be giving less even though he is actually offering more by providing the pupil with the key principles which will—in time— allow him to answer any question for himself.

Here we also encounter the solution to another potential problem: the question of how, if the address to Pausanias was really esoteric, it could ever have got into circulation and become public. There are, to begin with, some practical considerations relevant to this issue which we need to bear in mind—such as that even for Empedocles himself there need have been no incongruity whatever in preserving an esoteric doctrine in writing,[31] and that the technique of formally

---

[30] On the theoretical side cf. e.g. R. Otto, *Mysticism East and West* (New York, 1932), 31-2, R. Guénon, *The Great Triad* (Cambridge, 1991), 125; and for two examples of the practical aspect, L. Pauwels, *Monsieur Gurdjieff* (Paris, 1954), I. Tweedie, *Daughter of Fire* (Grass Valley, Calif., 1986).

[31] On the limits of literacy down to beyond Empedocles' time cf. e.g. West (1978), 60, with Ch. 9 and the refs. in n. 37. In a broader perspective, see Kingsley (1992), 341 with n. 13. Note also Brashear's comments, 46-7.

addressing a poem to an individual while really assuming a wider audience to begin with was a traditional device in ancient literature. But behind these considerations lies the basic fact that the esoteric nature of the poem, its implicitness, would automatically protect it. Romantic notions of an esoteric text as a document containing earth-shattering statements that need locking away from the profane are naïve and vastly oversimplistic. The fact is that hardly anyone would recognize such a text for what it is, let alone know how to use it; and there is no need here to quote the saying by Jesus and the prophets about people 'seeing and not seeing, hearing and neither hearing nor understanding'. As Empedocles himself said at the start of his address to Pausanias: for people satisfied with their own understanding and their own perception of the world, the esoteric teaching he has to offer 'is incapable of being perceived or heard or understood'.[32]

[32] B2.7–8.

24
From Empedocles to the Sufis: 'The Pythagorean Leaven'
*

At first sight it may seem astonishing that mysteries, embedded in riddling secrecy, and philosophy, which on the contrary raises—or at any rate should raise—everything into the clear light of rationality, could have anything at all to do with each other.

So begins a recent paper on the overlap between philosophy and mysteries in the ancient world. Its author goes on to acknowledge the significance as well as the existence of this overlap in the period of late antiquity, and traces the origins of such an apparently paradoxical phenomenon back to one particular person: Plato.[1] Yet again one finds Plato endowed with the attributes of a Prometheus or Palamedes: if a late antique idea or phenomenon can be traced back to him, then he must have originated it.

There are two presuppositions lying behind the view generally held nowadays on the relation between ancient philosophy and mystery. The first is that the Greek mysteries were concerned with perpetuating obscurities, enigmas, and secrecy whereas Hellenic philosophers—and Presocratic philosophers in particular—were concerned on the contrary with an ideal of clarity and 'rationality'. In other words mythology and theology had posed the riddles, and the philosophers viewed it as their function to unravel them.[2] In practice, however, things turn out not to be so simple. As far as Empedocles in particular is concerned, we have seen that he was primarily a creator of

---

[1] H. Dörrie in M. J. Vermaseren (ed.), *Die orientalischen Religionen im Römerreich* (Leiden, 1981), 341–5. For mystery-language in Plato see C. Riedweg, *Mysterienterminologie bei Platon, Philon und Klemens von Alexandrien* (Berlin, 1987).

[2] Cf. e.g. Dörrie, op. cit. 342–3; G. G. Stroumsa in S. Biderman and B.-A. Scharfstein (eds.), *Interpretation in Religion* (Leiden, 1992), 230–3.

riddles, not a solver of them; and it has been the failure to appreciate this point which has caused such fundamental misunderstandings about the nature and scope of his teaching.

The second, more subtle, presupposition is that the two categories of ancient philosophy and mysteries are so distinct that any overlap between the one and the other is bound to be purely formal and superficial rather than anything essential. References to 'the adoption by the philosophers of mystic terminology'[3] are always carefully phrased, regardless of the identity of the philosophers in question, so as to maintain the assumption that the philosophical enterprise is altogether different from the concerns of mysteries and mystics. As we are repeatedly told, 'Presocratic natural philosophy' and mysticism are diametrically opposed.[4] In the case of Empedocles— or, rather, in the specific case of his cosmological poem—this perspective is justified by appealing to his use of observation of the world around him: a factor assumed to establish his credentials as a bona fide philosopher and even scientist. But, as we have seen, this is only how Empedocles happens to appear to *our* perception. From a historical point of view it is imperative that in the first instance we situate his concern with observation not in some theoretical sequence of Presocratic 'philosophical evolution' but in the context of Empedocles' own role as a magician—along with all the involvement in mystery and mysteries which that entails.

As was noted earlier, the implications of questioning in this way the scope and usefulness of the category of 'philosophy' extend far beyond Empedocles. The study of the ancient world in its many aspects is riddled with assumptions about the nature of philosophy which persistently distort and misrepresent. So, to mention just one example, the writer of a recent book on Gnosticism states in passing that 'The "philosophical" Hermetica are of course not real philosophy' because they 'claim to be based not on observation and reason but on revelation'.[5] In fact, however, the Hermetica present much the same kind of mixture of observation of the natural world combined with a guiding awareness of divine revelation which

---

[3] Seaford 253.
[4] So e.g. H. Leisegang, *Der Heilige Geist*, i (Leipzig, 1919), 255–7.
[5] B. Layton, *The Gnostic Scriptures* (London, 1987), 447.

we find in Empedocles: when we look closely at Empedocles' poetry, and at the Hermetica, it becomes clear that what was considered important by the authors in both cases was an inner sense of revelation capable of pointing to the real nature and significance of the things outwardly observed. The revelation–observation dichotomy is a false one; and if the Hermetica are to be dismissed as 'not real philosophy', then the same must logically be done with Empedocles as well.

This comparison between the Hermetica and Empedocles is not inappropriate: we saw earlier that there are specific historical links, as well as purely formal analogies, between them. The overall problem we are faced with here, as so often elsewhere, is one of terminological distinctions and value judgements which have severed phenomena that are both formally and historically related, forced phenomena that are fundamentally heterogeneous to coalesce, and established dubious continuities while overlooking the more meaningful and profound ones simply because these run across established fields of scholarly specialization. On the one hand Empedocles' supposedly rational, scientific, and philosophical traits are of a highly questionable standing if defined from a traditional Aristotelian perspective. On the other hand a much later figure such as Bolus of Mendes has been dismissed as a mere occultist, devoid of any 'philosophical' capacity,[6] in spite of the fact that more recent writers have repeatedly drawn attention to the systematic and even rigorous logic which appears to have informed his writings.[7] What evidently needs to be grasped here, behind the over-simplistic use or abuse of terminology, is that there may be not just one but many different types of logic—and even many different types of rationality.

In the last resort, Aristotle and Theophrastus were as ill-equipped to understand Empedocles as anyone in their situation could be. Their disapproval of his poetic style, their impatient exploitation of it, and their arrogant misunderstandings yielded results which in hindsight were only to be expected. When we turn from general considerations to specifics we find, not surprisingly, that they misrepresented him on every single main point we have considered: his

[6] Kroll 231.
[7] Needham, v/4. 329 with n. d; Wilson 23.

equation of elements with divinities, his astronomy, his oracular or enigmatic style, and the overall function of his teaching within a context of revelation. The insistent assumption by modern scholars that we can look to Aristotle and Theophrastus as embodying 'the ancient tradition' and the ultimate 'ancient authority'[8] for the interpretation of Empedocles is doubly misleading. First, it is misleading because on close analysis this 'authority' turns out to rest on no solid foundation. Second, as we have also seen, when assessing Empedocles' historical affinities we need to take into account as well another tradition of an essentially different kind. That other tradition was composed of a stream of influences which passed down from southern Italy and Sicily into the world of Graeco-Roman Egypt, and specifically into the Hermetica and magical papyri. As soon as this tradition is taken into account, points which have scarcely if ever been noticed assume their proper significance. For example, it is hardly a coincidence that the Hermetic *Korē Kosmou*—a text which has striking affinities with Empedoclean and early Pythagorean doctrine—presents the four elements as personified divinities, 'each the ruler and commander of its own domain'.[9] And it is no coincidence, either, that a section of the Paris magical papyrus—another document showing the profoundest affinities with basic aspects of Empedocles' teaching—contains a number of features each of which points separately to Empedocles but which, together, present a virtual summary of Empedoclean themes and concerns. Here we have a ritual for regeneration and immortalization that has significant analogies both with Empedocles and with the 'Orphic' gold plates, an initial prayer to the four elements as immortal, personified beings, and a description of those divine elements as existing simultaneously out in the universe and inside one which takes us to the heart of Empedocles' theory of perception:

I request immortality for my 'child' and for him alone, initiate into this power of ours ... Spirit of spirit, origin of the spirit inside me; fire, god-given contribution to the mixture of things blended in me, origin of the fire inside me; water of water, origin of the water

---

[8] Bollack, iii. 170; Guthrie, ii. 144–5.
[9] *Korē Kosmou* 63 (NF iv. 20.25–6). Cf. Emp. B17.28–9; Ch. 19 with n. 37; Ch. 21.

inside me; substance of earth, origin of the terrestrial substance inside me ...[10]

*

Apart from Hermetism and the magical papyri, there is a third tradition to bear in mind as well: a tradition very closely related to them both. Long ago Diels pointed to analogies between the mystery-based teaching of Empedocles and early Pythagoreans on the one hand, and of the Graeco-Egyptian alchemists on the other.[11] In fact several times already we have noted points of intersection between Empedocles, or Pythagoreanism, and alchemical literature which reveal not just formal similarities but specific lines of continuity. And there is one other point on which Empedocles and the alchemists were in fundamental agreement. This was the need to present their doctrines in the form of riddles. It is more than just a curiosity to note that, in adopting this practice, Greek alchemists found themselves appealing to the authority of Presocratic philosophers—just as Arab and Persian writers were later to find themselves appealing to Empedocles for exactly the same reason.[12]

The parallelism on this point between alchemical and Islamic tradition calls for a few words. Earlier (Ch. 5) we looked at the Latin *Turba philosophorum* and its Arabic prototype, the *Muṣḥaf al-jamāʿa*. We saw that—via the intermediary of Greek alchemy—these works preserved genuine items of information about the Presocratics, including Empedocles: a significant fact in itself. But we also saw how, apart from the purely Gnostic and alchemical ideas they attributed to Empedocles, they supplemented their ultimate sources of information by restoring key themes in Empedocles' original teaching which the

---

[10] *PGM* IV.476–95, reading μύστηι for μυσται with Dieterich, Merkelbach. Cf. Emp. B109; Dieterich (1891), 57–8, (1923), 54–5; and on the Empedoclean principle of like perceiving like, Sext. *Math.* 1.302–3 with Altmann 5–6.

[11] Diels 434–5.

[12] Zos. Pan. 114.22–3, 116.7–9; Georgius Syncellus, *Ecloga chronographica* 297.28–298.1 Mosshammer; al-ʿĀmirī, *Kitāb al-amad ʿalā 'l-abad* 3.2 (Stern 328, 333, Rowson 70–1); Shihāb al-Dīn Yaḥyā al-Suhrawardī, *Kitāb ḥikmat al-išrāq* 10.13–16, trans. Corbin (1986*b*), 88 (cf. Scott, iv. 264–5); Quṭb al-Dīn al-Shīrāzī in al-Suhrawardī, *Kitāb ḥikmat al-išrāq* (Tehran edn., 1315/1936), 17, trans. Corbin (1986*b*), 243; Morris 36–7. Note also Plot. 4.8.1.17–22 with *Theol. Arist.* 1.30–3 (23.6–13 Badawī, Lewis 227).

mainstream Greek philosophic tradition had played down, misunderstood, or just ignored.

The *Muṣḥaf* and *Turba* were, themselves, to become among the most influential of all western alchemical texts.[13] But this should not be allowed to obscure the fact that they are by no means the only evidence indicating familiarity with Empedocles in Arab alchemical circles. There is also the famous body of alchemical texts attributed to Jābir ibn Ḥayyān: texts which, in spite of the attribution, were clearly not written by one and the same historical person. In their surviving form they can sometimes be dated to the late ninth or tenth century, and one of them—concerned with the alchemical principle of the 'balance'—was probably produced at much the same time as the *Muṣḥaf al-jamā'a*; at one point it refers to views on alchemical procedure held by the *ṭā'ifat anbadaqlīs*, the 'sect' or 'followers' of Empedocles.[14]

There may seem very little to be inferred from such a reference. And yet this is not the only mention in Arabic literature of a *ṭā'ifat anbadaqlīs*: of people who evidently took Empedocles as their authority and in some basic way identified with him. Writing about Empedocles in the late tenth century, the Persian writer al-'Āmirī notes that

a group of the Bāṭiniyya (*wa-ṭā'ifa min al-bāṭiniyya*) regard themselves as followers of his wisdom and hold him superior to all others, claiming that he has enigmatic allusions (*rumūz*) which can very rarely be comprehended.[15]

---

[13] For the *Muṣḥaf* see Plessner (1954), 332, Sezgin, iv. 60, 65, 92–3; for the *Turba*, Kingsley (1994c), n. 59.

[14] *Kitāb al-aḥjār 'alā ra'y balīnās*, ed. Kraus (1935), 187.13. It will be noted that in the list of authorities cited here the word *ṭā'ifa* is used only with regard to Empedocles and—in the next line—Pythagoras: cf. Kraus (1943), 45 n. 5, 46 n. 2. Sezgin's statement (iv. 49) that these followers of Empedocles are 'probably Neoplatonists' bears no relation whatever to the Jābir text itself. For use of the word *ṭā'ifa* elsewhere in the Jābir corpus as term of reference for an alchemical 'group' or 'circle' cf. Kraus (1935), 71.11; (1943), 125 n. 3. On the dating of the *Kitāb al-aḥjār 'alā ra'y balīnās*, = Kraus (1942), 78–9 § 308, see ibid. 75–6 with Needham, v/4. 393; on the dating of the *Muṣḥaf al-jamā'a*, Kingsley (1994c). For the authorship and dating of the Jābir corpus as a whole cf.—apart from Kraus's classic studies, (1930), (1942)—Plessner, *ZDMG* 115 (1965), 23–35 and *Ambix*, 19 (1972), 212–13; Needham, v/4. 390–7; S. H. Nasr, *An Introduction to Islamic Cosmological Doctrines* (Cambridge, Mass., 1964), 14 n. 31; P. Lory, *Jābir ibn Hayyân: Dix traités d'alchimie* (Paris, 1983), 34–62 and in Corbin (1986a), 18–22.

[15] *Kitāb al-amad 'alā 'l-abad* 3.2 (Stern 328, 333; Rowson 70–1).

This time we are given some information about the identity of the Empedocleans in question. They are Bāṭinites: a word which, certainly in a tenth-century writer, alludes in the first instance to Shi'ite Muslims—and in particular to the Ismā'īlī branch of the Shi'ites, who were in fact well known for assimilating Greek and other pre-Islamic traditions into their schemes of prophecy and revelation.[16]

This double mention of followers of Empedocles, one in an alchemical and the other in an Ismā'īlī context, is plainly no coincidence. There was, as we know from the Jābir corpus in particular, a very close connection between Arab alchemy and the Ismā'īlīs.[17] Absorption of Graeco-Egyptian alchemical traditions as reprocessed by Islamic writers was, in fact, just one aspect of the Ismā'īlī tendency to adopt and 'appropriate' for themselves early Islamic sources famous for perpetuating traditions of the Greeks.[18] And in the particular case of Arab alchemy, with its deep roots in Islamic as well as pre-Islamic Egypt, it is important to note the specific concern of Ismā'īlīs to justify the founding of their Fāṭimid dynasty in Egypt during the early tenth century by discovering their own beliefs and ideas already prefigured in existing Arab traditions—even at the cost of inverting the historical sequence of events.[19] At every

---

[16] For *bāṭiniyya* and Ismā'īlīs in their relation to Greek philosophical traditions see Peters's summary, 174–9, and in general, Corbin (1983). For their particular interest in Empedocles see Appendix III. Some of this evidence for Ismā'īlī interest in Empedocles has already been noted for its relevance to the al-'Āmirī passage by Rowson (208), while Stern (326) had earlier argued that the prime reference of al-'Āmirī's *bāṭiniyya*—literally a denomination for those who adhere to the *bāṭin*: the 'inner' or 'esoteric'—must be to Ismā'īlism. However, in spite of the undeniably close links between Ismā'īlīs and Sufis—for which cf. e.g. Corbin (1964), 139–41, 149–51—Stern's appeal to Sufism as a possible secondary connotation of al-'Āmirī's term must be excluded: the word was only extended to include Sufis at a much later time (M. G. S. Hodgson, *EI*² i. 1099b), and in this respect al-Ghazālī's use of the term *bāṭiniyya* in his onslaught on Ismā'īlism at the end of the 11th c. is particularly significant (cf. Peters 178–9, with further refs.).

[17] Cf. esp. Kraus (1930) and (1942), pp. xlviii–lvii; Corbin (1964), 184–8 and (1986a), 147–219; Daftary 88.

[18] Regarding this tendency cf. e.g. Corbin's comments, (1983), 136, on the Ismā'īlī appropriation of Qusṭā ibn Lūqā. Note also the Ismā'īlī adoption of material from the ps.-Ammonius doxography: Rowson 46–7, Rudolph (1989), 13, 23–5, 117–18, 130, 137. For general comments on this concern of Ismā'īlism with 'assimilating the entire encyclopedia of the Greek sciences', 'supplanting Islamic law by means of its esoteric teaching', and 'substituting the lights of Greek science and philosophy for the revelation of the Koran', cf. Kraus (1942), p. xlix.

[19] Kingsley (1994c), § III. On the famous *ḥadīth* referred to in the Jābir corpus about

possible level, the Ismāʿīlīs were interested in embracing, assimilating, and perpetuating.

The existence of Empedocleans spanning the worlds of alchemy and Ismāʿīlism is an important aid in its own right to our understanding of the use made of Greek traditions in the Islamic world. But it is also important in helping us to understand one very particular phenomenon. That phenomenon is the so-called 'pseudo-Empedocles' literature—known to us directly or indirectly from Arabic, Persian, Hebrew, and Latin sources—which was to exert an immense influence on medieval mysticism, theology, and philosophy both inside and outside Europe.[20] In this respect it is important for two reasons. First, there are clear links between the pseudo-Empedocles material and alchemy: connections which have been neglected in modern scholarship because of the tendency to overestimate the primacy of the apparently Neoplatonic elements in this 'Empedocles' literature while playing down those aspects of the literature which are plainly irreducible to any form of Platonism. There are, in particular, correspondences between the Empedocles of the *Turba philosophorum*—a text showing no traces of Neoplatonism whatsoever—and the Arabic 'pseudo-Empedocles' which imply an extensive backdrop of alchemical, Hermetic, and Gnostic ideas attributed to Empedocles well before the late ninth century AD.[21] And, second, there are

the sun rising in the West, and on the relation between Fāṭimid Egypt and Persia in the development of Ismāʿīlī doctrine, cf. ibid., n. 76.

[20] See e.g. S. Munk, *Mélanges de philosophie juive et arabe* (Paris, 1857), 241–5; A. Lasson, *Giordano Bruno: Von der Ursache, dem Princip und dem Einen* (Berlin, 1872), 150–1; Kaufmann 1–63; Nagy 307–20, 325–44; S. M. Stern, *EI²* i. 483–4; Schlanger 58, 60, 76–88; Asín Palacios; Wilcox 62–78; Rudolph (1989), 130–2.

[21] So e.g. the mention by 'Empedocles' of a 'lower fire' in the *Ghāyat al-ḥakīm* (*Picatrix* 299.1) corresponds to *Turba* 53.17–18 Plessner, which refers elliptically to an upper as well as a lower fire (the lower being the central fire). For other points of correspondence between the *Turba* and the *Ghāyat al-ḥakīm* see Plessner 16 n. 36, 81 n. 184, 109 n. 272, 121–2. Also overlooked is the correspondence between the idea of shell within shell, *cortex* within *cortex*, in the *Turba* passage (52.12–53.16) and the central theme—on which cf. R. Arnaldez, *EI²* iii. 869–70, Rudolph (1989), 156—of shell within shell, *qišr* within *qišr*, in the 'pseudo-Empedocles' literature. For *cortex* = *qišr* cf. Altheim and Stiehl 16 n. 17, and for the overall theme of shell within shell around a central point (as in *Turba* 52.11–53.20: above, Ch. 5 with n. 22) both in 'pseudo-Empedocles' and, subsequently, in Cabbala cf. Duhem, v. 120–2; also Altmann 177–9. On the Hermetic and Gnostic background to the idea see Rudolph (1989), 156.

equally clear links between the pseudo-Empedocles material and Ismāʿīlism. On the one hand certain striking features in 'pseudo-Empedocles', particularly as preserved for us by al-Shahrastānī, bear the unmistakable hallmarks of later Ismāʿīlī doctrine; on the other, when the evidence is submitted to close analysis an interesting pattern emerges of this pseudo-Empedocles material being repeatedly copied, used, and elaborated by Ismāʿīlī writers.[22] Alchemists, Ismāʿīlīs, Empedocleans, and medieval Empedoclean literature are all intimately interrelated.

What relevance, it will be asked, could all this have to the historical Empedocles? The answer provided by western scholarship as a whole during the nineteenth and twentieth centuries has been simple: to call this oriental and medieval Empedocles 'pseudo-Empedocles', claim that by the time this 'pseudo-Empedocles' came into being in the Arab world 'the authentic tradition of the genuine ancient Greek ideas had been severed and lost', and add remarks to the effect that nothing of any value or interest for our understanding of the ancient world could, anyway, be hoped for from the 'Arab race'.[23] And yet it is important to appreciate that things are not so clear-cut. Certainly the Arabic Empedocles contains a generous amount of the Gnostic, Hermetic, and Platonic ideas so characteristic of late classical and early medieval mysticism. But it also contains elements which very clearly correspond to, and derive from, the historical Empedocles as known from the surviving fragments of his poetry.[24] As for the transmission of Empedoclean ideas being severed by the time it reached the Arab world, we have already seen from the *Turba philosophorum* that this was far from being the case. And, what is perhaps even more important, we have repeatedly seen examples of the way that Arab writers—both in and out of alchemical tradition—managed intuitively to reconstruct aspects of Empedoclean or Pythagorean teaching which even by the time of Plato and Aristotle had been either misinterpreted or simply ignored.[25]

---

[22] See Appendix III.
[23] Cf. e.g. Kaufmann 13–56 (pseudo-Empedocles); ibid. 8 for the quotation about loss of 'the authentic tradition'; Karsten 76 with Nagy 326.
[24] Cf. e.g. Kaufmann 54 with n. 1; R. Arnaldez, *EI*² iii. 869–70.
[25] Above, Chs. 5, 14.

Here it will be worth giving just one more example of this last phenomenon: an especially suggestive example, considering that it has to do with one of the earliest surviving instances of Empedocles appearing in the Arab world. The *Theology of Aristotle* was evidently produced in the al-Kindī circle at Baghdad during or around the early ninth century. Although essentially an Arabic translation of extracts from Plotinus' *Enneads*, it is far more than just that: the author expands and illustrates Plotinus' original statements, supplements and modifies them at will. The opening section of the *Theology* covers ground which, in the original *Enneads*, contains a brief reference by Plotinus to Empedocles. Plotinus has very little to say about him: basically only that, according to Empedocles' teaching, if souls make some mistake they are punished by falling into incarnation and that he himself—Empedocles— was just one of those souls, banished from the realm of the divine. But the writer of the *Theology* goes on to say things about Empedocles which reach well beyond what we find in Plotinus:

It was as a fugitive from the anger of God that he too came to this world, for when he came down to this world he came as a help (*ghiyāth*) to those souls whose minds have become contaminated and mixed. And he became like a madman, calling out to people at the top of his voice and urging them to reject this realm and what is in it and go back to their own original, sublime, and noble world.[26]

There are several important features in this passage: features which raise questions about the interpretation of Empedocles in the Greek, the medieval, and the modern world. But here it will be enough just to comment on the idea, which has no correspondence whatsoever in Plotinus, of Empedocles as a *ghiyāth* or 'help' of souls. The idea is significant from a number of different aspects. First, the word *ghiyāth*—like the word *ghawth*, which has the same root and the same meaning—was to become a more or less technical term in Sufism as a way of describing the role of a prophet or saint, but above all as a way of describing the function of the 'Pole': the supreme 'shaikh of

---

[26] *Theol. Arist.* 1.31 (23.7–10 Badawī, Lewis 227); Plot. 4.8.1.17–20. On the background, date, and scope of the *Theology* see Zimmermann, with Genequand (1987–8), 15 and n. 55; C. D'Ancona Costa, *Medioevo*, 17 (1991), 83–134.

shaikhs' who is the spiritual guide and teacher of his time. In this respect, the introduction to the *Theology* prefigures the way that a writer such as the twelfth-century Persian mystic al-Suhrawardī was to paint Empedocles—along with Pythagoras—in the colours of a Sufi.[27] But at the same time, this characterization of Empedocles as a 'help' of souls also prefigures the important place reserved for the role of prophet and prophecy in the doctrine later ascribed to 'pseudo-Empedocles'.[28] And finally, the picture here of Empedocles as a teacher of asceticism, urging people to purify themselves so that they can return at last to the spiritual world, will continue to be invoked down at least to the seventeenth century in Persian traditions about certain Presocratics—Empedocles included—exemplifying the path of asceticism and spiritual struggle.[29]

In short, Henry Corbin was perfectly justified when he portrayed the figure of Empedocles in Islam as a 'hierophant and prophet' with 'the moral physiognomy of a Sufi'.[30] This is a point with major implications of its own, especially in allowing us to assess the evidence for connections between 'pseudo-Empedocles' and the famous Spanish mystic of the late ninth and early tenth century, Ibn Masarra: connections which, on the basis of a passage in Ṣāʿid al-Andalusī, were universally acknowledged until Samuel Stern reversed the scholarly consensus with his attack on Ṣāʿid's reliability.[31] In fact, closer analysis indicates that Stern's cynical dismissal of Ṣāʿid is—as

---

[27] So e.g. al-Suhrawardī, *Kitāb ḥikmat al-išrāq* 5.12–15 = Corbin (1986b), 80, on Empedocles as one of the *afrād*. On the term *fard/afrād* in classical Sufism see M. Chodkiewicz, *Le Sceau des saints* (Paris, 1986), 73–4, 133–43; for the term *ghawth* cf. e.g. ʿAbd al-Razzāq Kāšānī, *Iṣṭilāḥāt al-ṣūfiyya*, ed. M. K. I. Jaʿfar (Cairo, 1981), 167.14–15, Nicholson 130, Massignon 133, 199. On the Sufi idea of the 'Pole' or *quṭb*, and its traditional origins in the teaching of Dhū 'l-Nūn, see J. S. Trimingham, *The Sufi Orders in Islam* (Oxford, 1971), 163.

[28] Cf. esp. al-Shahrastānī, ii. 263.3–13 Cureton = ii. 71.1–10 Kīlānī. Note also Corbin's comments (1971–2), ii. 70 n. 88, and compare al-ʿĀmirī's explicit statement that Pythagoras received his knowledge 'from the niche of prophecy' (*min miškāt al-nubuwwah*: *Kitāb al-amad ʿalā 'l-abad* 3.3, Rowson 70–1, 209).

[29] ʿAbd al-Razzāq Lāhījī, *Guzīda-yi gawhar-i murād*, ed. S. Muwaḥḥid (Tehran, 1985), 320.8. Cf. also the passages from Ṣāʿid al-Andalusī, Ibn al-Qifṭī, al-Shahrastānī, and al-Shahrazūrī collected by Asín Palacios, 47 with n. 16, 56–7, 162.

[30] (1964), 305–8.

[31] Ṣāʿid al-Andalusī, *Kitāb ṭabaqāt al-umam*, 21 Cheikho (28.13–16 al-ʿUlūm, Blachère 59); Stern 325–8, followed e.g. by B. Radtke, *Der Islam*, 58 (1981), 330–1, P. E. Walker, *JAOS* 103 (1983), 761–2, Rudolph (1989), 132, Addas 57–8 and in *LMS* 912–14.

with the results of his rather too influential passion for debunking elsewhere—highly implausible and, at the very least, over-simplistic.[32] And yet the crucial point which has been missed is, to judge from the surviving evidence, the remarkable overlap of interests between the 'pseudo-Empedocles' tradition and Ibn Masarra: concern, in both cases, with the role of prophecy, with asceticism, with the purification and reascent of the soul, and with presenting these teachings in an enigmatic and riddling form.[33] To claim that there can be no common ground between the Arabic Empedocles and Ibn Masarra because the first was a philosopher, not a Sufi, and the second 'a Sufi, not a philosopher'[34] is to create distinctions between *taṣawwuf*—or mystical Sufism—and *falsafa*—or Greek philosophy—which certainly operated with considerable force later on but which have a very dubious validity when applied to the early tenth century; and it is to present an idea of 'Empedocles' as a rational philosopher which is as false for the tenth century AD as it was for the fifth century BC. For of course in his 'moral physiognomy' this Empedocles not only resembled a Sufi but also resembled the historical Empedocles—himself a spiritual guide and a prophet, not just in the limited sense of a diviner or seer but in the fuller sense of someone able to reveal

---

[32] To denounce Ṣāʿid's statement that Ibn Masarra was deeply indebted to the philosophy of 'Empedocles' as no more than an 'unfortunate idea', 'a superficial conclusion' 'thoughtlessly interpolated' (Stern 326) is, in the absence of supporting evidence, purely speculative—especially when we consider that Ṣāʿid, who lived in the century after Ibn Masarra, was also from Andalusia. Rowson (207) has followed Stern in claiming that, by mentioning Ibn Masarra when elaborating on al-ʿĀmirī's earlier statements about followers of Empedocles, Ṣāʿid was 'merely interpolating comments to supply local color and Islamic relevance—as he does with his note on Abū 'l-Hudhayl'. But what Rowson does not mention is that, in this note on Abū 'l-Hudhail, Ṣāʿid provides specific information which turns out to be highly accurate and reliable (ibid. 227–31). For the unfortunate influence exerted on subsequent scholarship in another area by Stern's debunkings, cf. R. Boase in *LMS* 458–73.

[33] For the evidence relating to Ibn Masarra cf. Morris, esp. 32 n. 45, 33, 36–8, now to be supplemented further with the texts edited by M. K. I. Jaʿfar in *Min qaḍāyā 'l-fikr al-islāmī* (Cairo, 1978), 311–60.

[34] Stern 327. It will also be noted that Stern's further assertion (ibid.) that 'there were Neoplatonic philosophers in Spain, Muslim as well as Jewish—but they had nothing to do with Ibn Masarra and his school', is fundamentally undermined by the Arabic texts of Ibn Masarra which have been discovered and published in the meantime: texts in which he refers repeatedly, and most respectfully, to 'philosophers' with a very obviously Neoplatonic (or rather Gnostic, Hermetic, and Platonic) background. Cf. *Kitāb al-ḥurūf*, ed. Jaʿfar, op. cit. 315, 330–1.

to people the origin and predicament of their souls and show them the way back to their spiritual source. This prophetic and guiding function was a crucial aspect of his teaching which Empedocles no doubt inherited from a common fund of ancient Near Eastern and Mediterranean traditions but which, for philosophical as well as theological reasons, Greek pagans and Christians alike were never at all happy to acknowledge.[35]

There would seem also to be a second aspect to this idea of Empedocles as *ghiyāth*. Quite apart from the theological associations of this particular term both in Sufism and in Islamic culture as a whole,[36] the notion of Empedocles 'coming down to this world as a help to souls whose minds have become contaminated' has very specific overtones strongly suggestive of Hermetism and Manichaeism. In fact the idea of a saviour figure descending into this world as a 'help' ($βοήθεια$), so as to remind incarnated souls of their origin and help them return to their spiritual home by urging them to reject the world of the senses, already plays a pronounced role in the earliest of the Hermetica.[37] This idea then became embodied in the technical vocabulary of Manichaeism: the help ($βοήθεια$) sent down by the transcendent God to souls trapped and deluded in this world was personified in the figure of the divine envoy or saviour known as 'the helper' (*p-boēthos*) or, more specifically, 'the helper of souls' (*p-boēthos nm-psychaue*)—an expression which passed from Manichaeism into Persian Sufism to become *ghiyāth al-nufūs*, 'help of souls', in the work of al-Suhrawardī.[38] As it happens, these striking similarities to

[35] Emp. B112–30, 135–47; for the historical and geographical background cf. esp. Grottanelli 649–70 plus Burkert's further comments, (1983), 115–19.
[36] For its popular use with direct reference to Allāh see Lane, vi. 2306.
[37] *CH* 1.22 (cf. 24.6); note also the language in *Korē Kosmou* 64 (NF iv. 21.5), *Asclepius* 37 (NF ii. 348.1–6). For Gnostic literature cf. esp. Nag Hammadi Codices I 90.26–7, II 92.1–2, V 55.15–16, 59.24, VIII 46.27–30, 47.18, IX 28.28, and the further refs. in F. Siegert, *Nag-Hammadi-Register* (Tübingen, 1982), 227. Not so close in concept or terminology to the *Theology* passage, but plainly influenced by Hermetic and Gnostic as well as specifically Pythagorean sources, is the idea of the divine man being 'sent down' to earth for the benefit of humanity which occurs in Atticus, Iamblichus, and Syrianus (O'Meara 36–9 with n. 22, 88, 125–6; cf. Burkert 141–6).
[38] Cf. esp. *A Manichaean Psalm-Book*, ed. C. R. C. Allberry, ii (Stuttgart, 1938), 92.24, 96.9–10, 98.31, 137.30, 204.1, 206.7–11, 209.22–210.6, 218.26–8; L. J. R. Ort, *Mani: A Religio-Historical Description of his Personality* (Leiden, 1967), 256, Puech, ii. 225, 234, M. Scopello, *L'Exégèse de l'âme* (Leiden, 1985), 127. For the expression *ghiyāth al-nufūs* cf. al-Suhrawardī, *Kitāb al-talwīḥāt* 55 (*OMM* i. 70.4); Corbin 119, with his further

Hermetism and Manichaeism in our passage from the *Theology* are hardly accidental: recent research has already brought into the open the background of Hermetic and Manichaean ideas which inspired the al-Kindī circle in general and, in particular, the additions to the text of Plotinus contained in the Arabic *Theology*.[39] In other words, here at the very first stage in the penetration of Empedocles into the Arab and Persian world we would seem to encounter a characteristically Hermetic Empedocles. The existence of an ultimately Hermetic background to the 'pseudo-Empedocles' material has in fact already been suspected on a number of occasions;[40] and this is an issue which obviously assumes a very particular significance in the light of the evidence considered earlier for crucial aspects of the *historical* Empedocles' teaching being preserved and perpetuated in Hermetic tradition.

'Pseudo-Empedocles' can, of course, only be 'pseudo' in relation to a genuine Empedocles. Orientalists are quite specific about what that genuine Empedocles is: it is 'the authentic Empedocles' preserved for us in Aristotelian tradition.[41] However, as we have repeatedly seen, the Empedocles presented to us in Greek philosophical—and specifically Aristotelian—tradition is in many fundamental respects a false or pseudo-Empedocles, while in these very same respects the

---

comments on the passage, *Itinéraire d'un enseignement* (Tehran, 1993), 37–8; Böwering 53. On al-Suhrawardī's indebtedness to Manichaeism see Corbin, *OMM* ii. 51–5.

[39] Genequand (1987–8). Genequand well notes the parallels in Arab Hermetism to another statement ascribed to Empedocles in the introduction to the *Theology*: the one comparing the body to 'rust' (ibid. 6–7; *Theol. Arist.* 1.38, 24.7–8 Badawī, Lewis 229); but as far as the Greek sources for this idea are concerned, more relevant than the text from Porphyry cited by Genequand (*Ad Marcellam* 13; note also Plot. 4.7.10.42–52) are Basilides *ap.* Clem. *Str.* 4.12.88.5 (cf. Asín Palacios 70–1) and *CH* 14.7. Genequand does not mention the passage in the *Theology* which we have been considering.

[40] Massignon in Festugière (1950), 387, 392; S. H. Nasr, *Islamic Studies* (Beirut, 1967), 72; F. E. Peters, *Allah's Commonwealth* (New York, 1973), 284, 357; Wilcox 65; and on the substratum of Hermetic and Gnostic ideas in the so-called 'Ammonius doxography' cf. Genequand's comments, (1991), 945–7. Plessner, *Studia Islamica*, 2 (1954), 48 n. 3, of course states the obvious in pointing out the impossibility of distinguishing clearly between Hermetic and Gnostic doctrines: cf. Fowden 112–14, 172–4, 202–4, Kingsley (1993b), and n. 37 above. For Empedocles and Gnosticism see Ch. 22 with n. 25; for Empedocles and Manichaeism, P. Alfaric's remarks, *Les Écritures manichéennes*, ii (Paris, 1919), 198–9.

[41] See e.g. S. M. Stern, *EI*² i. 483.

so-called 'pseudo-Empedocles' manages to preserve or restore genuine features of the historical Empedocles' teaching. Here we have not just some superficial irony but, to return to a recurring theme of this book, a telling symptom of the major flaws in western scholarship's treatment and assessment of Empedocles. Naturally this is not to suggest that the Empedocles literature in Arabic is intrinsically any more reliable a source for our understanding of the historical Empedocles than the Aristotelian and later Greek traditions of interpretation. But it is to emphasize the fact that just as we cannot automatically turn to 'pseudo-Empedocles' for clarification of certain aspects of the original Empedocles' teaching, so we cannot turn to the Greek philosophical traditions of interpretation either. Each single piece of evidence needs to be assessed on its own merits, without any preconceptions about the accuracy of Aristotelian—or Stoic, or Neoplatonic—interpretations of earlier philosophers.

It would be wrong to underestimate the importance, or the scope, of the reassessment and revision required. In this book the focus of attention has quite deliberately been placed on Empedocles' theory of the elements and on the exact way that he himself introduces it at the start of his poem to Pausanias: to use his own language, the elements represented the very roots of his system, and conclusions arrived at on such a basis will inevitably be relevant to the system as a whole. There is no need to repeat here how wrong the Greek philosophical interpretations of Empedocles' element theory have turned out to be; and as far as the other aspects of Empedocles' teaching are concerned, it is poor reasoning to assume that the end result will prove to be any different—or the modern consensus of opinion any closer to the mark. Just as each item of his teaching has to be approached in its own right and in a fresh light, so, also, must each item of Parmenides' teaching, not to mention further issues such as the relation between Empedocles' teaching and Parmenides' or any other Presocratic's. To retreat to the safe ground of the idea that 'Presocratic philosophy' is a recognizable tradition of rational discourse is to prejudice the entire issue. 'Rationality' is a blanket term which, in spite of its apparent definiteness, tends to obscure more than it clarifies. And as for the theory of a 'Presocratic discourse', it is questionable whether those philosophers who supposedly reacted and responded to the views of earlier Presocratics even knew of their existence, let alone of their work; while in those cases where they did, their responses were no doubt far subtler and more complex than is usually supposed.[42]

Shifting the boundaries for our understanding of Empedocles to such an extent—and, in the process, marginalizing the usefulness of Aristotle or Theophrastus as means of approach to his teaching—may still seem to some a rather radical procedure. Ultimately, it is the shift that must justify itself; but here it will be worth helping to put things in their proper perspective by addressing some historical pieces of evidence which Empedoclean scholarship has completely overlooked. It is normally assumed that the first real step in questioning Aristotle's accuracy as an interpreter of the Presocratics was taken by Harold Cherniss in the 1930s.[43] What is not mentioned is that in 1581 Francesco Patrizi published a four-volume work which dealt extensively, and in considerable detail, with Aristotle's treatment of the Presocratics from a philological and historical point of view, which radically criticized Aristotle for wilfully misrepresenting his predecessors and examined his motives for doing so, and which tackled the further task of re-evaluating the teachings of the Presocratics—independent of Aristotle's interpretations—on the basis of the surviving fragments of their works. It is scarcely a coincidence, in the light of the connections we have already observed, that Patrizi had a particular interest in Hermetism and was in fact one of the first modern editors of the Hermetica.[44]

In a sense, Patrizi's philological onslaught on Aristotle was one very particular manifestation of the Renaissance tendency to view the name of Aristotle as 'the synonym of intellectual oppression'.[45] But it is important to appreciate that this general

---

[42] Cf. West's comments (1971), 99, 218–19.

[43] See the Introduction, with n. 6.

[44] For the work on Aristotle see Franciscus Patritius, *Discussionum Peripateticarum tomi quattuor* (Basle, 1581), plus M. Muccillo's paper 'La storia della filosofia presocratica nelle "Discussiones Peripateticae" di Francesco Patrizi da Cherso', *La Cultura*, 13 (1975), 48–105 (90–3 on Empedocles). For Patrizi's edition of the Hermetica see Scott, i. 36–40; F. van Lamoen, *Hermes Trismegistus, Pater philosophorum* (Amsterdam, 1990), 80–7.

[45] T. Whittaker, *The Neo-Platonists*² (Cambridge, 1928), 195.

From Empedocles to the Sufis 387

attitude had roots extending much further into the past—and into the Arab and Persian world. Four hundred years earlier the Persian writer al-Suhrawardī, founder of the *Ishrāqī* line of Sufis, had already adopted a less analytic but also more subtle approach. For him the teaching of the great Greek philosophers before Aristotle, especially Empedocles and Pythagoras, was one that combined the highest mystic and intuitive grasp of reality with the ability to argue for their teaching when necessary; but with Aristotle this balanced synthesis was destroyed. Instead, Aristotle substituted an exclusive concern with reasoning and rationalism which 'absorbed' later philosophers (*šaghalahum*) and gave rise to an increasingly sterile occupation with mere intellectualism and argument. Thanks to him, as time went on 'the traces of the paths of the ancient sages disappeared'; their ability to discover and communicate the reality that lay behind their teachings was lost; 'their directions were either effaced, or corrupted and distorted'.[46]

Once again, it is no coincidence that al-Suhrawardī saw his own work as falling very firmly in the line of Hermetic tradition.[47] And his ideas are in fact only one aspect of a larger historical picture. On the one hand, his perception of Aristotle as responsible for losing the thread of understanding of Presocratic philosophy and 'setting himself up in opposition to them' (*va aristū muḥālafat karde bā īšān*) was perpetuated by the *Ishrāqī* tradition down to and beyond the seventeenth century.[48] But on the other hand, these views of his were not new to him either. Two hundred years before him, in the early tenth century, the great Persian physician al-Rāzī had already radically criticized Aristotle for 'corrupting philosophy' (*fasada al-falsafa*) by departing from the tradition of the

---

[46] ... *wa-inṭamasat adillatuhum wa-ḥurrifa 'alaihim ghairuhā*: *Ḥikmat al-išrāq* 5.19–6.6 = Corbin (1986b), 80–1, with Quṭb al-Dīn al-Shīrāzī's summary (ibid. 237). Cf. also al-Suhrawardī, op. cit. 10.13–11.11 (ibid. 88–90); Nasr 61 and in M. M. Sharif (ed.), *A History of Muslim Philosophy*, i (Wiesbaden, 1963), 376. It will be noted that al-Suhrawardī maintained this estimate of Aristotle in spite of his mistaken belief that the so-called *Theology of Aristotle* was a work of his (Corbin 304 n. 2), and in spite of the attribution of alchemical and related doctrines to Aristotle in the Arab world: Kraus (1943), 45 n. 6.

[47] Scott, iv. 264 with n. 4; Nasr 59–63; Corbin, *L'Archange empourpré* (Paris, 1976), pp. xv–xvi.

[48] 'Abd al-Razzāq Lāhījī (died 1662), *Guzīda-yi gawhar-i murād*, ed. S. Muwaḥḥid (Tehran, 1985), 320.4–6 (cf. Corbin 171).

Presocratics—especially Empedocles and Pythagoras.[49] Al-Rāzī was an alchemist, and held ideas commonly referred to as Pythagorean.[50] The same basic idea of Aristotle corrupting philosophy can be traced back, in our Greek sources, as far as Numenius of Apamea in the second century: to the man who wanted to rescue Plato from the distorted picture of him presented by Aristotle and others through emphasizing, instead, his indebtedness to Pythagorean tradition.[51] Needless to say, such views of Aristotle as distorting, covering over, or corrupting the teachings of his predecessors have no absolute value of their own: Numenius' view of early Greek philosophy was tinted with Platonism, and the Persian writers in particular had no access to the poetry of Empedocles in the way that we do. But what is striking and significant about these views is that they fall into a pattern which corresponds remarkably to the conclusions drawn throughout the course of this book on the basis of direct examination of the earliest Greek texts.

Al-Suhrawardī, writing in the early twelfth century, happens to be very specific about what he considered the source of his information on the Presocratics. For him there was essentially only one true wisdom, or 'eternal leaven' (*al-ḥamīrat al-azaliyya*), which at a certain point had divided into two branches: an eastern and a western. In the East, it had been transmitted down to his time primarily by the great Persian sages. In the West, it had appeared in the figures of rare individuals such as Empedocles and Pythagoras; and there—in spite of being covered over by Aristotelian rationalism—it had been preserved and eventually transmitted back, via Egypt, to the East. He also explains when and where this transmission of what he describes as 'the Pythagorean leaven' back to the East occurred. 'The Pythagorean leaven fell to the share of the brother of Akhmīm, and from him came down to the traveller from Tustar and his party' (*fa-ḥamīrat al-faithāghūriyyīn waqaʿat ilā aḫī iḫmīm wa-minhu nazalat ilā sayyār tustar wa-šīʿatihi*).

[49] Ṣāʿid al-Andalusī, *Kitāb ṭabaqāt al-umam*, 33 Cheikho (42.13–18 al-ʿUlūm, Blachère 75); Asín Palacios 8 with n. 18.

[50] J. Ruska, *Der Islam*, 22 (1935), 281–319; Needham, v/4. 398; S. Pines, *Beiträge zur islamischen Atomenlehre* (Berlin, 1936), 82 n. 3, Peters 171.

[51] Cf. Leemans 33, Puech, i. 45–7, O'Meara 12 and n. 14 (Numenius and Atticus); ibid. 96–7, 100 (Iamblichus), 122–4 (Syrianus). For Numenius see Ch. 19 with n. 49, Ch. 20 with n. 40; for Atticus, Iamblichus, and Syrianus, above, n. 37.

The 'brother of Akhmīm' refers to Dhū 'l-Nūn al-Miṣrī, and 'the traveller from Tustar' to Sahl al-Tustarī.[52] Akhmīm is, of course, a place-name we have already encountered. Apart from being, as Arab and Persian writers remembered, a major centre of Hermetism in the Graeco-Egyptian world, it was also the place of composition of the Arabic prototype of the *Turba philosophorum*: an alchemical work preserving elements of Empedoclean doctrine within the literary framework of an assembly of sages presided over by Pythagoras.[53] But that is not all. Dhū 'l-Nūn, known as 'head of the Sufis' because of the immense influence he exerted on the subsequent development of Sufism, is strongly connected in medieval sources with alchemy and Hermetism. These connections have proved a major source of embarrassment for those interested in maintaining the purely Islamic nature of Sufism and denying its links with previous, non-Arab traditions; but their historical nature can be—and since the start of this century has been—established.[54] It seems not to have been noticed, however, that there is one witness to Dhū 'l-Nūn's involvement with alchemical tradition which pre-dates the mid-tenth-century writers (Ibn Umail and al-Mas'ūdī) who are usually considered our earliest sources of information. That is the alchemist 'Uthmān Ibn Suwaid, from Dhū 'l-Nūn's home town of Akhmīm. Writing little more, and perhaps even less, than a generation after Dhū 'l-Nūn himself, either in the late ninth or the very early tenth century, his list of published works—all of them on the subject of alchemy, as their titles clearly show—includes a 'Book of Refutation of the Accusation Against Dhū 'l-Nūn al-Miṣrī' (*Kitāb ṣarf al-tawahhum 'an dhī-al-nūn al-miṣrī*). It was this same Ibn Suwaid who, as mentioned earlier, was almost certainly the author of the *Muṣḥaf al-jamā'a*: the Arabic

---

[52] *Kitāb al-maṣārī' wa-'l-muṭāraḥāt* 223 (*OMM* i. 502.13–503.6). Cf. Corbin, *OMM* i, pp. xli–xlii, *OMM* ii, Prolegomena 23 n. 49, and (1971–2), ii. 35–6 with n. 42; Nasr 62; Böwering 52; also H. Ziai in C. E. Butterworth (ed.), *The Political Aspects of Islamic Philosophy: Essays in Honor of Muhsin S. Mahdi* (Cambridge, Mass., 1992), 326–7, 334.

[53] Above, Ch. 5; Kingsley (1994c), § III. For Akhmīm (Greek Panopolis), Hermes Trismegistus, and Hermetism cf. Fowden 27, 173–4; Asín Palacios 165; D. Pingree, *The Thousands of Abū Ma'shar* (London, 1968), 14–15.

[54] Cf. esp. Nicholson's careful discussion, *JRAS* (1906), 309–20; also Massignon 201–7, M. Smith, *EI*[2] ii. 242, R. T. Wallis, *Neoplatonism* (London, 1972), 165, Needham, v/4. 397, 401, Böwering 53–4.

prototype of the *Turba philosophorum*.⁵⁵ The particular connection here between Dhū 'l-Nūn and Ibn Suwaid is highly indicative of the links between early Sufism and Arab alchemy, and in more ways than one. For example, one of the other works attributed to Ibn Suwaid—*Kitāb al-kibrīt al-aḥmar*, 'Book of the Red Sulphur'—points not only to the central importance of the notion of red sulphur in alchemy but also to the use of this very same book title in subsequent Sufi tradition.⁵⁶

As for 'the traveller from Tustar and his party' whom al-Suhrawardī names after Dhū 'l-Nūn, the historical nature of the teacher–disciple relationship ascribed in the medieval sources to Dhū 'l-Nūn and Sahl al-Tustarī must be considered established.⁵⁷ And, as Gerhard Böwering has shown, we must take very seriously the reports of 'a line of secret gnostic teaching' passing through Dhū 'l-Nūn, al-Tustarī, and continuing into various subsequent chains of Islamic Sufi tradition.⁵⁸ This same guiding theme of secrecy and esotericism is, as we have seen, one that also runs back to the historical Empedocles; and it is clearly futile to try to understand Empedocles himself—or the repercussions of his teaching on later individuals, either in the West or in the East—without

⁵⁵ Ibn al-Nadīm 359.4–5 = Dodge, ii. 865; Kingsley (1994c), § III.
⁵⁶ Ch. 5 with the refs. in nn. 24, 48. Any idea of Islamic Sufism being responsible for 'spiritualizing' such alchemical symbolism by interpreting it mystically and allegorically must be countered by the basic observation that early Sufis such as Dhū 'l-Nūn clearly stood in the same line of tradition as Zosimus from Panopolis (i.e. Akhmīm), who already testifies over 500 years earlier to the radical spiritualizing or introversion of alchemical procedure. Cf. esp. Zos. Pan. 120.12–122.2 with Fowden 122–3 (although Fowden's statement that according to this passage 'the procedures of conventional alchemy are strictly preparatory to the purification and perfection of the soul' is wide of the mark: Zosimus is stating that the procedures of conventional alchemy are to be reinterpreted as, themselves, symbolizing the purification and perfection of the soul); Green 90–1; and note also Needham's comments, v/4. 397, 401. On links between early Sufism and alchemy see further Massignon 154–5, 205; Kraus (1942), pp. xl, xliii; P. Nwyia, *Exégèse coranique et langage mystique* (Beirut, 1970), 157–8; Corbin (1971–2), i. 92.
⁵⁷ Böwering 43, 50–8, 66: they probably met at Mecca, but possibly also in Egypt. For al-Tustarī's own relation to alchemy cf. ibid. 54–5.
⁵⁸ Ibid. 53–4. It will be noted that Ibn Masarra in particular, apart from having certain affinities with the teachings of Dhū 'l-Nūn (Asín Palacios 40, 86–8 with n. 36 and *Picatrix* 297.21, 128, 174; Morris 15 n. 22, 21, Appendix, p. xv), was—as Asín Palacios suspected, and as has now been confirmed—especially indebted to Sahl al-Tustarī. See Asín Palacios 88 and 127 with D. Gril in M. Chodkiewicz (ed.), *Les Illuminations de la Mecque* (Paris, 1988), 427–8. For Ibn Masarra and the Arabic Empedocles see above, with n. 32.

giving this esoteric dimension of his doctrine the emphasis and study it deserves. The task of reassessing Empedocles' own teaching in this light, not to mention its subsequent channels of influence, is a vast and complex one. But no one said it would be easy.

APPENDIX I

Parmenides and Babylon

(See pp. 54-5)

*

PARALLELS between Parmenides' proem and the Gilgamesh epic have already been pointed out by Burkert (1969), 18-19, and the probabilities of a Near Eastern background to other features of his poetry have also been noted: cf. e.g. Pfeiffer 124-30, Kingsley (1990), 255-6 with n. 66. It is clearly relevant here that we know—from internal evidence as well as from the testimony of Herodotus—that the religious traditions and cult practices in Parmenides' home town of Velia were of specifically Anatolian origin: see Ch. 15 n. 28. For the dissemination of Babylonian mythology—and, to be more precise, of motifs relating to the Gilgamesh epic—into Anatolia see Ch. 15 n. 20.

What has not so far been adequately emphasized, however, is the fact that not just the motif of a descent to the great Goddess in the underworld (Burkert, loc. cit.; cf. Ch. 17 n. 6) but also Parmenides' detailed imagery of guarded doors, gates, bolts, bars, keys, and locks preventing or providing access to the underworld (B1.11-17) is a central feature of Babylonian cosmic mythology. What is more, the gateways in question are routinely described in Babylonian sources as lying on the paths of the sun and moon—just as Parmenides' guarded gates stand on 'the paths of night and day' (B1.11). Cf. esp. Heimpel 132-43; W. Horowitz, 'Mesopotamian Cosmic Geography', Ph.D. thesis (Birmingham, 1986), 385-7, 405-7 (for Parmenides' 'bonds' or πείρατα, B8.26, 10.7, and *passim*, cf. ibid. 384-5). But most important of all is Parmenides' description— immediately before his mention of the 'gates at the paths of night and day'—of the 'daughters of the Sun' (Ἡλιάδες κοῦραι) departing from and returning to the 'house of Night' (δώματα Νυκτός, B1.8-11). Certainly there is a link here with Hesiod's mention of a 'house of Night' (Νυκτὸς οἰκία, *Th.* 744; Pellikaan-Engel 26, 51-2), and yet there are equally certain links between Hesiod's creation myth and Babylonian cosmology: cf. P. Walcot, *Hesiod and the Near East* (Cardiff, 1966), West (1966), 2, 22-4, 28-30, 212-13, 337, 379-81, and above, Ch. 14 with n. 37. More to the point, however, is the fact that the expression 'House of Night', *bīt mūši*, is itself attested in

I. Parmenides and Babylon

Babylonian cult and religion; that it was cited specifically in the context of rituals aimed at balancing the lengths of day and night at the winter and summer solstices; and that according to Babylonian tradition the 'house' or 'temple' of night is approached and entered at certain specified times by none other than the 'daughters of Esagil'— that is, the daughters of the god of the Esagil temple, Marduk, here associated with the sun. Cf. BM 34035.1–8, erratically transcribed by Livingstone (1986), 256 (the names of the daughters are Şilluştab and Katunna). The surviving *copy* of the text was written in 137 BC; but it clearly preserves much older traditions, and the scribe states explicitly in the colophon (ibid. 259) that he has simply transcribed the text from an 'ancient long tablet'. For Esagil and Marduk cf. E. Unger, *Reallexikon der Assyriologie*, i. 353–9, W. Sommerfeld, ibid. vii. 366–8, and for Marduk and the sun, VAT 8917 rev. 5, ed. Livingstone (1986), 82–3, 90–1 = (1989), 101. The solar connotation of Esagil in this particular case is confirmed by BM 34035.8, 'Esagil is the temple of day' (*esagil bīt ūmu šū*; cf. also Unger, op. cit. 355b, 357b, on the 'Sun-gates' of Esagil).

For the continuing indebtedness to and acquaintance with Babylonian traditions—astronomical, astrological, and theological—in the Academy of Plato during the century after Parmenides, see Kingsley (1995c).

APPENDIX II

Nergal and Heracles

(See p. 275)

*

FOR Nergal = Akkadian Erra = Erragal see J. J. M. Roberts, *The Earliest Semitic Pantheon* (Baltimore, 1972), 22, 29, 44, 107 n. 355; for the Nergal–Heracles etymology, M. K. Schretter, *Alter Orient und Hellas* (Innsbruck, 1974), 170–1, Dalley 63–6; and for new evidence regarding the spelling and pronunciation of the name Nergal, J. Eidem in D. Charpin and F. Joannès (eds.), *Marchands, diplomates et empereurs: Études sur la civilisation mésopotamienne offertes à Paul Garelli* (Paris, 1991), 195, 205. Fundamental problems with the traditional Greek etymology of Heracles as 'Hera's glory' remain, in spite of the overly imaginative defence of it by P. Kretschmer, *Glotta*, 8 (1917), 121–9, and W. Pötscher, *Emerita*, 39 (1971), 169–84: cf. H. Usener, *Die Sintfluthsagen* (Bonn, 1899), 58–60, J. Zwicker, *RE* viii. 523–8, Burkert (1979), 179 n. 17. But what is generally forgotten by both attackers and defenders of the traditional etymology is the persistent Greek habit of adapting foreign names by modifying their vowel structure so as to give them a seeming—and as a rule transparently inadequate—etymology which is *secondary* to their original one: see Kingsley (1993*b*), 11–15, (1995*b*), § 2, and above, Ch. 22 with n. 9.

On the non-Greek character and origin of Heracles see E. Kalinka, *Klio*, 22 (1929), 258 with, for Heracles' bow, Botta and Flandin (as in Ch. 17 n. 87), locc. citt.; also E. Schwyzer, *Griechische Grammatik*, i (Munich, 1939), 61–2. As for W. G. Lambert's objections to the Nergal–Heracles etymology, *Levant*, 23 (1991), 185, they either miss the point or—in the case of his main argument, the supposed absence of Nergal from Syria—are contradicted by the Nergal coinage from Tarsus which portrays him complete with lion, bow, and club. See Burkert (1985), 432 n. 21, (1979), 82 and n. 17; Dalley 65, plus ibid. 62, 64, for the further evidence of Nergal in Syria; and regarding the equation of Heracles and Nergal in Syria see H. Seyrig, *Syria*, 24 (1944–5), 62–80. For Nergal as god of pestilence and plagues cf. ibid. 71–2; E. von Weiher, *Der babylonische Gott Nergal* (Neukirchen–Vluyn, 1971), 83–7; Burkert (1979), 82 (with, for Heracles, above, Ch. 17 nn. 85, 88); Dalley 62; also Jourdain-Annequin 169.

APPENDIX III
Empedocles and the Ismāʿīlīs
(See pp. 377, 379)

*

ONE example of distinctively Ismāʿīlī doctrine in the Arabic Empedocles is the cyclical doctrine of prophecy: al-Shahrastānī, ii. 263.3–13 Cureton = ii. 71.1–10 Kīlānī. For the Ismāʿīlī teaching see Corbin (1983), *passim*; Daftary 63, 66–7, 234, and esp. 87 on the notion of *dāʾirat al-nubuwwa*. Another is the notion of soul within soul within soul—with the lower soul serving as shell or body for the higher soul (al-Shahrastānī, ii. 262.1–5 Cureton = ii. 70.3–6 Kīlānī)—which recurs exactly in the 13th-century *Taṣawwurāt* by Naṣīr al-Dīn al-Ṭūsī. This parallel with al-Ṭūsī has been pointed out by Corbin (1983), 55 n. 98, although his statement that the term 'shell' (*qišr*) in al-Shahrastānī has been replaced by the term 'body' (*jasad*) in al-Ṭūsī is inaccurate: al-Sharastānī's Empedocles already uses the term 'body', *jasad*, as well (ii. 262.4 Cureton = ii. 70.5 Kīlānī; Altheim and Stiehl 5, Altmann and Stern 184).

The most striking instances of 'Empedocles' being used by Ismāʿīlīs are the cases of Nasafī—cf. Rudolph (1989), 137—and al-Sijistānī (Rowson 208, Daftary 239), both from the 10th century. For Ismāʿīlī use of the so-called Ammonius doxography see Ch. 24 n. 18; for use of the text by Abū Ḥātim al-Rāzī (early 10th century) in particular, Rudolph (1989), 13, 23–30, 219 n. 61, 220 n. 70, H. Daiber in *Proceedings of the 1st International Congress on Democritus* (Xanthi, 1984), ii. 260. On the Ismāʿīlī affiliations of al-Shahrastānī, our single most important witness for Empedocles in the Arab world, see esp. W. Madelung in A. Dietrich (ed.), *Akten des VII. Kongresses für Arabistik und Islamwissenschaft* (Göttingen, 1976), 250–9; also C. Baffioni, *Sulle tracce di sofia* (Naples, 1990), 366 and n. 759; Green 166. For Ismāʿīlī use of the *Theology of Aristotle* cf. Zimmermann 129, 196–208; and for the repeated mention of Empedocles by Judah Halevi (*Kuzari* 4.25, 5.14), see S. Pines, *Jerusalem Studies in Arabic and Islam*, 2 (1980), 160–243. Worth noting as well, in view of the parallels between his writings and 'pseudo-Empedocles' (Altmann and Stern 184), is the fact that Isaac Israeli was closely connected with the court of the first Fāṭimids in Egypt: see Ch. 24 with n. 19.

APPENDIX III
Empedocles and the Ismāʿīlīs
(See pp. 377, 379)

*

ONE example of distinctively Ismāʿīlī doctrine in the Arabic Empedocles is the cyclical doctrine of prophecy: al-Shahrastānī, ii. 263.3–13 Cureton = ii. 71.1–10 Kīlānī. For the Ismāʿīlī teaching see Corbin (1983), *passim*; Daftary 63, 66–7, 234, and esp. 87 on the notion of *dā'irat al-nubuwwa*. Another is the notion of soul within soul within soul—with the lower soul serving as shell or body for the higher soul (al-Shahrastānī, ii. 262.1–5 Cureton = ii. 70.3–6 Kīlānī)—which recurs exactly in the 13th-century *Taṣawwurāt* by Naṣīr al-Dīn al-Ṭūsī. This parallel with al-Ṭūsī has been pointed out by Corbin (1983), 55 n. 98, although his statement that the term 'shell' (*qišr*) in al-Shahrastānī has been replaced by the term 'body' (*jasad*) in al-Ṭūsī is inaccurate: al-Sharastānī's Empedocles already uses the term 'body', *jasad*, as well (ii. 262.4 Cureton = ii. 70.5 Kīlānī; Altheim and Stiehl 5, Altmann and Stern 184).

The most striking instances of 'Empedocles' being used by Ismāʿīlīs are the cases of Nasafī—cf. Rudolph (1989), 137—and al-Sijistānī (Rowson 208, Daftary 239), both from the 10th century. For Ismāʿīlī use of the so-called Ammonius doxography see Ch. 24 n. 18; for use of the text by Abū Ḥātim al-Rāzī (early 10th century) in particular, Rudolph (1989), 13, 23–30, 219 n. 61, 220 n. 70, H. Daiber in *Proceedings of the 1st International Congress on Democritus* (Xanthi, 1984), ii. 260. On the Ismāʿīlī affiliations of al-Shahrastānī, our single most important witness for Empedocles in the Arab world, see esp. W. Madelung in A. Dietrich (ed.), *Akten des VII. Kongresses für Arabistik und Islamwissenschaft* (Göttingen, 1976), 250–9; also C. Baffioni, *Sulle tracce di sofia* (Naples, 1990), 366 and n. 759; Green 166. For Ismāʿīlī use of the *Theology of Aristotle* cf. Zimmermann 129, 196–208; and for the repeated mention of Empedocles by Judah Halevi (*Kuzari* 4.25, 5.14), see S. Pines, *Jerusalem Studies in Arabic and Islam*, 2 (1980), 160–243. Worth noting as well, in view of the parallels between his writings and 'pseudo-Empedocles' (Altmann and Stern 184), is the fact that Isaac Israeli was closely connected with the court of the first Fāṭimids in Egypt: see Ch. 24 with n. 19.

ABBREVIATIONS

*

For modern works cited by the name of the author only, see the Bibliography.

Ael. *VH*	Aelian, *Varia historia*
Aesch.	Aeschylus. *Ag.* = *Agamemnon*; *Eum.* = *Eumenides*; *Pr.* = *Prometheus Bound*. Fragments and testimonia, ed. S. Radt (*TrGF* iii)
AJP	*American Journal of Philology*
Alex. *Met.*	Alexander of Aphrodisias, *In Aristotelis Metaphysica commentaria*, ed. M. Hayduck (*CAG* i)
ANRW	*Aufstieg und Niedergang der römischen Welt* (Berlin, 1972–)
Ap. *Mir.*	Apollonius, *Mirabilia*
Apul. *Met.*	Apuleius, *Metamorphoses*
Arist.	Aristotle. *Cael.* = *De caelo;* *GC* = *De generatione et corruptione*; *Met.* = *Metaphysics*; *Meteor.* = *Meteorology*; *Ph.* = *Physics*; *Rh.* = *Rhetoric*. Fragments, ed. V. Rose (Leipzig, 1886); O. Gigon's Berlin, 1987 edition has, as a rule, not been used because of its errors and omissions
Aristox.	Aristoxenus, ed. F. Wehrli (Basle, 1945)
Ascl. *Met.*	Asclepius, *In Aristotelis Metaphysicorum libros A–Z commentaria*, ed. M. Hayduck (*CAG* vi/2)
BM	British Museum
BR	M. Berthelot and C.-É. Ruelle, *Collection des anciens alchimistes grecs* (Paris, 1887–8)
CAG	*Commentaria in Aristotelem graeca* (Berlin, 1882–1909)
Calcid.	*Timaeus a Calcidio translatus commentarioque instructus*, ed. J. H. Waszink (Leiden, 1962)
CH	*Corpus Hermeticum*, ed. A. D. Nock and A.-J. Festugière (NF i–ii)
Cic.	Cicero. *Div.* = *De divinatione;* *Fam.* = *Epistulae ad familiares*; *ND* = *De natura deorum*; *Or.* = *De oratore*; *Tusc.* = *Tusculanae disputationes*
Claud. *Rapt.*	Claudian, *De raptu Proserpinae*
Clem.	Clement of Alexandria, ed. O. Stählin (Leipzig,

	1906–9; Berlin, 1970–85). *Pr.* = *Protrepticus*; *Str.* = *Stromateis*
CMAG	J. Bidez, F. Cumont, *et al.*, *Catalogue des manuscrits alchimiques grecs* (Brussels, 1924–32)
CQ	*Classical Quarterly*
CR	*Classical Review*
CRAI	*Comptes rendus de l'Académie des Inscriptions et Belles-Lettres*
Dam. *Phd.*	Damascius, *In Phaedonem*, ed. L. G. Westerink (Amsterdam, 1977)
Dem.	*The Homeric Hymn to Demeter*, ed. N. J. Richardson (Oxford, 1974; repr. 1979)
Diod.	Diodorus Siculus, *Bibliotheca historica*
DK	H. Diels, *Die Fragmente der Vorsokratiker*^{6}, ed. W. Kranz (Berlin, 1951–2)
D.L.	Diogenes Laertius, *Vitae philosophorum*
Dox.	H. Diels, *Doxographici Graeci* (Berlin, 1879)
EA	*Epigraphica Anatolica*
EH	*Fondation Hardt, Entretiens sur l'antiquité classique*
EI^2	*The Encyclopaedia of Islam*^{2} (Leiden, 1960–)
Emp.	Empedocles (DK 31)
Eur.	Euripides. *IT* = *Iphigenia Taurica*; *Or.* = *Orestes*; *Phoen.* = *Phoenissae*. Fragments, ed. A. Nauck (Leipzig, 1926^2)
Eus. *PE*	Eusebius, *Praeparatio evangelica*
Eust.	Eustathius. *Il.* = *Ad Iliadem*; *Od.* = *Ad Odysseam*
FGrH	F. Jacoby, *Die Fragmente der griechischen Historiker* (Berlin and Leiden, 1923–58)
FHSG	W. W. Fortenbaugh, P. M. Huby, R. W. Sharples, D. Gutas, *et al.*, *Theophrastus of Eresus: Sources for his Life, Writings, Thought and Influence* (Leiden, 1992)
Fic. *Op.*	Marsilio Ficino, *Opera* (Basle, 1576)
fr(s).	fragment(s)
GMPT	*The Greek Magical Papyri in Translation*, ed. H. D. Betz (Chicago, 1992^2)
GRBS	*Greek, Roman, and Byzantine Studies*
Hdt.	Herodotus
Heracl. *Alleg.*	Heraclitus, *Homeric Allegories*
Heraclid.	Heraclides of Pontus, ed. F. Wehrli (Basle and Stuttgart, 1969^2)
Heraclit.	Heraclitus of Ephesus (DK 22)
Hes.	Hesiod. *Op.* = *Works and Days*; *Th.* = *Theogony*

Abbreviations

Hipp. *Ref.*	Hippolytus, *Refutatio omnium haeresium*, ed. M. Marcovich (Berlin, 1986)
HSCP	*Harvard Studies in Classical Philology*
HTR	*Harvard Theological Review*
Iam.	Iamblichus. *Pr.* = *Protrepticus*; *VP* = *De vita Pythagorica*
Ibn al-Nadīm	Ibn al-Nadīm, *Kitāb al-fihrist*, ed. G. Flügel (Leipzig, 1871–2); trans. B. Dodge (New York, 1970)
Il.	Homer, *Iliad*
JAOS	*Journal of the American Oriental Society*
JHS	*Journal of Hellenic Studies*
JNES	*Journal of Near Eastern Studies*
JRAS	*Journal of the Royal Asiatic Society*
JWCI	*Journal of the Warburg and Courtauld Institutes*
KRS	G. S. Kirk, J. E. Raven, and M. Schofield, *The Presocratic Philosophers*² (Cambridge, 1983)
LIMC	*Lexicon iconographicum mythologiae classicae* (Zurich, 1981–)
LMS	S. K. Jayussi (ed.), *The Legacy of Muslim Spain* (Leiden, 1992)
LSJ	H. G. Liddell, R. Scott, and H. S. Jones, *A Greek–English Lexicon* (Oxford, 1925–40)
LSS	M. H. Jameson, D. R. Jordan, and R. D. Kotansky, *A lex sacra from Selinous* (Durham, NC, 1993)
Luc.	Lucian. *Alex.* = *Alexander*; *DM* = *Dialogues of the Dead*
Lucr.	Lucretius, *De rerum natura*
Lyd. *Mens.*	Lydus, *De mensibus*, ed. R. Wünsch (Leipzig, 1898)
Macr.	Macrobius. *Sat.* = *Saturnalia*; *Somn.* = *In Somnium Scipionis*
MagH	C. A. Faraone and D. Obbink (eds.), *Magika hiera* (New York, 1991)
MEFRA	*Mélanges d'archéologie et d'histoire de l'École Française de Rome*
MH	*Museum Helveticum*
MT	R. Merkelbach and M. Totti, *Abrasax* (Opladen, 1990–2)
NBHL	G. Awetik'ean, X. Siwrmēlean, and M. Awgerean, *Nor Baṙgirk' Haykazean Lezui* (Venice, 1836–7)

NF	A. D. Nock and A.-J. Festugière, *Hermès Trismégiste* (Paris, 1946–54)
NGG	*Nachrichten von der königlichen Gesellschaft der Wissenschaften zu Göttingen, Philologisch-historische Klasse*
Od.	Homer, *Odyssey*
OF	O. Kern, *Orphicorum fragmenta* (Berlin, 1922)
Olymp.	Olympiodorus. *Phd.* = *In Phaedonem*, ed. L. G. Westerink (Amsterdam, 1976)
OMG	*Orfismo in Magna Grecia* (Atti del quattordicesimo convegno di studi sulla Magna Grecia; Naples, 1975)
OMM	Shihāb al-Dīn Yaḥyā al-Suhrawardī, *Opera metaphysica et mystica*, ed. H. Corbin (i: Istanbul, 1945; ii: Tehran, 1952)
Ov. *Met.*	Ovid, *Metamorphoses*
Pap. Derv.	Derveni Papyrus, cited according to the column numbering announced at Princeton in 1993 by K. Tsantsanoglou; the edition published in *ZPE* 47 (1982) can be consulted by deducting 4 from these new column numbers
Parm.	Parmenides (DK 28)
Paus.	Pausanias, *Description of Greece*
PG	J.-P. Migne, *Patrologiae cursus completus*, series graeca (Paris, 1857–1936)
PGM (or PGM[2])	K. Preisendanz and A. Henrichs, *Papyri graecae magicae* (Stuttgart, 1973–4)
Philo	Philo of Alexandria. *Prov.* = *De providentia*, ed. J.-B. Aucher, *Philonis Iudaei sermones tres hactenus inediti* (Venice, 1822), 1–121; *QG* = *Quaestiones in Genesin*, partial edn. by J. Paramelle (Geneva, 1984)
Philostr. *VA*	Philostratus, *Vita Apollonii*
Pi.	Pindar. *Isthm.*, *Nem.*, *Ol.*, *Pyth.* = *Isthmian*, *Nemean*, *Olympian*, *Pythian Odes*. Fragments, ed. H. Maehler (Leipzig, 1989)
Picatrix	*'Picatrix': Das Ziel des Weisen von Pseudo-Maǧrīṭī*, trans. H. Ritter and M. Plessner (London, 1962)
Pl.	Plato. *Crat.* = *Cratylus*; *Crit.* = *Critias*; *Gorg.* = *Gorgias*; *Phd.* = *Phaedo*; *Phdr.* = *Phaedrus*; *Phlb.* = *Philebus*; *Rep.* = *Republic*; *Tht.* = *Theaetetus*; *Tim.* = *Timaeus*
Plin. *HN*	Pliny the Elder, *Naturalis historia*

Plot.	Plotinus, *Enneads*, ed. P. Henry and H.-R. Schwyzer (Paris and Brussels, 1951–73)
Plut.	Plutarch. *De fac.* = *De facie in orbe lunae*; *De gen.* = *De genio Socratis*; *De Isid.* = *De Iside et Osiride*; *De sera* = *De sera numinis vindicta*; *Frig.* = *De primo frigido*; *Marc.* = *Marcellus*; *Qu. conv.* = *Quaestiones convivales*; *Qu. Plat.* = *Quaestiones Platonicae*. Ps.-Plut. *Placita* = ps.-Plutarch, *Placita philosophorum*, ed. J. Mau (Leipzig, 1971; the final figure in refs. is the paragraph number)
PMG	D. L. Page, *Poetae melici graeci* (Oxford, 1962)
Porph.	Porphyry. *VP* = *Vita Pythagorae*
PP	*La Parola del Passato*
Procl.	Proclus. *Eucl.* = *In Euclidem commentarius*, ed. G. Friedlein (Leipzig, 1873); *Rep.* = *In Platonis Rempublicam commentarii*, ed. W. Kroll (Leipzig, 1899–1901); *Th. Pl.* = *In Platonis theologiam*, ed. H. D. Saffrey and L. G. Westerink (Paris, 1968–); *Tim.* = *In Platonis Timaeum commentaria*, ed. E. Diehl (Leipzig, 1903–6)
RE	G. Wissowa, W. Kroll, *et al.*, *Paulys Realencyclopädie der classischen Altertumswissenschaft* (Stuttgart and Munich, 1894–1978). Suppl. = Supplementband
REA	*Revue des études anciennes*
REG	*Revue des études grecques*
RFIC	*Rivista di filologia e di istruzione classica*
RhM	*Rheinisches Museum*
RHR	*Revue de l'histoire des religions*
Ṣā'id al-Andalusī	Ṣā'id al-Andalusī, *Kitāb ṭabaqāt al-umam*, ed. L. Cheikho (Beirut, 1912); ed. M. B. al-'Ulūm (Najuf, 1967); trans. R. Blachère (Paris, 1935)
Schol.	scholiast; scholion
Sen.	Seneca. *Herc. Oet.* = *Hercules Oetaeus*; *QN* = *Quaestiones naturales*
Sext. Math.	Sextus Empiricus, *Adversus mathematicos*
SH	H. Lloyd-Jones and P. Parsons, *Supplementum hellenisticum* (Berlin, 1983)
al-Shahrastānī	al-Shahrastānī, *Al-milal wa-'l-niḥal*, ed. W. Cureton (London, 1846); ed. M. S. Kīlānī (Cairo, 1961)
SIG[3]	W. Dittenberger, *Sylloge inscriptionum graecarum*[3] (Leipzig, 1915–24)
Sil.	Silius Italicus, *Punica*

Simpl.	Simplicius. *Cael.* = *In Aristotelis De caelo commentaria*, ed. J. L. Heiberg (*CAG* vii); *Ph.* = *In Aristotelis Physica commentaria*, ed. H. Diels (*CAG* ix–x)
SO	*Symbolae Osloenses*
Soph.	Sophocles. *OC* = *Oedipus Coloneus*; *Tr.* = *Trachiniae*. Fragments, ed. S. Radt (*TrGF* iv)
Speus.	Speusippus, ed. L. Tarán (Leiden, 1981)
Stob.	Ioannes Stobaeus, ed. C. Wachsmuth and O. Hense (Berlin, 1882–95). Cited by volume, page, and line
al-Suhrawardī	Shihāb al-Dīn Yaḥyā al-Suhrawardī, *Kitāb ḥikmat al-išrāq*, ed. H. Corbin (*OMM* ii. 1–260)
Suppl. Mag.	R. W. Daniel and F. Maltomini, *Supplementum magicum* (Opladen, 1990–1)
SVF	H. von Arnim, *Stoicorum veterum fragmenta* (Leipzig, 1903–5)
TAPA	*Transactions of the American Philological Association*
Tert.	Tertullian. *An.* = *De anima*, ed. J. H. Waszink (Amsterdam, 1947)
TGL	K. Tsantsanoglou and G. M. Parássoglou, 'Two Gold Lamellae from Thessaly', *Hellenika*, 38 (1987), 3–16
Theocr.	Theocritus, *Idylls*
Theol. Arist.	*Theology of Aristotle*, ed. 'A. Badawī, *Aflūṭin 'ind al-'arab* (Cairo, 1955), 8–164; trans. G. L. Lewis in *Plotini opera*, ed. P. Henry and H.-R. Schwyzer, ii (Paris and Brussels, 1959)
Theophr.	Theophrastus. *CP* = *De causis plantarum*; *Met.* = *Metaphysics*
TrGF	B. Snell, R. Kannicht, and S. Radt, *Tragicorum graecorum fragmenta* (Göttingen, 1971–)
Turba	*Turba philosophorum*, ed. J. Ruska (Berlin, 1931); partial edn. in Plessner 38–83
Tzetz.	Tzetzes. *Alleg. Il.* = *Allegoriae Iliadis*, ed. J. F. Boissonade (Paris, 1851)
VAT	Vorderasiatisches Museum, Berlin
Virg.	Virgil. *Aen.* = *Aeneid*; *G.* = *Georgics*
Vitr.	Vitruvius, *De architectura*
ZDMG	*Zeitschrift der Deutschen Morgenländischen Gesellschaft*
Zos. Pan.	*Zosimo di Panopoli: Visioni e risvegli*, ed. A. Tonelli (Milan, 1988)
ZPE	*Zeitschrift für Papyrologie und Epigraphik*

BIBLIOGRAPHY

*

Papers are usually cited in their latest published form. Reprints of books are only mentioned if they contain corrections or supplementary material.

ADAM, J.: *The Republic of Plato* (Cambridge, 1902; 1963).
ADDAS, C.: *Quest for the Red Sulphur: The Life of Ibn 'Arabī* (Cambridge, 1993).
ALTHEIM, F., and STIEHL, R.: 'New Fragments of Greek Philosophers', *East and West*, 12 (1961), 3–18.
ALTMANN, A.: *Studies in Religious Philosophy and Mysticism* (London, 1969).
—— and STERN, S. M.: *Isaac Israeli* (Oxford, 1958).
ANDÒ, V.: 'Nestis o l'elemento acqua in Empedocle', *Kokalos*, 28–9 (1982–3), 31–51.
ANNAS, J.: 'Plato's Myths of Judgement', *Phronesis*, 27 (1982), 119–43.
ASÍN PALACIOS, M.: *The Mystical Philosophy of Ibn Masarra and his Followers* (Leiden, 1978); orig. pub. as *Abenmasarra y su escuela* (Madrid, 1914).
AUSTIN, R. G.: *P. Vergili Maronis Aeneidos Liber Sextus* (Oxford, 1977).
BAYET, J.: *Les Origines de l'Hercule romain* (Paris, 1926).
BECK, R.: 'Thus Spake Not Zarathuštra: Zoroastrian Pseudepigrapha of the Greco-Roman World', in M. Boyce and F. Grenet, *A History of Zoroastrianism*, iii (Leiden, 1991), 491–565.
BELOCH, J.: *Campanien* (Berlin, 1879).
BEN, N. VAN DER: *The Proem of Empedocles' Peri Physios* (Amsterdam, 1975).
BERNAND, A.: *Sorciers grecs* (Paris, 1991).
BETZ, H. D.: 'Fragments from a Catabasis Ritual in a Greek Magical Papyrus', *History of Religions*, 19 (1980), 287–95.
BIDEZ, J.: *La Biographie d'Empédocle* (Ghent, 1894).
—— and CUMONT, F.: *Les Mages hellénisés* (Paris, 1938).
BIELER, L.: Θεῖος ἀνήρ (Vienna, 1935–6).
BIGNONE, E.: *Empedocle* (Turin, 1916).
BLUCK, R. S.: *Plato's Meno* (Cambridge, 1964).
BODRERO, E.: *Il principio fondamentale del sistema di Empedocle* (Rome, 1904).

BOECKH, A.: *Philolaos des Pythagoreers Lehren nebst den Bruchstücken seines Werkes* (Berlin, 1819).
BOLLACK, J.: *Empédocle* (Paris, 1965–9).
BOLTON, J. D. P.: *Aristeas of Proconnesus* (Oxford, 1962).
BONNER, C.: *Studies in Magical Amulets* (Ann Arbor, Mich., 1950).
BOSTOCK, D.: *Plato's Phaedo* (Oxford, 1986).
BOUSSET, W.: *Die Himmelsreise der Seele* (Darmstadt, 1971) (1st pub. 1901).
BOWEN, A. C.: 'The Foundations of Early Pythagorean Harmonic Science: Archytas Fragment 1', *Ancient Philosophy*, 2 (1982), 79–104.
BÖWERING, G.: *The Mystical Vision of Existence in Classical Islam* (Berlin, 1980).
BOYANCÉ, P.: *Le Culte des Muses chez les philosophes grecs*² (Paris, 1972).
BRASHEAR, W. M.: *A Mithraic Catechism from Egypt* (Vienna, 1992).
BRISSON, L.: *Platon: Les Mots et les mythes* (Paris, 1982).
—— (1987): 'Proclus et l'Orphisme', in J. Pépin and H. D. Saffrey (eds.), *Proclus: Lecteur et interprète des anciens* (Paris), 43–104.
BROWN, P.: *Society and the Holy in Late Antiquity* (London, 1982).
BUFFIÈRE, F.: *Les Mythes d'Homère et la pensée grecque* (Paris, 1956).
BURGER, R.: *The 'Phaedo': A Platonic Labyrinth* (New Haven, Conn., 1984).
BURKERT, W.: *Lore and Science in Ancient Pythagoreanism* (Cambridge, Mass., 1972).
—— (1961): 'Hellenistische Pseudopythagorica', *Philologus*, 105: 16–43, 226–46.
—— (1962): 'ΓΟΗΣ. Zum griechischen "Schamanismus"', *RhM* 105: 36–55.
—— (1968): 'Orpheus und die Vorsokratiker', *Antike und Abendland*, 14: 93–114.
—— (1969): 'Das Proömium des Parmenides und die Katabasis des Pythagoras', *Phronesis*, 14: 1–30.
—— (1977): 'Orphism and Bacchic Mysteries', in W. Wuellner (ed.), *The Center for Hermeneutical Studies, Protocol of the 28th Colloquy* (Berkeley, Calif.), 1–8.
—— (1979): *Structure and History in Greek Mythology and Ritual* (Berkeley, Calif.).
—— (1982): 'Craft Versus Sect: The Problem of Orphics and Pythagoreans', in B. F. Meyer and E. P. Sanders (eds.), *Jewish and Christian Self-Definition*, iii (London), 1–22, 183–9.
—— (1983): 'Itinerant Diviners and Magicians: A Neglected Element in Cultural Contacts', in R. Hägg (ed.), *The Greek Renaissance of the Eighth Century B.C.: Tradition and Innovation* (Stockholm), 115–19.

—— (1985): *Greek Religion* (Oxford).
—— (1987): *Ancient Mystery Cults* (Cambridge, Mass.).
—— (1989): contribution to A. C. Cassio and D. Musti (eds.), *Tra Sicilia e Magna Grecia* (Atti del Convegno Napoli, 19–20 Mar. 1987; Pisa), 209–11.
—— (1992): *The Orientalizing Revolution: Near Eastern Influence on Greek Culture in the Early Archaic Age* (Cambridge, Mass.).
BURNET, J.: *Plato's Phaedo* (Oxford, 1911).
—— (1892): *Early Greek Philosophy*[1] (London).
—— (1930): *Early Greek Philosophy*[4] (London).
CALDER, W. M. III: *The Inscription from Temple G at Selinus* (Durham, NC, 1963).
CAMERON, A.: *The Pythagorean Background of the Theory of Recollection* (Menasha, Wis., 1938).
CASSIO, A. C.: 'Nicomachus of Gerasa and the Dialect of Archytas, Fr. 1', *CQ* 38 (1988), 135–9.
CAVEN, B.: *Dionysius I* (New Haven, Conn., 1990).
CHERNISS, H. (1935): *Aristotle's Criticism of Presocratic Philosophy* (Baltimore).
—— (1944): *Aristotle's Criticism of Plato and the Academy*, i (Baltimore).
CIACERI, E.: *Culti e miti nella storia dell'antica Sicilia* (Catania, 1911).
CLARK, R. J.: *Catabasis: Vergil and the Wisdom-Tradition* (Amsterdam, 1979).
COLE, S. G.: 'New Evidence for the Mysteries of Dionysos', *GRBS* 21 (1980), 223–38.
COOK, A. B.: *Zeus* (Cambridge, 1914–40).
CORBIN, H.: *Spiritual Body and Celestial Earth* (Princeton, NJ. 1977).
—— (1964): *Histoire de la philosophie islamique*, i (Paris).
—— (1971–2): *En Islam iranien* (Paris).
—— (1978): *The Man of Light in Iranian Sufism* (Boulder, Colo.).
—— (1983): *Cyclical Time and Ismaili Gnosis* (London).
—— (1986*a*): *Alchimie comme art hiératique* (Paris).
—— (1986*b*): *Sohravardi: Le Livre de la sagesse orientale* (Lagrasse).
CROON, J. H.: *The Herdsman of the Dead* (Utrecht, 1952).
CULIANU, I. P.: *Psychanodia*, i (Leiden, 1983).
CUMONT, F.: *Lux perpetua* (Paris, 1949).
—— (1942): *Recherches sur le symbolisme funéraire des Romains* (Paris).
DAFTARY, F.: *The Ismāʿīlīs: Their History and Doctrines* (Cambridge, 1990).
DAIBER, H.: *Aetius Arabus* (Wiesbaden, 1980).
DALLEY, S.: 'Near Eastern Patron Deities of Mining and Smelting in the Late Bronze and Early Iron Ages', *Report of the Department of Antiquities, Cyprus* (1987), 61–6.

DANFORTH, L. M.: *Firewalking and Religious Healing* (Princeton, 1989).
DELATTE, A.: 'Faba Pythagorae cognata', *Serta Leodiensia* (Liège, 1930), 33–57.
—— (1915): *Études sur la littérature pythagoricienne* (Paris).
—— (1927): *Anecdota Atheniensia*, i (Liège).
—— (1934): *Les Conceptions de l'enthousiasme chez les philosophes présocratiques* (Paris).
DELCOURT, M.: *Héphaistos* (Paris, 1957).
DELLA CASA, A.: *Nigidio Figulo* (Rome, 1962).
DETIENNE, M.: 'Héraclès, héros pythagoricien', *RHR* 158 (1960), 19–53.
—— (1962): *Homère, Hésiode et Pythagore* (Brussels).
—— (1963): *La Notion de Daimôn dans le Pythagorisme ancien* (Paris).
—— (1973): *Les Maîtres de vérité dans la Grèce archaïque*[2] (Paris).
DEUBNER, L.: *De incubatione* (Leipzig, 1900).
DEWAILLY, M.: *Les Statuettes aux parures du sanctuaire de la Malophoros à Sélinonte* (Naples, 1992).
DICKS, D. R.: *Early Greek Astronomy to Aristotle* (London, 1970).
DIELS, H.: *Kleine Schriften zur Geschichte der antiken Philosophie* (Darmstadt, 1969).
—— (1901): *Poetarum philosophorum fragmenta* (Berlin).
—— (1920): *Antike Technik*[2] (Leipzig).
DIETERICH, A.: *Nekyia* (Leipzig, 1893).
—— (1891): *Abraxas* (Leipzig).
—— (1911): *Kleine Schriften* (Leipzig).
—— (1923): *Eine Mithrasliturgie*[3] (Leipzig).
DILLON, J.: *The Middle Platonists* (London, 1977).
DILTHEY, K.: 'Ueber die von E. Miller herausgegebenen griechischen Hymnen', *RhM* 27 (1872), 375–419.
DODDS, E. R.: *The Greeks and the Irrational* (Berkeley, Calif., 1951).
—— (1959): *Plato: Gorgias* (Oxford).
—— (1960): *Euripides: Bacchae*[2] (Oxford).
—— (1965): *Pagan and Christian in an Age of Anxiety* (Cambridge).
DORTER, K.: *Plato's 'Phaedo': An Interpretation* (Toronto, 1982).
DUBOIS, L.: *Inscriptions grecques dialectales de Sicile* (Rome, 1989).
DUHEM, P.: *Le Système du monde: Histoire des doctrines cosmologiques de Platon à Copernic* (Paris, 1913–59).
EDSMAN, C.-M.: *Ignis divinus* (Lund, 1949).
EISLER, R.: *Weltenmantel und Himmelszelt* (Munich, 1910).
EITREM, S.: *Opferritus und Voropfer der Griechen und Römer* (Christiania, 1915).
ELIADE, M.: *Shamanism: Archaic Techniques of Ecstasy* (Princeton, NJ, 1964).

ESSEN, C. C. VAN: *Did Orphic Influence on Etruscan Tomb Paintings Exist?* (Amsterdam, 1927).
FABBRI, M., and TROTTA, A.: *Una scuola-collegio di età augustea* (Rome, 1989).
FARAONE, C. A.: *Talismans and Trojan Horses* (New York, 1992).
FARNELL, L. R.: *The Cults of the Greek States* (Oxford, 1896–1909).
FAURE, P.: *Alexandre* (Paris, 1985).
FERRARI, G. R. F.: *Listening to the Cicadas* (Cambridge, 1987).
FESTUGIÈRE, A.-J. (1949): *La Révélation d'Hermès Trismégiste*, ii (Paris).
—— (1950): *La Révélation d'Hermès Trismégiste*, i² (Paris).
—— (1967): *Hermétisme et mystique païenne* (Paris).
—— (1972): *Études de religion grecque et hellénistique* (Paris).
—— (1981): *L'Idéal religieux des grecs et l'Évangile*² (Paris).
FEYERABEND, B.: 'Zur Wegmetaphorik beim Goldblättchen aus Hipponion und dem Proömium des Parmenides', *RhM* 127 (1984), 1–22.
FIEDLER, W.: *Antiker Wetterzauber* (Stuttgart, 1931).
FINAMORE, J. F.: *Iamblichus and the Theory of the Vehicle of the Soul* (Chico, Calif., 1985).
FLASHAR, H.: 'Empedokles, Frgm. B111 und seine Stellung im Lehrgedicht Über die Natur', *Actes de la XIIe conférence internationale d'études classiques, Eirene* (Bucharest and Amsterdam, 1975), 547–51.
FOWDEN, G.: *The Egyptian Hermes* (Cambridge, 1986).
FRANK, E.: *Plato und die sogenannten Pythagoreer* (Halle, 1923).
FRASER, P. M.: *Ptolemaic Alexandria* (Oxford, 1972).
FRAZER, J. G. (1911): *Taboo and the Perils of the Soul* (= *The Golden Bough*³, Part II; London).
—— (1919): *Adonis Attis Osiris* (= *The Golden Bough*³, Part IV), i (London).
FREEMAN, E.: *History of Sicily* (Oxford, 1891–4).
FRIEDLÄNDER, P.: *Plato* (New York and London, 1958–69).
FRITZ, K. VON (1940): *Pythagorean Politics in Southern Italy* (New York).
—— (1973): 'Philolaos', *RE* Suppl. xiii. 453–84.
FRUTIGER, P.: *Les Mythes de Platon* (Paris, 1930).
FURLEY, D. J.: *Cosmic Problems* (Cambridge, 1989).
GALLAVOTTI, C.: *Empedocle: Poema fisico e lustrale* (Milan, 1975).
GANSCHINIETZ, R.: 'Katabasis', *RE* x (1919), 2359–449.
—— (1913): *Hippolytos' Capitel gegen die Magier* (Leipzig).
GEMELLI MARCIANO, L.: *Le metamorfosi della tradizione: Mutamenti di significato e neologismi nel Peri Physeos di Empedocle* (Bari, 1990).
GENEQUAND, C. (1987–8): 'Platonism and Hermetism in al-Kindī's

Fī al-nafs', *Zeitschrift für Geschichte der arabisch-islamischen Wissenschaften*, 4: 1–18.

—— (1991): Review of Rudolph (1989), *Bibliotheca Orientalis*, 48: 944–7.

GERNET, L.: *Anthropologie de la Grèce antique* (Paris, 1968).

GILBERT, O.: *Die meteorologischen Theorien des griechischen Altertums* (Leipzig, 1907).

GOTTSCHALK, H. B.: *Heraclides of Pontus* (Oxford, 1980).

GOW, A. S. F.: *Theocritus* (Cambridge, 1950).

GRAF, F.: 'Textes orphiques et rituel bacchique: A propos des lamelles de Pélinna', in P. Borgeaud (ed.), *Orphisme et Orphée* (Geneva, 1991), 87–102.

—— (1974): *Eleusis und die orphische Dichtung Athens in vorhellenistischer Zeit* (Berlin).

GREEN, T. M.: *The City of the Moon God: Religious Traditions of Harran* (Leiden, 1992).

GRESE, W. C.: *Corpus Hermeticum XIII and Early Christian Literature* (Leiden, 1979).

GRIFFITH, R. D.: 'Pelops and Sicily: The Myth of Pindar', *JHS* 109 (1989), 171–3.

GRIFFITHS, J. G.: *Apuleius of Madauros: The Isis-Book* (Leiden, 1975).

GRONAU, K.: *Poseidonios* (Leipzig, 1914).

GRONINGEN, B. A. VAN: *La Composition littéraire archaïque grecque*² (Amsterdam, 1960).

—— (1956): 'Le Fragment 111 d'Empédocle', *Classica et Mediaevalia*, 17: 47–61.

GROTTANELLI, C.: 'Healers and Saviours of the Eastern Mediterranean in Pre-Classical Times', in U. Bianchi and M. J. Vermaseren (eds.), *La soteriologia dei culti orientali nell'Impero Romano* (Leiden, 1982), 649–70.

GRUPPE, O.: 'Herakles', *RE* Suppl. iii (1918), 910–1121.

GUTHRIE, W. K. C.: *A History of Greek Philosophy* (Cambridge, 1962–81).

—— (1952): *Orpheus and Greek Religion*² (London).

HACKFORTH, R.: *Plato's Phaedo* (Cambridge, 1955).

HALLEUX, R.: *Le Problème des métaux dans la science antique* (Paris, 1974).

—— (1969): 'Lapis-lazuli, azurite ou pâte de verre?, à propos de *kuwano* et *kuwanowoko* dans les tablettes mycéniennes', *Studi micenei ed egeo-anatolici*, 9: 47–66.

—— (1981): *Les Alchimistes grecs*, i (Paris).

HARDIE, C.: 'The Crater of Avernus as a Cult-Site', in Austin 279–86.

HARRISON, J. E.: *Prolegomena to the Study of Greek Religion*³ (Cambridge, 1922).
HARVA, U.: *Die religiösen Vorstellungen der altaischen Völker* (Helsinki, 1938).
HARWARD, J.: *The Platonic Epistles* (Cambridge, 1932).
HAUSSLEITER, J.: *Der Vegetarismus in der Antike* (Berlin, 1935).
HEATH, T. L.: *Aristarchus of Samos* (Oxford, 1913).
HEIM, R.: *Incantamenta magica graeca latina* (Leipzig, 1892).
HEIMPEL, W.: 'The Sun at Night and the Doors of Heaven in Babylonian Texts', *Journal of Cuneiform Studies*, 38 (1986), 127–51.
HENRICHS, A. (1982): 'Changing Dionysiac Identities', in B. F. Meyer and E. P. Sanders (eds.), *Jewish and Christian Self-Definition*, iii (London), 137–60, 213–36.
HOPFNER, T.: *Griechisch-ägyptischer Offenbarungszauber*² (Amsterdam, 1974–90).
HUFFMAN, C. A.: *Philolaus of Croton: Pythagorean and Presocratic* (Cambridge, 1993).
HUIZINGA, J.: *Homo ludens* (London, 1949).
IMBRAGUGLIA, G., *et al.*: *Index Empedocleus* (Genoa, 1991).
INWOOD, B.: *The Poem of Empedocles* (Toronto, 1992).
JAEGER, W.: *Aristotle*² (Oxford, 1948).
JANKO, R.: 'Forgetfulness in the Golden Tablets of Memory', *CQ* 34 (1984), 89–100.
JORDAN, D. R.: 'A Love Charm with Verses', *ZPE* 72 (1988), 245–59.
JOURDAIN-ANNEQUIN, C.: *Héraclès aux portes du soir* (Paris, 1989).
JUNG, C. G. (1968a): *Aion*² (Collected Works, ix/2; London).
—— (1968b): *Alchemical Studies* (CW xiii; London).
—— (1970): *Mysterium coniunctionis*² (CW xiv; London).
KAHN, C. H.: *Anaximander and the Origins of Greek Cosmology* (New York, 1960).
—— (1971): 'Religion and Natural Philosophy in Empedocles' Doctrine of the Soul', in J. Anton and G. Kustas (eds.), *Essays in Ancient Greek Philosophy* (New York), 3–38 (1st pub. 1960).
KARSTEN, S.: *Empedoclis Agrigentini carminum reliquiae* (Amsterdam, 1838).
KAUFMANN, D.: *Studien über Salomon ibn Gabirol* (Budapest, 1899).
KEDAR, B.: 'Netherworld', *Encyclopaedia Judaica*, xii (1971), 997–8.
KERÉNYI, K.: *Dionysos* (London, 1976).
KERSCHENSTEINER, J.: *Platon und der Orient* (Stuttgart, 1945).
—— (1962): *Kosmos: Quellenkritische Untersuchungen zu den Vorsokratikern* (Munich).
KINGSLEY, P. (1990): 'The Greek Origin of the Sixth-Century

Dating of Zoroaster', *Bulletin of the School of Oriental and African Studies*, 53: 245–65.
—— (1992): 'Ezekiel by the Grand Canal: Between Jewish and Babylonian Tradition', *JRAS*[3] 2: 339–46.
—— (1993a): 'Empedocles in Armenian', *Revue des études arméniennes*, 24: 47–57.
—— (1993b): 'Poimandres: The Etymology of the Name and the Origins of the Hermetica', *JWCI* 56: 1–24.
—— (1994a): 'Empedocles and his Interpreters: The Four-Element Doxography', *Phronesis*, 39: 235–54.
—— (1994b): 'Empedocles' Sun', *CQ* 44: 316–24.
—— (1994c): 'From Pythagoras to the *Turba philosophorum*: Egypt and Pythagorean Tradition', *JWCI* 57: 1–13.
—— (1994d): 'Greeks, Shamans and Magi', *Studia Iranica*, 23: 187–98.
—— (1994e): 'The Christian Aristotle: Theological Interpretation and Interpolation in Medieval Versions of *On the Heavens*', *Le Muséon*, 107: 195–205.
—— (1994f): Review of Huffman, *CR* 44: 294–6.
—— (1995a): 'Notes on Air: Some Questions of Meaning in Empedocles and Anaxagoras', *CQ* 45.
—— (1995b): 'Artillery and Prophecy: Sicily in the Reign of Dionysius I', *Prometheus*, 21.
—— (1995c): 'Meetings with Magi: Iranian Themes among the Greeks, from Xanthus of Lydia to Plato's Academy', *JRAS*[3] 5.
—— (1995d): 'Empedocles' Two Poems', *Hermes*, 122.
KIRK, G. S.: *Heraclitus: The Cosmic Fragments* (Cambridge, 1954).
KOHLSCHITTER, S. A.: 'The Interpretation of Empedocles in the Tradition of Middle- and Neoplatonism', Ph.D. thesis (Dublin, 1991).
KRAFFT, F.: *Dynamische und statische Betrachtungsweise in der antiken Mechanik* (Wiesbaden, 1970).
KRANZ, W.: *Empedokles: Antike Gestalt und romantische Neuschöpfung* (Zurich, 1949).
KRAUS, P. (1930): 'Dschābir ibn Ḥajjān und die Isma'īlijja', *Dritter Jahresbericht des Forschungs-Instituts für Geschichte der Naturwissenschaften in Berlin* (Berlin), 23–42.
—— (1935): *Jābir ibn Ḥayyān: Essai sur l'histoire des idées scientifiques dans l'Islam*, i: *Textes choisis* (Paris).
—— (1942): *Jābir ibn Ḥayyān: Contribution à l'histoire des idées scientifiques dans l'Islam*, i (Cairo).
—— (1943): *Jābir ibn Ḥayyān: Contribution à l'histoire des idées scientifiques dans l'Islam*, ii (Cairo).

Bibliography

KROLL, W.: 'Bolos und Demokritos', *Hermes*, 69 (1934), 228–32.
KUHRT, A.: 'Survey of Written Sources Available for the History of Babylonia Under the Later Achaemenids', in H. W. A. M. Sancisi-Weerdenburg (ed.), *Achaemenid History*, i (Leiden, 1987), 147–57.
KUSTER, B.: *De tribus carminibus papyri Parisinae magicae* (Königsberg, 1911).
LABARBE, J.: *L'Homère de Platon* (Liège, 1949).
LAÍN ENTRALGO, P.: *The Therapy of the Word in Classical Antiquity* (New Haven, Conn., 1970).
LANE, E. W.: *An Arabic–English Lexicon* (London, 1863–93).
LEEMANS, E.-A.: *Studie over den wijsgeer Numenius van Apamea* (Brussels, 1937).
LEFKOWITZ, M.: *Lives of the Greek Poets* (London, 1981).
LEHRS, K.: *De Aristarchi studiis Homericis*[3] (Leipzig, 1882).
LETROUIT, J. (1995): 'Chronologie des alchimistes grecs', in S. Matton (ed.), *Alchimie: Art, histoire et mythes. Actes du premier Colloque international de la Société d'Étude de l'Histoire de l'Alchimie* (Paris).
LEWY, H.: *Chaldaean Oracles and Theurgy*[2] (Paris, 1978).
LINDSAY, J.: *The Origins of Alchemy in Graeco-Roman Egypt* (London, 1970).
LINFORTH, I. M.: *The Arts of Orpheus* (Berkeley, 1941).
LIUZZI, D.: *Nigidio Figulo* (Lecce, 1983).
LIVINGSTONE, A. (1986): *Mystical and Mythological Explanatory Works of Assyrian and Babylonian Scholars* (Oxford).
—— (1989): *Court Poetry and Literary Miscellanea* (State Archives of Assyria, 3; Helsinki).
LLOYD, G. E. R.: 'Plato and Archytas in the Seventh Letter', *Phronesis*, 35 (1990), 159–74.
—— (1979): *Magic, Reason and Experience* (Cambridge).
—— (1983): *Science, Folklore and Ideology* (Cambridge).
LLOYD-JONES, H.: 'Pindar and the After-Life', *EH* xxxi (1984), 245–83.
LOBECK, C. A.: *Aglaophamus* (Königsberg, 1829).
LOICQ-BERGER, M.-P.: *Syracuse* (Brussels, 1967).
LONG, H. S.: *A Study of the Doctrine of Metempsychosis in Greece from Pythagoras to Plato* (Princeton, NJ, 1948).
MAASS, E.: *Orpheus* (Munich, 1895).
MCKAY, A. G.: *Ancient Campania*, i (Hamilton, Ont., 1972).
MALTEN, L.: 'Hephaistos', *RE* viii (1912), 311–66.
MANSFELD, J.: *Studies in Later Greek Philosophy and Gnosticism* (London, 1989); cited by chapter and page.

MANSFELD, J. (1990): *Studies in the Historiography of Greek Philosophy* (Assen).
—— (1992): *Heresiography in Context* (Leiden).
MARCOVICH, M.: *Heraclitus*, ed. maior (Mérida, 1967).
MARSDEN, E. W.: *Greek and Roman Artillery* (Oxford, 1969–71).
MARTINEZ, D. G.: *A Greek Love Charm from Egypt* (*P. Mich. 757*) (Atlanta, 1991).
MASSIGNON, L.: *Essai sur les origines du lexique technique de la mystique musulmane*[2] (Paris, 1954).
MÉAUTIS, G.: *Recherches sur le Pythagorisme* (Neuchâtel, 1922).
MEULI, K.: *Gesammelte Schriften* (Basle, 1975).
MILLERD, C. E.: *On the Interpretation of Empedocles* (Chicago, 1908).
MOMIGLIANO, A.: *The Development of Greek Biography*[2] (Cambridge, Mass., 1993).
MORRIS, J. W.: 'Ibn Masarra: A Reconsideration of the Primary Sources', unpub. thesis (Oberlin College, Ohio, 1973).
MORRISON, J. S.: 'An Introductory Chapter in the History of Greek Education', *Durham University Journal*, 41 (1949), 55–63.
MOURELATOS, A. P. D.: *The Route of Parmenides* (New Haven, Conn., 1970).
NAGY, A.: 'Di alcuni scritti attribuiti ad Empedocle', *Rendiconti della Reale Accademia dei Lincei, Classe di scienze morali, storiche e filologiche*[5], 10 (1901), 307–20, 325–44.
NASR, S. H.: *Three Muslim Sages* (Cambridge, Mass., 1964).
NEEDHAM, J.: *Science and Civilisation in China* (Cambridge, 1954–84).
NEUGEBAUER, O.: *Astronomy and History* (New York, 1983).
NICHOLSON, R. A.: *Studies in Islamic Mysticism* (Cambridge, 1921).
NILSSON, M. P.: *Opuscula selecta* (Lund, 1951–60).
NISSEN, H.: *Italische Landeskunde* (Berlin, 1883–1902).
NOCK, A. D.: *Essays on Religion and the Ancient World* (Oxford, 1972).
—— (1926): *Sallustius: Concerning the Gods and the Universe* (Cambridge).
NORDEN, E.: *Agnostos theos* (Leipzig, 1913).
O'BRIEN, D.: *Empedocles' Cosmic Cycle* (Cambridge, 1969).
—— (1981): *Pour interpréter Empédocle* (Paris and Leiden).
ODER, E.: 'Beiträge zur Geschichte der Landwirthschaft bei den Griechen', *RhM* 45 (1890), 58–99, 212–22.
OLERUD, A.: *L'Idée de macrocosmos et de microcosmos dans le Timée de Platon* (Uppsala, 1951).
OLIVIERI, A.: *Lamellae aureae orphicae* (Bonn, 1915).
O'MEARA, D. J.: *Pythagoras Revived: Mathematics and Philosophy in Late Antiquity* (Oxford, 1989).

ONIANS, R. B.: *The Origins of European Thought*² (Cambridge, 1954).
OSBORNE, C.: *Rethinking Early Greek Philosophy* (London, 1987).
PACE, B.: *Arte e civiltà della Sicilia antica* (Milan, 1935–49).
PARKE, H. W.: *Sibyls and Sibylline Prophecy in Classical Antiquity* (London, 1988).
PARMENTIER, L.: *Recherches sur le traité d'Isis et d'Osiris de Plutarque* (Brussels, 1913).
PELLIKAAN-ENGEL, M. E.: *Hesiod and Parmenides: A New View on their Cosmologies and on Parmenides' Proem* (Amsterdam, 1974).
PETERS, F. E.: *Aristotle and the Arabs* (New York, 1968).
PFEIFFER, E.: *Studien zum antiken Sternglauben* (Leipzig, 1916).
PHILIP, J. A.: *Pythagoras and Early Pythagoreanism* (Toronto, 1966).
PHILLIPS, C. R. III: 'The Sociology of Religious Knowledge in the Roman Empire', *ANRW* ii.16.3 (1986), 2677–773.
PLESSNER, M.: *Vorsokratische Philosophie und griechische Alchemie* (Wiesbaden, 1975).
—— (1954): 'The Place of the *Turba Philosophorum* in the Development of Alchemy', *Isis*, 45: 331–8.
PUECH, H.-C.: *En quête de la Gnose* (Paris, 1978).
RATHMANN, W.: *Quaestiones Pythagoreae Orphicae Empedocleae* (Halle, 1933).
RAVEN, J. E.: 'Plants and Plant Lore in Ancient Greece', *Annales Musei Goulandris*, 8 (1990), 129–80.
REICKE, B.: *The Disobedient Spirits and Christian Baptism* (Copenhagen, 1946).
REITZENSTEIN, R. (1906): *Hellenistische Wundererzählungen* (Leipzig).
—— (1927): *Die hellenistischen Mysterienreligionen*³ (Leipzig).
REYNDERS, B.: *Lexique comparé du texte grec et des versions latine, arménienne et syriaque de l''Adversus Haereses' de Saint Irénée* (Louvain, 1954).
RICHARDSON, N. J.: *The Homeric Hymn to Demeter* (Oxford, 1974; repr. 1979).
—— (1975): 'Homeric Professors in the Age of the Sophists', *Proceedings of the Cambridge Philological Society*, 201 (1975), 65–81.
RIGINOS, A. S.: *Platonica* (Leiden, 1976).
ROHDE, E.: *Psyche* (Eng. trans., London, 1925).
ROSCHER, W. H.: *Über Selene und Verwandtes* (Leipzig, 1890).
ROSTAGNI, A.: *Scritti minori* (Turin, 1955–6).
ROWSON, K. E.: *A Muslim Philosopher on the Soul and its Fate* (New Haven, Conn., 1988).
RUDOLPH, U.: 'Christliche Theologie und vorsokratische Lehren in der *Turba philosophorum*', *Oriens*, 32 (1990), 97–123.
—— (1989): *Die Doxographie des Pseudo-Ammonios* (Stuttgart).

RUSKA, J.: *Turba philosophorum* (Berlin, 1931).
SACHS, E.: *Die fünf platonischen Körper* (Berlin, 1917).
SAFFREY, H. D.: 'Les Néoplatoniciens et les Oracles Chaldaïques', *Revue des études augustiniennes*, 27 (1981), 209–25.
SANTILLANA, G. DE: *Reflections on Men and Ideas* (Cambridge, Mass., 1968).
SCHIBLI, H. S.: *Pherekydes of Syros* (Oxford, 1990).
SCHLANGER, J.: *La Philosophie de Salomon ibn Gabirol* (Leiden, 1968).
SCHMIDT, M.: *Die Erklärungen zum Weltbild Homers und zur Kultur der Heroenzeit in den bT-Scholien zur Ilias* (Munich, 1976).
SCOTT, W.: *Hermetica* (Oxford, 1924–36).
SEAFORD, R.: 'Dionysiac Drama and the Dionysiac Mysteries', *CQ* 31 (1981), 252–75.
SEDLEY, D.: 'Teleology and Myth in the *Phaedo*', in J. J. Cleary and D. C. Shartin (eds.), *Proceedings of the Boston Area Colloquium in Ancient Philosophy*, v (Lanham, Md., 1991), 359–83.
SEZGIN, F.: *Geschichte des arabischen Schrifttums* (Leiden, 1967–).
SFAMENI GASPARRO, G.: *I culti orientali in Sicilia* (Leiden, 1973).
SHAW, G. (1985): 'Theurgy: Rituals of Unification in the Neoplatonism of Iamblichus', *Traditio*, 41: 1–28.
—— (1993): 'The Geometry of Grace: A Pythagorean Approach to Theurgy', in H. J. Blumenthal and E. G. Clark (eds.), *The Divine Iamblichus: Philosopher and Man of Gods* (Bristol), 116–37.
SMITH, M.: 'Prolegomena to a Discussion of Aretalogies, Divine Men, the Gospels and Jesus', *Journal of Biblical Literature*, 90 (1971), 174–99.
—— (1973): *Clement of Alexandria and a Secret Gospel of Mark* (Cambridge, Mass.).
SOLMSEN, F.: 'Chaos and "Apeiron"', *Studi italiani di filologia classica*, 24 (1950), 235–48.
SPOERRI, W.: *Späthellenistische Berichte über Welt, Kultur und Götter* (Basle, 1959).
STEIN, H.: *Empedoclis Agrigentini fragmenta* (Bonn, 1852).
STERN, S. M.: 'Ibn Masarra, Follower of Pseudo-Empedocles—An Illusion', *Actas, 4 Congresso de Estudos Árabes e Islâmicos, Coimbra–Lisboa 1968* (Leiden, 1971), 325–37 = *Medieval Arabic and Hebrew Thought* (London, 1983), ch. v.
STEWART, J. A.: *The Myths of Plato* (London, 1905).
STURZ, F. W.: *Empedocles Agrigentinus* (Leipzig, 1805).
TARÁN, L. (1975): *Academica: Plato, Philip of Opus, and the Pseudo-Platonic 'Epinomis'* (Philadelphia).
—— (1981): *Speusippus of Athens* (Leiden).
TAYLOR, A. E.: *A Commentary on Plato's Timaeus* (Oxford, 1928).

THESLEFF, H. (1961): *An Introduction to the Pythagorean Writings of the Hellenistic Period* (Åbo).
—— (1965): *The Pythagorean Texts of the Hellenistic Period* (Åbo).
THOMAS, H. W.: *Ἐπέκεινα: Untersuchungen über das Ueberlieferungsgut in den Jenseitsmythen Platons* (Würzburg, 1938).
TRAGLIA, A.: *Studi sulla lingua di Empedocle* (Bari, 1952).
VERMASEREN, M. J.: *Liber in Deum* (Leiden, 1976).
VIANO, C.: 'Olympiodore l'alchimiste et les présocratiques: Une doxographie de l'unité', in S. Matton (ed.), *Alchimie: Art, histoire et mythes. Actes du premier Colloque international de la Société d'Étude de l'Histoire de l'Alchimie* (Paris).
VLASTOS, G.: *Socrates* (Cambridge, 1991).
VODRASKA, S. L.: 'Pseudo-Aristotle *De causis proprietatum et elementorum*', Ph.D. thesis (London, 1969).
VOLLGRAFF, C. W.: '"Ἔριφος ἐς γάλ' ἔπετον": Over den oorsprong der dionysische mysteriën', *Mededeelingen der koninklijke Akademie van Wetenschappen Amsterdam*, 57 (1924), 19–53.
WASSERSTROM, S. M.: 'The Magical Texts in the Cairo Genizah', in J. Blau and S. C. Reif (eds.), *Genizah Research After Ninety Years* (Cambridge, 1992), 160–6.
WEHRLI, F.: *Herakleides Pontikos*[2] (Basle and Stuttgart, 1969).
WEIDLICH, T.: *Die Sympathie in der antiken Litteratur* (Stuttgart, 1894).
WEINREICH, O.: *Ausgewählte Schriften* (Amsterdam, 1969–79).
WELLMANN, M.: *Die Georgika des Demokritos* (Berlin, 1921).
—— (1934): *Marcellus von Side als Arzt und die Koiraniden des Hermes Trismegistos* (Leipzig).
WEST, M. L.: *The Orphic Poems* (Oxford, 1983).
—— (1966): *Hesiod: Theogony* (Oxford).
—— (1971): *Early Greek Philosophy and the Orient* (Oxford).
—— (1975): 'Zum neuen Goldblättchen aus Hipponion', *ZPE* 18: 229–36.
—— (1978): *Hesiod: Works and Days* (Oxford).
—— (1982): 'Cosmology in the Greek Tragedians', *Balkan and Asia Minor Studies* (Tokai University, Tokyo) 8: 1–13.
WESTERINK, L. G.: *The Greek Commentaries on Plato's Phaedo* (Amsterdam, 1976–7).
WHITTAKER, J.: 'The Value of Indirect Tradition in the Establishment of Greek Philosophical Texts or the Art of Misquotation', in J. N. Grant (ed.), *Editing Greek and Latin Texts* (New York, 1989), 63–95.
WIETEN, J. H.: *De tribus laminis aureis quae in sepulcris Thurinis sunt inventae* (Amsterdam, 1915).
WILAMOWITZ-MOELLENDORFF, U. VON: *Platon*[2] (Berlin, 1920).

WILAMOWITZ-MOELLENDORFF, U. VON (1909): *Euripides: Herakles*² (Berlin).
—— (1931–2): *Der Glaube der Hellenen* (Berlin).
—— (1935): *Kleine Schriften*, i (Berlin).
WILCOX, J. C.: *The Transmission and Influence of Qusta ibn Luqa's 'On the Difference Between Spirit and the Soul'* (Ann Arbor, Mich., 1987).
WILSON, C. A.: *Philosophers, Iōsis and Water of Life* (Leeds, 1984).
WIND, E.: *Pagan Mysteries in the Renaissance*² (London, 1968).
WOODHEAD, W. D.: *Etymologizing in Greek Literature from Homer to Philo Judaeus* (Toronto, 1928).
WORTMANN, D. (1968a): 'Neue magische Texte', *Bonner Jahrbücher*, 168: 56–111.
—— (1968b): 'Die Sandale der Hekate–Persephone–Selene', *ZPE* 2: 155–60.
WRIGHT, M. R.: *Empedocles: The Extant Fragments* (New Haven, Conn., 1981).
WUILLEUMIER, P.: *Tarente* (Paris, 1939).
ZANDEE, J.: *Death as an Enemy According to Ancient Egyptian Conceptions* (Leiden, 1960).
ZASLAVSKY, R.: *Platonic Myth and Platonic Writing* (Lanham, Md., 1981).
ZELLER, E.: *Die Philosophie der Griechen*, ed. W. Nestle, i⁶ (Leipzig, 1919–20), ii/1⁴ (1889).
ZHMUD', L. J.: '"All is Number"?', *Phronesis*, 34 (1989), 270–92.
ZIMMERMANN, F. W.: 'The Origins of the So-Called *Theology of Aristotle*', in J. Kraye, W. F. Ryan, and C. B. Schmitt (eds.), *Pseudo-Aristotle in the Middle Ages* (London, 1986), 110–240.
ZUNTZ, G.: *Persephone: Three Essays on Religion and Thought in Magna Graecia* (Oxford, 1971).
—— (1976): 'Die Goldlamelle von Hipponion', *Wiener Studien*, 89: 129–51.

THESLEFF, H. (1961): *An Introduction to the Pythagorean Writings of the Hellenistic Period* (Åbo).
—— (1965): *The Pythagorean Texts of the Hellenistic Period* (Åbo).
THOMAS, H. W.: Ἐπέκεινα: *Untersuchungen über das Ueberlieferungsgut in den Jenseitsmythen Platons* (Würzburg, 1938).
TRAGLIA, A.: *Studi sulla lingua di Empedocle* (Bari, 1952).
VERMASEREN, M. J.: *Liber in Deum* (Leiden, 1976).
VIANO, C.: 'Olympiodore l'alchimiste et les présocratiques: Une doxographie de l'unité', in S. Matton (ed.), *Alchimie: Art, histoire et mythes. Actes du premier Colloque international de la Société d'Étude de l'Histoire de l'Alchimie* (Paris).
VLASTOS, G.: *Socrates* (Cambridge, 1991).
VODRASKA, S. L.: 'Pseudo-Aristotle *De causis proprietatum et elementorum*', Ph.D. thesis (London, 1969).
VOLLGRAFF, C. W.: ' Ἔριφος ἐς γάλ' ἔπετον: Over den oorsprong der dionysische mysteriën', *Mededeelingen der koninklijke Akademie van Wetenschappen Amsterdam*, 57 (1924), 19–53.
WASSERSTROM, S. M.: 'The Magical Texts in the Cairo Genizah', in J. Blau and S. C. Reif (eds.), *Genizah Research After Ninety Years* (Cambridge, 1992), 160–6.
WEHRLI, F.: *Herakleides Pontikos*² (Basle and Stuttgart, 1969).
WEIDLICH, T.: *Die Sympathie in der antiken Litteratur* (Stuttgart, 1894).
WEINREICH, O.: *Ausgewählte Schriften* (Amsterdam, 1969–79).
WELLMANN, M.: *Die Georgika des Demokritos* (Berlin, 1921).
—— (1934): *Marcellus von Side als Arzt und die Koiraniden des Hermes Trismegistos* (Leipzig).
WEST, M. L.: *The Orphic Poems* (Oxford, 1983).
—— (1966): *Hesiod: Theogony* (Oxford).
—— (1971): *Early Greek Philosophy and the Orient* (Oxford).
—— (1975): 'Zum neuen Goldblättchen aus Hipponion', *ZPE* 18: 229–36.
—— (1978): *Hesiod: Works and Days* (Oxford).
—— (1982): 'Cosmology in the Greek Tragedians', *Balkan and Asia Minor Studies* (Tokai University, Tokyo) 8: 1–13.
WESTERINK, L. G.: *The Greek Commentaries on Plato's Phaedo* (Amsterdam, 1976–7).
WHITTAKER, J.: 'The Value of Indirect Tradition in the Establishment of Greek Philosophical Texts or the Art of Misquotation', in J. N. Grant (ed.), *Editing Greek and Latin Texts* (New York, 1989), 63–95.
WIETEN, J. H.: *De tribus laminis aureis quae in sepulcris Thurinis sunt inventae* (Amsterdam, 1915).
WILAMOWITZ-MOELLENDORFF, U. VON: *Platon*² (Berlin, 1920).

WILAMOWITZ-MOELLENDORFF, U. VON (1909): *Euripides: Herakles*² (Berlin).
—— (1931–2): *Der Glaube der Hellenen* (Berlin).
—— (1935): *Kleine Schriften*, i (Berlin).
WILCOX, J. C.: *The Transmission and Influence of Qusta ibn Luqa's 'On the Difference Between Spirit and the Soul'* (Ann Arbor, Mich., 1987).
WILSON, C. A.: *Philosophers, Iōsis and Water of Life* (Leeds, 1984).
WIND, E.: *Pagan Mysteries in the Renaissance*² (London, 1968).
WOODHEAD, W. D.: *Etymologizing in Greek Literature from Homer to Philo Judaeus* (Toronto, 1928).
WORTMANN, D. (1968a): 'Neue magische Texte', *Bonner Jahrbücher*, 168: 56–111.
—— (1968b): 'Die Sandale der Hekate–Persephone–Selene', *ZPE* 2: 155–60.
WRIGHT, M. R.: *Empedocles: The Extant Fragments* (New Haven, Conn., 1981).
WUILLEUMIER, P.: *Tarente* (Paris, 1939).
ZANDEE, J.: *Death as an Enemy According to Ancient Egyptian Conceptions* (Leiden, 1960).
ZASLAVSKY, R.: *Platonic Myth and Platonic Writing* (Lanham, Md., 1981).
ZELLER, E.: *Die Philosophie der Griechen*, ed. W. Nestle, i⁶ (Leipzig, 1919–20), ii/1⁴ (1889).
ZHMUD', L. J.: '"All is Number"?', *Phronesis*, 34 (1989), 270–92.
ZIMMERMANN, F. W.: 'The Origins of the So-Called *Theology of Aristotle*', in J. Kraye, W. F. Ryan, and C. B. Schmitt (eds.), *Pseudo-Aristotle in the Middle Ages* (London, 1986), 110–240.
ZUNTZ, G.: *Persephone: Three Essays on Religion and Thought in Magna Graecia* (Oxford, 1971).
—— (1976): 'Die Goldlamelle von Hipponion', *Wiener Studien*, 89: 129–51.

Index

Academy of Plato 111, 177–9, 196–202, 211, 256, 328–9, 393
Acheron 101, 102 n., 122–3
Acherusian Lake 80, 98, 118–19, 123, 125
Acragas viii, 1, 8, 34, 76, 224–5, 243, 257, 352, 357, 364–8
Adad 294 n.
Adranus viii, 76, 281–2, 349 n.
Aelian 342
aer 15–18, 20, 24–36, 44–6, 56, 64, 123–5
'Aëtius' 34 n.
Aidoneus 13–14
air (as element) 13–48, 56, 63–4, 123–5
aither 15–29, 31 n., 32 n., 35–6, 40–4, 49, 75, 91, 127 n., 218 n., 252 n., 255
Akhmīm ix, 10, 59, 60, 66–8, 119 n., 365 n., 388–90
Albertus Magnus 204–8, 212–13
alchemy 55–68, 205, 221, 229, 298 n., 326, 344, 356, 375–9, 387 n., 388–90
Alexander of Abonuteichus 294 n.
Alexander the Great 151, 276, 336
Alexandria ix, 60–2, 64–5, 102, 183, 242–4, 339
Alexis 348
allegory 26, 36–9, 102–8, 109 n., 113–16, 122–6, 139, 159, 165–70, 190, 192, 219, 324, 357, 390 n.
al-'Āmirī 375 n., 376–7, 381 n., 382 n.
Ammonius doxography 62 n., 377 n., 384 n., 395
Amphiaraus 284 n., 287
Anatolia 223 n., 225, 293–4, 331, 392
Anatolius 183–7, 189, 192, 198
Anaxagoras 18 n., 92 n.
Anaxilaus of Larissa 322 n., 332
Anaximander 61 n., 65 n.
Anaximenes 1 n., 61 n., 63 n.
Andania 353–4
Androtion 299 n.
anger 101–2, 104, 247–8
Aphrodite 75, 270–1
Apollo 136, 225 n., 261–2, 273 n., 274 n., 351 n.
Apollonius of Tyana 253 n., 276, 293 n., 319 n., 332

apotheosis 234–5, 253, 256–7, 274–5
Aquinas, St Thomas 204–8, 212–13
Archedemus 95 n., 160
Archytas 94–5, 144, 146–7, 149, 156–7, 160, 164, 203, 262, 324
Arignote 262 n.
Arignotus 332, 333 n.
Aristotle: on Empedocles 2–5, 20, 43–4, 53, 346–7, 360, 373–4, 384–8; and Neopythagoreanism 335–41; on Plato 106 n., 111, 127; on Platonists 177–80, 197–8; on Presocratics 3–4, 18, 48, 174, 208–10, 386–8; on Pythagoreanism 157 n., 172–82, 187–9, 195–200, 203–13, 342, 368; translations of 204–12
Aristoxenus 143–6, 149, 247, 284 n., 323–4
artillery, history of 145–58
astrology 173, 229, 288, 331, 393
Athenocentrism 156, 340
Athens 9, 103, 149–53, 156–8, 178, 198, 296, 323, 339–41, 346
Atticus 383 n., 388 n.
Attis, mysteries of 265–7
Avernus viii, 98–9, 134 n., 137 n.

Babylon 226, 304
Babylonian and Assyrian traditions 54–5, 173, 189, 191, 210–11, 223 n., 275, 294 n., 304, 331, 392–4
Bacchic mysteries 162 n., 164 n., 259–72, 290, 308–13; *see also* Dionysus
Baghdad 380
bāṭiniyya 376–7
Baubo 249 n.
beans 283 n., 284–6
Biton 145–6, 150, 153
Bolus of Mendes 293 n., 298, 299 n., 325–8, 335–41, 343, 345, 347, 373
Book of Enoch 211, 345 n.
breathing 142, 159 n.
Brontinus 159 n.
bronze 237–40

Caecilia Secundina 309, 311
Calcidius 200–1, 203–4
Calvin, John 211

Index

Camarina viii, 115, 143
Carthage 155, 312
caves and hollows 36–8, 66, 74, 82–4, 90, 93, 98, 100, 105–6, 133, 136, 281–2, 353
Chaldaean Oracles 124 n., 223 n., 303–4
chasm 79, 83, 126–8, 137 n., 138, 141, 282
Christian traditions 67, 119 n., 135 n., 193, 203–4, 207–13, 224 n., 252, 294 n., 297 n., 334, 383
Cicero 280–1, 284–6, 317, 324–5
Clinias 144–5
Cocytus 96–101, 120 n., 122
Copernicus 173
counter-earth 91–2, 172–4, 186–7
Crete 223 n., 224, 255 n., 256, 307, 331, 361
Croton viii, 143–4, 152 n., 276, 331
Cumae viii, 84 n., 98–101, 141, 153–5, 225 n.
Cynics 276, 341

Damascius 106 n., 120–5, 181–2, 193, 199
death 40, 55 n., 105, 222–3, 252 n.; and initiation 54–5 n., 251–2, 258, 264–9, 287 n., 291, 365 n.; and regeneration 223 n., 251–2, 259, 264–8, 287, 289, 291, 303, 313; return from 40–1, 136–7, 225–6; *see also* apotheosis, Hades, incarnation and reincarnation
Delphi 136–7
Demeter 74 n., 97–8, 115–16, 164, 225, 230, 240, 243, 245–6, 262 n., 270, 283, 307, 351–2
Democritus 288 n., 326–8, 337 n.
Derveni papyrus 116, 118, 122, 124, 126, 136 n., 164–5, 169 n.; and Empedocles 46 n., 47 n.
Dhū 'l-Nūn al-Miṣrī 59 n., 381 n., 388–90
Dicaearchia viii, 100, 137 n.
Dicaearchus 368
Diocles of Carystus 222 n., 337
Diodorus of Ephesus 273–6
Diodorus Siculus 86, 150, 224–5, 253 n.
Diogenes of Apollonia 33 n.
Diogenes Laertius 228, 237, 253–6, 272–4, 363–5
Dionysius I 150, 153
Dionysius II 152, 157
Dionysus 162 n., 164 n., 222 n., 259–71, 272 n., 276 n., 307, 312, 361; *see also* Bacchic mysteries
dodecahedron 93, 330 n.
doxographic tradition 3 n., 58, 61–7
dreams and dream oracles 134, 136–8, 225 n., 281–8, 326; *see also* incubation

earth: as element 13–16, 19–21, 24–5, 29–33, 44–7, 57–8, 63–6; 'other' 91–3, 186–7; position of 172–91; spherical 41 n., 89–90, 105–7; 'true' 90–3, 107
egg 57–8, 66 n., 67–8
Egypt: pre-Christian 54, 68 n., 173 n., 193 n., 242–4, 263–4, 268–9, 287, 298, 312, 325–6, 331–3, 343–6; Christian period 10, 59–60, 67–8, 119 n., 240–8, 286–8, 314, 326, 332–3, 339, 343–6, 365 n., 374–7, 389; Islamic 10, 59, 377–8, 388–90, 395
element(s): fifth 17 n., 122 n.; four 1, 13–68, 74–5, 120–4, 183–4, 218 n., 300–1, 348–50, 354–62, 374; transmutation of 29–30, 32–4
Eleusis 115–16, 299 n., 351, 353–4
Elijah 236 n.
Empedocles: and alchemy 56–68, 221, 298 n., 356, 375–9; clepsydra 27, 141–2; cosmic cycle 7, 142; dating 1; death 1, 135, 233–56, 272–83, 289–92; and Egypt 10, 56–68, 344–6, 365 n., 374–9, 388–90; four elements 1, 6, 13–48, 300–1, 348–62, 374, 385; a god 220, 223–4, 233–8, 255, 258–9, 272–7, 296–7, 319 n.; and gold plates 256–9, 272, 314, 351, 361; and Heracles 252–7, 272–6, 297 n.; and Hermetism 135, 221, 334, 344–6, 372–4, 378–9, 383–4; and Hesiod 26, 45, 78, 298 n., 299 n.; and Homer 26–8, 42–5, 52–3, 192, 223–4, 247; in the Islamic world 56–67, 375–91, 395; and Jesus 297 n., 370; legends 227–8, 233–56, 272–83, 289–92, 301, 314; Love and Strife 7, 20, 298–300, 338; and magic 217–32, 238–49, 289–91, 294, 296–304, 314–15, 339, 344–5, 359, 374; on papyrus 10, 67, 365 n.; and Plato 112–14, 134–5, 141–3, 158–9, 342–3; poems of 7–8, 297, 363–70; as poet 2, 5, 38, 42–5, 51–3, 296; *Purifications* 7, 36–9, 363–8; and Pythagoreanism 112–13, 135, 142, 158–9, 172, 180–1, 193, 217, 225, 240–8, 251–3, 276–7, 282–4, 290–6, 317–19, 328–30, 339, 342–7, 354–5, 366 n., 374–5, 379, 381, 387–9; and Sophists 247; teaching methods 38, 42–5, 219–21, 230–1, 330, 354, 360–76, 382
Empusa 240 n., 249 n.
Epicharmus 284 n.
Epimenides 92 n., 223 n., 284 n., 287 n.
epopteia 230 n., 299 n., 367–8
Eratosthenes 266
Esagil 393
esoteric transmission 221, 230–1, 277, 289–91, 322, 330–4, 344, 354, 359–70, 376, 377 n., 390–1
Etna viii, 35 n., 72–86, 100, 115, 133–5; cult activity 279–83; and Empedocles 135, 233–9, 248–56, 273–9, 282–3, 289–91, 301; and heaven 52–3, 78 n., 252 n., 255, 280; and Plato 94; and underworld 73–4, 77, 84 n., 239–40, 280
etymologizing in antiquity 15, 47 n., 96, 165 n., 167–70, 350; double etymology 350–1, 394

Ficino, Marsilio 56 n., 170 n., 247 n.
fire: in alchemy 55–8, 65–8; and anger 101–2, 104; central 55–6, 65–6, 172–213, 378 n.; dual aspect 74–8; as element 13–23, 25, 44–

Index 419

58, 68; and heroization 258; immortalizes 252–5; purifies 193 n., 252–5; subterranean 33–5, 51–8, 63–8, 71–80, 86, 93, 100–1, 112, 180–1; volcanic 34, 53, 56, 71–80, 86, 100, 180–1, 193, 206, 239–40; *see also* Etna, Hades
Furies 240 n., 249

Gaggera 243, 271, 306–8
Gehenna 193, 207, 210
Gela viii, 298 n.
Gerard of Cremona 207–12
al-Ghazālī, Abū Ḥāmid 377 n.
ghiyāth 380, 383
Gilgamesh 223 n., 392
Glaucus of Rhegium 272 n., 327 n.
Gnostic traditions 57, 66–7, 92 n., 302, 355–6, 375, 378–9, 382 n., 383 n., 384 n.
gold plates 256–72, 276–7, 290, 303, 308–14, 323 n., 351, 361, 374
Gorgias 113 n., 170 n., 220–1, 248 n.
Grail 135

Hades: as air 14, 16, 36–42; descent to 38, 40–1, 54 n., 74, 83 n., 115, 125, 133–8, 225–6, 249–52, 272 n., 282–3, 286–7, 289–92, 311 n., 392; as earth 14, 46–7; entrances to 73–4, 84 n., 97–8, 134, 138, 240, 252–3, 282–3; etymology of 47 n., 169 n.; as fire 14, 47–53, 68, 71–8, 193, 239–40, 248–9, 282–3; gates of 285 n.; and inversion 77, 186–7; location of 36–41, 46–7, 51, 105–6, 183–7, 190; and paradox 77; and Persephone 71 n., 74, 97, 348–58; and sun 39 n., 50–6
healing 220 n., 222–3, 229–30, 247–8, 327–8, 331, 335–9, 341–5
heaven: and *aither* 16–19, 23, 29, 32 n., 35, 40–2, 49; ascent to 135, 234–5, 244, 251–8, 275, 297 n.; *see also* Etna
Hecate 238–46, 249–50, 270–2, 283, 286, 289, 294, 302
hell, hell-fire 37, 41 n., 55–6, 66, 77, 193, 204–5, 207–12, 252
Hephaestus 20 n., 44–5, 74–8, 281
Hera 13–15, 26, 36, 44–8, 348–51, 355–7
Heraclea viii, 143–5, 155
Heracles 252–8, 262 n., 266, 269, 272 n., 273 n., 274–6, 281, 297 n., 394
Heraclides of Pontus 92, 99 n., 137 n., 234–7, 256, 278 n., 285 n., 297 n., 303 n.
Heraclitus (allegorist) 26 n., 47 n., 96 n.
Heraclitus of Ephesus 1 n., 18 n., 30, 147, 294, 331 n., 339 n.
Hermann of Carinthia 57 n.
Hermetic tradition 61 n., 135, 221, 290, 301 n., 321, 333–4, 343–6, 355 n., 372–5, 378–9, 382–4, 386–9
Herodorus 143, 253 n., 275 n.
Herodotus 225 n., 263–4, 392
heroization 256–8

Hesiod: *aer* 17 n., 26, 123; and Babylonia 392; and Empedocles 45, 78, 224, 298 n., 299 n.; and Orphic tradition 128–9; and Philolaus 188–90, 210, 212
Hiera viii, 76, 133, 134 n.
Himera viii, 34
Hippasus 330 n.
Hippocratic writings 16 n., 26, 220, 223 n., 224 n., 225 n., 229–30
Hippolytus of Rome 24–5 n., 58, 62–3, 65 n.
Hipponium viii, 257 n., 259–60, 263, 266, 310–11
Homer: *aer* and *aither* 15–16, 17 n., 20 n., 26, 36, 123–4; and Empedocles 26–8, 42–5, 52–3, 223–4, 247; and Greek culture 81, 191–2, 223; in magic 247–8; in Plato 79, 109, 120, 122 n., 126, 129–30, 165–6, 168 n.; and Pythagoreanism 109, 183–93, 198–9, 212, 308; word-play 15 n.
hot springs 32–5, 71–2, 74 n., 76 n., 83, 85, 112, 205–6, 357
Hybla Geleatis viii, 83 n., 133, 134 n., 281–2

Iamblichus: on Assyrians 304 n.; cosmology 39 n.; and Julian 132 n.; on Pythagoreanism 144–5, 149, 247, 285–6, 293 n., 305, 363 n., 366 n., 383 n., 388 n.
iatromantis 220 n.
Ibn Masarra 381–2, 390 n.
Ibn Sīnā 206
Ibn Suwaid 59, 66, 389–90
immortality and immortalization 135, 220, 223 n., 233–8, 244, 251–9, 264–8, 272, 275 n., 290, 297 n., 303, 313–14, 374
incantations and charms 222, 247–8, 300, 307–8, 342
incarnation and reincarnation 40, 160, 164–5, 257, 267 n., 286, 344–5, 366–9, 380
incense 279–81
incubation 225 n., 284–8, 331
Iranian traditions 223 n., 227, 304 n.; *see also* Persia
Isaac Israeli 395
Islamic traditions 31 n., 55–60, 66–7, 193, 209, 375–90, 395
Ismāʿīlīs 377–9, 395

Jābir ibn Ḥayyān 376–7
Jewish traditions 193, 211, 236 n., 242, 308 n., 378 n., 382 n., 395
Judah Halevi 395
Jung, Carl 57–8, 355–6

katabasis 38, 41; *see also* Hades
al-Kindī, Abū Yaʿqūb 380, 384
Korē Kosmou 301 n., 344–6, 374, 383 n.
kratēr viii, 82–4, 133–41, 147, 159–60, 244, 262, 282–3
Kronos 71 n., 355
Kyane 83 n., 97–9, 134 n., 353

Lāhījī, 'Abd al-Razzāq 381 n., 387 n.
lead tablets 239 n., 244 n., 245-6, 249 n., 269-72, 306-8
Lesbos 294 n.
lightning 255, 257-8
Lipara, Liparan islands viii, 76, 86, 100
Locri viii, 271
Lucan 242
Lucian 226-7, 249, 292, 332
Lucifer 77

macrocosm and microcosm 102, 142
Magi 226-7, 328 n.
magic 217-347; conservatism of 315; and healing 220, 222-6, 229-30, 247-8, 299, 335-45; and mysticism 313; and philosophy 217-32, 296-305, 335-47; and religion 229, 295, 305-16; and rhetoric 220, 248 n., 306; and science 219, 229-30, 295, 335-7; and theurgy 301-4
magical papyri 55, 221-2, 223 n., 238-49, 251, 266, 286-8, 290, 296, 307-16, 321 n., 326, 332-3, 345, 374; dating 240-1
magos 225-6
Manichaeism 383-4
Marduk 54 n., 210, 393
Mecca 390 n.
medicine: *see* healing, Hippocratic writings, magic, plants
Mediterranean 80, 85, 100, 105
Menestor 299 n.
metallurgical traditions 66, 224, 291
Metapontum viii, 143-4, 265
Miletus 150-3
milk 264-71, 276
Milo of Croton 253 n., 276
Milton, John 37, 74, 77
mist 15 n., 16, 26-30, 35, 64, 123-5
moon 26 n., 39 n., 50, 91-2, 136-8, 173, 187, 392
Motya viii, 150
Musaeus 116 n., 120 n., 136 n.
mystery terminology 230-1, 299, 367-8, 371-2
myth: and *logos* 80-1, 167 n.; and philosophy 113-14, 239, 301; rationalization of 237-8, 248, 250-1, 254-5, 293; and science 80-1, 85, 90-1, 173-4

Nag Hammadi ix, 67, 321, 383 n.
Naples viii, 98-101, 133-4, 154
Neoplatonists: and alchemy 60-1; on Empedocles 7, 36-40; and Orphic tradition 120-32, 140, 305; and Plato 38-9, 102-3, 121, 130-2, 305; and Pythagoreanism 102-3, 131-2, 181-2, 193-4, 209, 304-5; and theurgy 301-5
Neopythagoreanism 102-3, 283 n., 305 n., 317-47
Nergal 275, 394
Nestis 13, 43 n., 348-56

Night 54, 136-8, 272, 392
Nigidius Figulus 284 n., 317, 322 n., 324-5
Numenius 304-5, 328, 388

observation: and magic 295-6, 335, 337, 339, 372; and revelation 372-3
Ocean 101, 120 n., 123, 353
Olympiodorus (alchemist) 60-2, 64-5
Olympiodorus (Neoplatonist) 60, 61 n., 94 n., 113-14, 120-1, 130-1
oracles 134, 136-8, 220 n., 240, 252 n., 282-8, 303, 319, 326, 331 n., 343, 362, 363 n., 366 n.; *see also* riddles and obscurity
Orpheus: descent to Hades 115, 133, 135-8, 226, 282-3, 287; and shamanism 226; at Tarentum 162 n.; writings of 115-32, 142-3, 161, 261 n., 285 n.
Orphic tradition 67, 68 n., 115-44, 147-8, 159-65, 259-69, 308-9, 321 n., 355; and allegory 120-6; and Bacchic mysteries 259-64, 268-9; and Egypt 67-8, 241, 243-4, 263-4, 312; and Eleusis 115-16; and Hesiod 128-9; and magic 241; and Pythagoreanism 115, 125-6, 131-2, 139, 143-4, 159-64, 259-64, 282-3, 305; and Sicily 115-16, 141
Oxyrhynchus ix, 60 n.; lead tablet 239 n., 245-6, 269-72

Pachrates 332, 351 n.
Palici viii, 35 n., 83 n., 85-6, 115, 133, 134 n.
Pandolfus 57 n.
Pandora 78
Panopolis ix, 59-60, 67-8, 389 n., 390 n.; *see also* Akhmīm
paradosis 230, 367-8
Paris magical papyrus (*PGM* IV) 221-2, 223 n., 230 n., 238-47, 249 n., 251, 287, 301 n., 308 n., 313, 333 n., 374-5
Parmenides: *aither* 18, 19 n., 35 n.; and Empedocles 43, 51, 54, 354, 385; paradox in 51, 54; proem 18 n., 54-5, 252 n., 354, 392-3; and Pythagoreanism 89 n., 329-30
Pasikrateia 243
Patrizi, Francesco 386
Pausanias (disciple of Empedocles) 7, 219, 224, 234-6, 360-1, 364-70
Pausanias (geographer) 280-1, 354
Persephone and Aphrodite 270-1; descent to Hades 74, 83 n., 84 n.; and Dionysus 260-2, 268, 270-1, 312; *hagne* 243 n., 353-4; in magic 243-6, 307; mysteries 74 n., 95, 97-9, 115, 164, 225, 240, 245-52, 260-2, 267-71, 350-8; in Orphic tradition 115, 260-2, 267-9, 351, 356 n.; in Sicily 74, 83 n., 84 n., 95, 97-9, 115, 164, 225, 243, 271, 307, 350-3, 357-8; and water 97-8, 348-58; *see also* Hades
Persia: classical 9, 151, 152 n., 226-7; Islamic

Index

375–8, 381, 383–4, 387–9; *see also* Iranian traditions
persuasion 90–1 n., 306
Petelia viii, 244 n., 257 n., 309, 311
pharmaka 222–3
Pherecydes 47 n., 140 n., 188, 189 n.
Philicus 351 n., 352
Philistus 150, 281 n., 282
Philo of Alexandria 17 n., 20–3, 30 n., 31–4, 62, 64, 102 n.
Philolaus 19 n., 90 n., 92, 126 n., 144–5, 146 n., 149, 162 n., 172–213, 256, 262, 324
philosophy: and healing 229–30, 327, 335–45; and magic 217–32, 296–305, 336–7, 341–5; and mysteries 371–2; origins of the word 339
Phoenicia 243, 253, 281, 312
Picatrix (*Ghāyat al-ḥakīm*) 57 n., 378 n., 390 n.
Pindar 73–4, 79, 100, 164, 185, 193, 222 n., 257, 261 n., 357
plagues 273–6, 302, 307–8, 394
planets 50, 54 n., 92 n., 172–4, 189–91; invisible 91–2, 186
plants 222–4, 229, 240 n., 299–300, 321 n., 328, 335, 337, 341–2
Plato: *Critias* 179, 197, 201–2; dialectic and myth 80–1, 165–70; and Dionysius II 152, 157; and Empedocles 112–14, 134–5, 141–3, 158–9, 342–3; *Gorgias* 37, 94 n., 104–5, 107, 109 n., 113–14, 116–17, 165–70; and Homer 109, 122 n., 129, 165, 168 n.; irony in 167–70; *Laws* 91 n., 170, 330 n.; *Meno* 110 n., 160–5; myths of 79–171, 346; and Orphic tradition 115–32, 139–41, 159–65, 261 n.; *Phaedo* 74, 79–170, 181, 185, 192, 198, 203, 262 n., 263, 346, 356–7; *Phaedrus* 167, 261 n.; poetic allusions in 109, 114, 121–2, 128–30, 168; and Pythagoreanism 88–115, 131–2, 139, 142–7, 152, 156–64, 177–9, 196–203, 263, 328–30, 339–40, 346, 388; *Republic* 38, 106, 116, 117 n., 161, 165–7, 346; and science 80–1, 90–1, 147; in Sicily 81, 93–5, 157; *Timaeus* 81, 93, 135, 159, 160 n., 166 n., 177 n., 201–4, 302 n., 325 n., 330 n.
Plotinus 38–9, 57 n., 300, 302–3, 375 n., 380, 384
Plutarch 45, 146 n.; on Empedocles 28, 57 n., 64, 65 n., 72, 230 n., 365 n., 366 n.; myths of 39 n., 106 n., 135–41, 187 n., 262 n.; on Orphic rites 161; on Plato 178–80
'Pole' (*quṭb*) 380–1
Porphyry 26 n., 36 n., 38, 55 n., 239 n., 289 n., 293 n., 366 n., 384 n.
Posidonius 33 n., 139, 326 n.
Proclus: and alchemy 61 n.; on Empedocles 51–2, 72; on Orphic tradition 120, 122, 125 n., 131–2, 136, 261 n.; on Plato 39 n., 131–2; on Pythagoreanism 131, 181–2, 193, 199–201, 203

purity and purification 193 n., 220, 230, 231 n., 252–5, 273–6, 285–6, 301, 313 n., 366–8, 381–2, 390 n.
Pyriphlegethon 74, 80, 85, 96–104, 109 n., 118–19, 122–3
Pythagoras: in alchemy 56, 376 n., 389; and Democritus 288 n., 327 n., 328, 337 n.; and Empedocles 38, 43 n., 248, 292, 294, 319, 328, 342, 376 n., 381, 387–9; and healing 328, 331, 342; imitation of 294 n., 305 n.; and inspiration 318–19, 381 n.; from Ionia 84 n., 152, 225, 293–4, 331; in Italy 144, 152, 276; legends 248, 255 n., 258, 291–4, 325; and magic 287–8, 291, 294, 342; in Middle Ages 56, 204–5, 376 n., 381, 387–9; not named 163; and Pythagoreanism 276, 320, 326, 331, 333, 342; rationalizations of 248, 292–3; and reincarnation 286, 368; ritual and taboo 285–8, 291, 319, 342
Pythagoreanism: *akousmata* 126 n., 319, 324 n.; *akousmatikoi* and *mathematikoi* 323–4, 329; anonymity 163; and Bacchic religion 162 n., 164 n., 259–65; centres in Italy 94–5, 143–7, 265, 276; cosmic breathing 142; cosmology 172–203; creativity 92–3, 160, 182, 191–4, 199, 319, 328–34; Demeter and Persephone 98, 164, 240, 246, 262, 283; and divination 283–8, 318, 331 n.; and the East 152, 173, 189–91, 225, 293–4, 331; and Egypt 10, 67–8, 119 n., 242–8, 263–4, 287–8, 325–6, 331–3, 339, 343–6, 375, 388–9; and Heracles 253, 276; and Hermetism 135, 321, 333–4, 343–6, 383 n.; and magic 247–8, 251 n., 286–96, 317–18, 326, 332–3; and Neopythagoreanism 283 n., 317–47; number mysticism 157, 174, 177, 183, 292 n.; ritual, cult, and mysteries 98, 162 n., 163–4, 240, 246, 251, 262–4, 283–96, 303–5, 317–19, 324, 332, 335, 339; secrecy 38, 290, 330, 368; *symbola* 290, 305 n.; and warfare 145–7, 157–8; and women 160–4; *see also* Empedocles, Orphic tradition, Plato

Qusṭā ibn Lūqā 377 n.

al-Rāzī, Abū Ḥātim 395
al-Rāzī, Muḥammad ibn Zakariyyā 387–8
red sulphur 58, 66 n., 390
rhetoric 1, 220, 248 n., 306
Rhodes 224–5
riddles and obscurity 38, 43–5, 330, 360–3, 366 n., 371–2, 375–6, 382
rivers: diversion of 273–6; subterranean 73–4, 79–80, 82–7, 96–102, 105, 107, 118–23, 126–7, 135–43, 353 n., 356–7
Rome viii, 154, 242, 244, 317, 322 n., 325
roots and root-cutting 6, 13, 222 n., 230, 299, 337, 345, 355, 385

Sahl al-Tustarī 388–90
Ṣāʿid al-Andalusī 381–2, 388 n.

sandals 239 n., 287, 294 n.; bronze 233-41, 245-7, 250-1, 254, 270, 272, 289-92, 294, 301
Satyrus 220 n., 228
science: and magic 219, 229-30, 295, 335-9; modern 8-9; and mysticism 292 n.; and myth 80-1, 85, 90-1, 166 n., 173-4
secrecy 290, 330, 344, 354, 368, 371, 390
Selinus viii, 34-5, 243, 244 n., 246, 256, 269-74, 306-8
al-Shahrastānī 379, 381 n., 395
shamanism 40, 226-7, 252 n., 291 n.
Sicily: and the East 9-10, 152, 224-5, 243, 275, 293-4; and Egypt 240-9, 269-70, 314, 339, 343, 374; geography 34-5, 46, 71-87, 93-109, 112, 118, 133-5, 137 n., 141-2, 278-83, 356-8; magic 239 n., 242-9, 289, 306-8; religion and cult 74 n., 76-7, 95, 97-9, 115, 164, 224-5, 243, 246, 252-3, 270-83, 293-4, 306-8, 349-53, 357
Simplicius 20 n., 24 n., 176 n., 181-2, 194, 196-7, 200, 202 n., 203, 350, 365 n.
Sophron 242-3
Sostratus 276
Speusippus 177-80, 196-7
stars 19 n., 49-50, 53, 54 n., 56, 63 n., 91-2, 173-4
Stoicism 14, 15, 17-18, 26 n., 32 n., 34, 36, 41 n., 47, 124
Strabo 73, 82, 83 n., 86, 100, 225 n., 278-9, 356
Strongyle viii, 133
Styx 98-9, 120 n., 123
Sufism 66 n., 377 n., 380-3, 387-90
al-Suhrawardī, Shihāb al-Dīn Yaḥyā 375 n., 381, 383, 387-90
sun: in alchemy 55-8, 66-8; 'black' 56; daughters of 54, 392-3; in Empedocles 19, 20 n., 23, 25-6, 29, 32 n., 49-53; as fire 17-19, 20 n., 23, 25, 29, 32 n., 49-58, 68; and Hades 39 n., 50-1, 55; rising in the West 378 n.; and Saturn 54 n.; and seasons 63 n.; and underworld 49-58
Swinden, Tobias 212-13
symbola 238, 289-90, 301-3, 305 n.
sympathy and antipathy 296, 298-300, 335-8
Syracuse viii, 95, 97-9, 115, 133, 143, 146, 150, 152-5, 157, 164, 242, 352-3
Syria 275, 394; Syriac Christianity 208
Syrianus 383 n., 388 n.

ṭā'ifat anbadaqlīs 376
Tarentum viii, 95, 143-8, 151-7, 162 n., 164, 257-8, 262, 265, 299 n.
Tartarus 55; in Christianity 193, 203-4, 207, 210-12; and Etna 73-4, 82, 185; and Hades 74, 185-7, 190, 193; and Hecate 238, 270; in Hesiod 128, 188-90; in Homer 185, 189-90, 212; in *Phaedo* myth 79-80, 82, 93, 101, 118, 126-8, 141, 185, 193, 203-4; in Philolaus 185-95, 199, 203, 210, 212
Telchines 224-5
Theagenes of Rhegium 26
Thebes (Egyptian) ix, 241
Theocritus 242, 244
Theology of Aristotle 57 n., 375 n., 380-4, 387 n., 395
Theophrastus: and alchemy 61-2; and Bolus 336, 338; botany 299, 338; on Empedocles 14, 47-50, 346-7, 360, 373-4, 386; on Philolaus 172; and Plato 178-80; on Presocratics 3-4, 48, 50, 362 n.; on Speusippus 179-80
Thessaly gold plates 256, 259 n., 260-1, 264 n., 265 n., 266-8, 310 n., 312-13
theurgy 301-5, 313
Thurii viii, 257-8, 261 n., 263 n., 264-7, 271-2, 310, 313 n., 348, 351
Timaeus of Tauromenium 234, 253 n., 278 n., 330 n.
Timocreon 114, 168
Titans 25, 188, 199, 203, 210
Trophonius 252 n., 282 n.
Turba philosophorum 56-68, 375-9, 389-90
al-Ṭūsī, Naṣīr al-Dīn 395

underworld: *see* Hades

Velia (Elea) viii, 143, 225 n., 284 n., 323 n., 392
Vesuvius viii, 134 n., 137 n.
Vulcan 281

water (as element) 13, 25-7, 29-33, 56-7, 63-4, 348-58
William of Moerbeke 204, 207, 208
words: power of 220, 229-31, 247-8, 309-12; as seeds 230-1, 299, 362-3

Xanthus of Lydia 223 n., 227
Xenocrates 139
Xenophanes 33 n., 61 n., 63, 192 n., 368 n.
Xenophilus 149

Zeus 13-14, 24-5 n., 36, 42-7, 224, 255, 348-51, 355, 356 n., 357; home of 40 n., 42, 195-7, 255; prison of 187-9, 195-213; sons of 266
Zopyrus of Tarentum 143-60, 262, 282-3, 287, 299 n.
Zoroastrianism 193 n., 226, 304
Zosimus of Panopolis 59-60, 61 n., 67-8, 375 n., 390 n.